A BRIEF VIEW OF ASTRONOMY

Jay M. Pasachoff

Field Memorial Professor of Astronomy
Director of the Hopkins Observatory
Williams College
Williamstown, Massachusetts

SAUNDERS GOLDEN SUNBURST SERIES

SAUNDERS COLLEGE PUBLISHING
Philadelphia   New York   Chicago
San Francisco   Montreal   Toronto
London   Sydney   Tokyo   Mexico City
Rio de Janeiro   Madrid

Address orders to:
383 Madison Avenue
New York, NY 10017

Address editorial correspondence to:
West Washington Square
Philadelphia, PA 19105

Text Typeface: Baskerville
Compositor: Clarinda
Acquisitions Editor: John Vondeling
Developmental Editor: Margaret Mary Kerrigan
Project Editors: Joanne Fraser and Robin Bonner
Copyeditor: Robin Bonner
Art Director: Carol C. Bleistine
Art/Design Assistant: Virginia A. Bollard
Text Design: Edward A. Butler
Cover Design: Lawrence R. Didona
New Text Artwork: Linda Maugeri and Larry Ward
Production Manager: Tim Frelick
Assistant Production Manager: JoAnn Melody

Cover Credit: Halley's comet in 1910, photographed at the Lowell Observatory and transformed
to color equal-intensity regions at the National Optical Astronomy Observatories. *Back cover:*
The diamond-ring effect at the total solar eclipse of November 22/23, 1984, photographed in
Papua New Guinea by J. M. Pasachoff; image processed at the Institute for Astronomy of the
University of Hawaii in collaboration with John W. MacKenty.

**Library of Congress Cataloging in
Publication Data**

Pasachoff, Jay M.
   A brief view of astronomy.

   Bibliography: p.
   Includes index.

   1. Astronomy. I. Title.
QB43.2.P357 1985   520   85-8220

ISBN 0-03-058422-1

A BRIEF VIEW OF ASTRONOMY   ISBN 0–03–058422–1

6—032—98765432

CBS COLLEGE PUBLISHING
Saunders College Publishing
Holt, Rinehart and Winston
The Dryden Press

PREFACE

We live in a golden age of astronomy: Our view of the universe and of the fascinating objects in it has been changed and enhanced in recent years by observational and theoretical discoveries. And new studies continue.

It is a pleasure for me to be able to describe the exciting state of astronomy to you. But one can't cover the whole universe in a few months. I have therefore had to pick and choose, while trying to cover a wide range and to convey the spirit of contemporary astronomy and of the scientists working in it. My mix includes both much basic astronomy and many of the exciting topics now at the forefront. Massive black holes in the centers of galaxies and quasars, observations of our galaxy from the Infrared Astronomical Satellite, spacecraft views of Halley's Comet and Uranus, and new ideas about the first fraction of a second of the universe's existence are among the topics covered. I hope that the treatment here, necessarily limited in scope, will whet your appetite enough to stimulate you to take another astronomy course or to read some of the books about astronomy available (some of which are listed at the end of the book) and new articles as they come out. You will also note that this book is almost entirely lacking in math; to understand astronomy more deeply, you may want to look in my other textbooks, which include some basic mathematical development. The discussions of modern space research and of new discoveries with ground-based telescopes give a more accurate feeling of contemporary astronomy than would more emphasis on historical results, but an understanding of the historical and mathematical foundations of astronomy are important too.

In writing this book, I share the goals of a commission of the Association of American Colleges, which reported in 1984 on the college curriculum that: "A person who understands what science is recognizes that scientific concepts are created by acts of human intelligence and imagination; comprehends the distinction between observation and inference and between the occasional role of accidental discovery in scientific investigation and the deliberate strategy of forming and testing hypotheses; understands how theories are formed, tested, validated, and accorded provisional acceptance; and discriminates between conclusions that rest on unverified assertion and those that are developed from the application of scientific reasoning."

I have provided aids to make the book easy to read and to study from. New vocabulary is italicized in the text, listed in the key words at the end of each chapter, and defined in the glossary. The index provides further aid in finding explanations. End-of-chapter questions cover a range of material, and include some that are straightforward to answer from the text and others that require more thought. All sections are numbered for ready reference. Appendices provide some standard and recent information on planets, stars, constellations, and non-stellar objects. The exceptionally beautiful new star maps by Wil Tirion will help you find your way around the sky when you go outside to observe the stars. Be sure you do.

Note to professors: Astronomy is too wide-ranging to be placed in so few pages without taking Procrustean steps. I have tried to keep a broad array of material, since no part of astronomy is inherently more important than any other. Those of you who feel that certain topics must be gone into more deeply may find my text *Contemporary Astronomy* (3rd edition, 1985) more appropriate.

New and exciting material is distributed through the book, so that students find particularly interesting topics at regular intervals. The book leads up to the topic that students and professors perhaps find most interesting: cosmology and the future of the universe. Earlier, the search for life elsewhere in the universe—which is of widespread interest, though it is out of the mainstream of astronomy—closes the section on the planets. Distributed throughout the book are sections describing what it is like to carry out actual astronomical research: there are sections on observing with a ground-based optical telescope, working on an eclipse expedition, using a telescope on a satellite, and using a radio telescope.

Acknowledgments: The publishers and I have always placed a heavy premium on accuracy for my books, and have made certain that the manuscript and proof have been read not only by students for clarity and style but also by professional astronomers for scientific comments. As a result, you will find that the statements in this book, brief as they are, are authoritative.

We would particularly like to thank the readers of the manuscript of this new brief version, including James L. Regas (California State University at Chico), Robert N. Zitter (Southern Illinois University at Carbondale), James Pierce (Mankato State University), Thomas H. Robertson (Ball State University), Tom Bullock (West Valley College), Stephen P. Lattanzio (Orange Coast College), Leo Connolly (Southeast Missouri State University), Roy W. Clark (Middle Tennessee State University), George Carlson (Citrus Community College), Theodore Spickler (West Liberty State College) and Louis Winkler (Pennsylvania State University).

We also thank the many student readers who have commented on my books. We are glad that the books have proved so enjoyable, and we appreciate the feedback.

We remain grateful to the various readers of other versions of my texts: Thomas T. Arny, James G. Baker, Laszlo Baksay, Bruce E. Bohannan, Kenneth Brecher, Bernard Burke, Clark Chapman, James W. Christy, Martin Cohen, Peter Conti, Lawrence Cram, Dale Cruikshank, Morris Davis, Raymond Davis, Jr., Dennis di Cicco, Marek Demianski, Gerard de Vaucouleurs, Richard B. Dunn, John A. Eddy, James L. Elliot, Farouk El-Baz, David S. Evans, J. Donald Fernie, William R. Forman, Peter V. Foukal, George D. Gatewood, John E. Gaustad, Tom Gehrels, Riccardo Giacconi, Owen Gingerich, Stephen T. Gottesman, Jonathan E. Grindlay, Alan R. Guth, Ian Halliday, U. O. Herrmann, James Houck, Robert F. Howard, John Huchra, H. W. Ibser, Christine Jones, Bernard J. T. Jones, Agris Kalnajs, Robert Kirshner, David E. Koltenbah, Jerome Kristian, Edwin C. Krupp, Karl F. Kuhn, Marc L. Kutner, Karen B. Kwitter, John Lathrop, Lawrence S. Lerner, Jeffrey L. Linsky, Sarah Lee Lippincott, Bruce Margon, Brian Marsden, Janet Mattei, R. Newton Mayall, Everett Mendelsohn, George K. Miley, Freeman D. Miller, Alan T. Moffet, William R. Moomaw, David D. Morrison, R. Edward Nather, David Park, Carl B. Pilcher, James B. Pollack, B. E. Powell, Edward L. Robinson, Herbert Rood, Maarten Schmidt, David N. Schramm, Leon W. Schroeder, Richard L. Sears, P. Kenneth Seidelmann, Maurice Shapiro, Joseph I. Silk, Lewis E. Snyder, Theodore Spickler, Hyron Spinrad, Paul Steinhardt, Alan Stockton, Robert G. Strom, Jean Pierre

Swings, Eugene Tademaru, Gustav Tammann, Joseph H. Taylor, Jr., Joe S. Tenn, Yervant Terzian, David Theison, Laird Thompson, M. Nafi Toksöz, Juri Toomre, Kenneth D. Tucker, Brent Tully, Barry Turner, Peter van de Kamp, Joseph Veverka, Gerald J. Wasserburg, Leonid Weliachew, Ray Weymann, John A. Wheeler, J. Craig Wheeler, Ewen A. Whitaker, Reinhard A. Wobus, LeRoy A. Woodward, and Susan Wyckoff.

I appreciate the special assistance of David Malin of the Anglo-Australian Observatory and Jurrie van der Woude of the Jet Propulsion Laboratory in obtaining many new pictures.

Nancy Pasachoff Kutner has done excellent work on the index.

I thank many people at Saunders College Publishing for their efforts on my books. John J. Vondeling and Lloyd Black merit special thanks for their continued support. Joanne Fraser, Robin Bonner, and Margaret Mary Kerrigan worked hard on many aspects of production. New artwork was drawn by Linda Maugeri and Larry Ward. Many of the drawings were executed from my sketches by George Kelvin of Science Graphics for the first edition of *Contemporary Astronomy*.

I appreciate the capable editorial work of Susan Welsch over the past several years. Martine Westermann has ably taken on various aspects of production. I also thank Karen Kowitz for her important assistance.

Various members of my family have provided vital and valuable editorial services, in addition to their general support. My father contributed so much to my books over the years. I also appreciate the work of my mother and of my wife, Naomi. Now that our daughters Eloise and Deborah are ten and eight, respectively, they are taking a bigger share of proofreading alongside their own studies and compositions. Eric Kutner has also helped with proofreading.

I am grateful to Sidney C. Wolff, Donald N. B. Hall, Dale P. Cruikshank, Donald A. Landman, and others at the Institute for Astronomy of the University of Hawaii for their hospitality during my sabbatical leave and for the opportunity to use their magnificent telescopes and other facilities.

At the request of the donor, it should be mentioned that the Field Memorial Professorship of Astronomy at Williams College was established in 1865 "In Memoriam" of Harriet Louisa Dudley Field. It is "devoted to the study, teaching, and advancement of the Science of Astronomy."

Ancillary Materials: The *Teacher's Guide and Test Bank for A Brief View of Astronomy* contains possible syllabi, lab exercises, tests, answers to all questions in this book, and lists films, tapes, videodisks, and other audio and visual aids for use as supplementary materials. The publishers are also making available selected artwork and photographs in the format of 100 overhead-projection transparencies. For information about them, contact your local Saunders Field Representative, with a copy of any letters to me. Telephone numbers for the regional field offices are: Eastern—(212) 599-8435; Central—(312) 323-0205; and Western—(415) 692-6386. Inquiries made through the central marketing office in New York are often considerably delayed.

If any problems arise with delivery of texts or of ancillary materials, I can often help. Please don't hesitate to write or telephone me.

To Students: Astronomy is very varied, and you are sure to find some parts that interest you more than others. Do try to get an overview from this text, and then go on to do additional reading (suggestions are provided at the end of the book). To get the most out of this text, you must read each chapter more than once. After you read each chapter the first time, you should go carefully through the list of key words, trying to identify or define each one. If you cannot do so, look up the word in the glossary, and also find the definition that appears with the word the first time it was used in the chapter. The index will help you find these references. Next, read through the chapter again especially carefully. Finally, answer the questions at the end of the chapter.

I am extremely grateful to all the individuals named above for their assistance. Of course, it is I who have put this all together, and I alone am responsible for it. I would appreciate hearing from readers with suggestions for improved presentation of topics, with comments about specific points that need clarification, with typographical or other errors, or just to tell me how you like your astronomy course. I invite readers to write me c/o Williams College, Hopkins Observatory, Williamstown, Massachusetts 01267. I promise a personal response to each writer.

JAY M. PASACHOFF

Field Editors: Susan Welsch, Martine Westermann

TABLE OF CONTENTS

Auxiliary material will include a Teacher's Guide with Test Bank, overhead transparencies, and computer material.

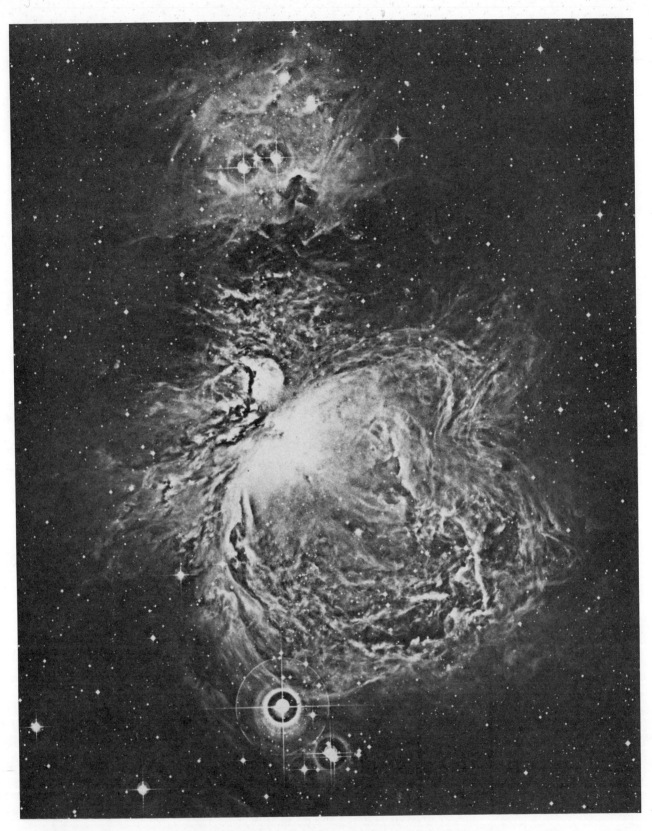

The Great Nebula in Orion, printed with a special technique that brings out small-scale structures.

PART I   Observing the Universe

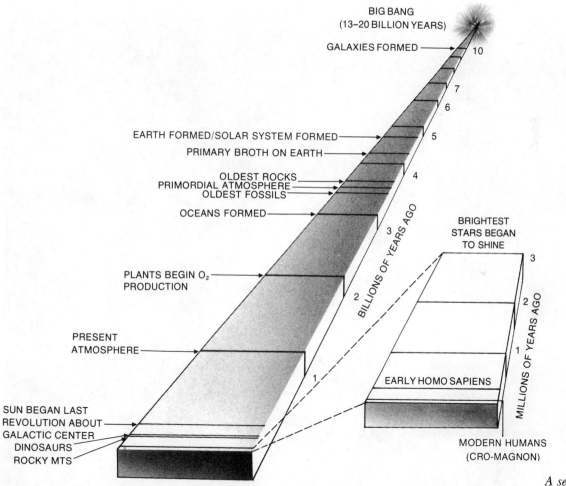

BIG BANG
(13–20 BILLION YEARS)

GALAXIES FORMED

10

7

6

EARTH FORMED/SOLAR SYSTEM FORMED

5

PRIMARY BROTH ON EARTH

OLDEST ROCKS
PRIMORDIAL ATMOSPHERE
OLDEST FOSSILS

4

OCEANS FORMED

BILLIONS OF YEARS AGO

3

BRIGHTEST
STARS BEGAN
TO SHINE

3

PLANTS BEGIN O_2
PRODUCTION

2

2

MILLIONS OF YEARS AGO

1

PRESENT
ATMOSPHERE

1

EARLY HOMO SAPIENS

SUN BEGAN LAST
REVOLUTION ABOUT
GALACTIC CENTER
DINOSAURS
ROCKY MTS

MODERN HUMANS
(CRO-MAGNON)

A sense of time.

As we shall study, astronomers have deduced that the universe began about 20 billion years ago. Let us consider that the time between the origin of the universe and the year 2000 is one day. If the universe began at midnight, then it wasn't until 4 P.M. that the earth formed; the first fossils date from 10 P.M. The first humans appeared only 2 seconds ago, and it is only 3/1000 second since Columbus discovered America. The year 2000 will arrive in only 1/10,000 second.

Still, the sun should shine for another 8 hours; an astronomical time scale is much greater than the time scale of our daily lives. Astronomers use a wide range of technology and theories to find out about the universe, what is in it, and what its future will be. This book surveys what we have found, how we look, and how we interpret and evaluate the results.

Chapter 1

The Whirlpool Galaxy, M51, in the constellation Canes Venatici.

THE UNIVERSE: AN OVERVIEW

The universe is a place of great variety—after all, it has everything in it! Some of the things astronomers study are of a size and scale that we humans can easily comprehend: the planets, for instance. Most astronomical objects, however, are so large and so far away that our minds have trouble grasping their sizes and distances.

Moreover, astronomers study the very small in addition to the very large. Most everything we know comes from the study of energy travelling through space in the form of "radiation"—of which light and radio waves are examples. The radiation we receive from distant bodies is emitted by atoms, which are much too small to see with the unaided eye. Also, the properties of the large astronomical objects are often determined by changes that take place on a minuscule scale—that of atoms or their nuclear cores. And recent studies of the universe's earliest fraction of a second are linked with studies of the still smaller particles that make up the nuclei. Thus the astronomer must be an expert in the study of the tiniest as well as the largest objects.

Such a variety of objects at very different distances from us or with very different properties often must be studied with widely differing techniques. Clearly, a certain type of equipment is required to analyze the properties of solid particles like Martian soil (which we have analyzed with equipment in a spacecraft sitting on the Martian surface). Another type of equipment and another method are required to study the light or radio waves from a gaseous body like a quasar deep in space in order to interpret its composition and behavior. However, one unifying method does link much of astronomy. With this method, astronomers analyze components of the light or other radiation that we receive from distant objects and study these components in detail. Of course, astronomers also make images—pictures—of, for example, planets, the sun, and galaxies. They may also study the variation of the amount of light coming from a star or a galaxy over time, or send spacecraft to visit the moon and planets.

The explosion of astronomical research in the last few decades has been fueled by our new ability to study radiation other than light—gamma rays, x-rays, ultraviolet radiation, infrared radiation, and radio waves. Astronomers' use of their new abilities to study such radiation is a major theme of this book. *Radiation* is energy carried in electric and magnetic fields; in a vacuum, it always moves at a huge constant speed known as "the speed of light." We can think of radiation as waves; all types of radiation have similar properties except for the length of the waves. Although x-rays and visible light may be similar, our normal experiences tell us that very different techniques are necessary to study them. All the kinds of radiation together make up the *spectrum*. Breaking up sunlight into its component colors gives the "visible spectrum" (Color Plate 5).

Technically, certain atomic and subatomic particles moving through space—"alpha rays" and "beta rays," for example—are also known as "radiation." They do not usually concern us in astronomy, except when we study "cosmic rays" (Section 21.2).

1.1 A SENSE OF SCALE

Let us try to get a sense of scale of the universe, starting with sizes that are part of our experience and then expanding toward the infinitely large. Each diagram will show a square 100 times greater on a side.

Some of the units used in astronomy and the prefixes used with them are listed in Appendix 1.

We shall use the metric system, which is commonly used by scientists. The basic unit of length is the meter, which is equivalent to 39.37 inches, slightly more than a yard. Prefixes are used (Appendix 1) in conjunction with the word "meter," abbreviated "m," to define new units. The most frequently used prefixes are "milli-," meaning 1/1000, "centi-," meaning 1/100, and "kilo-," meaning 1000 times. Thus 1 millimeter is 1/1000 of a meter, or about .04 inch, and a kilometer is 1000 meters, or about 5/8 mile. We will keep track of the powers of 10 by which we multiply 1 m by writing the number of tens we multiply together as an exponent; 1000 m, for example, is 10^3m.

The distance travelled by an object is equal to the rate at which the object is travelling (its velocity) times the time spent travelling ($d = vt$).

We can also keep track of distance in units that are based on the length of time that it takes light to travel. The speed of light is, according to Einstein's special theory of relativity, the greatest speed that is physically attainable. Light travels at 300,000 km/s (186,000 miles/s), fast enough to circle the earth 7 times in a single second. Even at that fantastic speed, we shall

(continued on page 8)

BOX 1.1
Scientific Notation

$10^0 = 1$
$10^1 = 10$
$10^2 = 100$
$10^3 = 1000$

Note that anything to the zeroth power is 1 and anything to the first power is itself.

In astronomy we often find ourselves writing numbers that have strings of zeros attached, so we use what is called either *scientific notation* or *exponential notation,* to simplify our writing chores. Scientific notation helps prevent making mistakes when copying long strings of numbers, and aids astronomers in making calculations.

In scientific notation, which we use in Figures 1–4 to 1–14, we merely count the number of zeros, and write the result as a superscript to the number 10. Thus the number 100,000,000, a 1 followed by 8 zeros, is written 10^8. The superscript is called the *exponent.* We also say that "10 is raised to the **eighth power.**" When a number is not a power of 10, we divide it into two parts: a number between 1 and 10, and a power of 10. Thus the number 3645 is written as 3.645×10^3. The exponent shows how many places the decimal point was moved to the left.

We can represent numbers between zero and one by using negative exponents. A minus sign in the exponent of a number means that the number is actually one divided by what the quantity would be if the exponent were positive. Thus $10^{-2} = 1/10^2$. One can compute the exponent that follows the minus sign by counting the number of places by which the decimal point has to be moved to the right until it is at the right of the first non-zero digit. Thus, for example, $0.000001435 = 1.435 \times 10^{-6}$, since the decimal point in the written-out number has to be moved six places to the right to come after the digit 1 in exponential notation. Positive exponents, similarly, are equal to the number of places that the decimal point has to be moved to the left to reach the right side of the first non-zero digit.

Let us begin our journey through space with a view of something 1 mm across. Here we see a velvety tree ant (Fig. 1–1), observed through a scanning electron microscope. Every step we take will show a region 100 times larger in diameter than the previous picture.

FIGURE 1–1 1 mm = 0.1 cm

A square 100 times larger on each side is 10 centimeters × 10 centimeters. (Since the area of a square is the length of a side squared, the area of a 10-cm square is 10,000 times the area of a 1-mm square.) The area encloses a flower (Fig. 1–2).

FIGURE 1–2 10 cm = 100 mm

As we move far enough away to see an area 10 meters on a side, we are seeing an area approximately that taken up by half a tennis court (Fig. 1–3).

FIGURE 1–3 10 m = 1000 cm

A square 100 times larger on each side is now 1 kilometer square, about 250 acres. An aerial view of several square blocks in New York City shows how big an area this is (Fig. 1–4).

FIGURE 1–4 1 km = 10^3 m

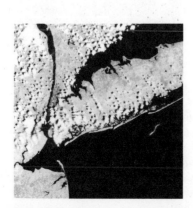

FIGURE 1–5 100 km = 10^5 m

The next square, 100 km on a side, encloses a major city, New York, and some of its suburbs. Note that though we are still bound to the limited area of the earth, the area we can see is increasing rapidly (Fig. 1–5).

A square 10,000 km on a side covers nearly the entire earth (Fig. 1–6).

FIGURE 1–6   10,000 km = 10^7 m

FIGURE 1–7 10^9 m = 3 lt sec

When we have receded 100 times farther, we see a square 100 times larger in diameter: 1 million kilometers across. It encloses the orbit of the moon around the earth (Fig. 1–7). We can measure with our wristwatches the amount of time that it takes light to travel this distance. If we were carrying on a conversation by radio with someone at this distance, there would be pauses of noticeable length after we finished speaking before we heard an answer. This is because radio waves, even at the speed of light, take over a second to travel that far. Astronauts on the moon have to get used to these pauses when speaking to earth. This photograph was taken by the Voyager 1 spacecraft en route to Jupiter and Saturn. Eastern Asia, the western Pacific Ocean, and part of the Arctic are on the illuminated portion of the earth (bottom).

When we look on from 100 times farther away still, we see an area 100 million kilometers across, ⅔ the distance from the earth to the sun. We can now see the sun and the two innermost planets in our field of view (Fig. 1–8).

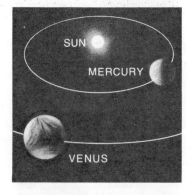

FIGURE 1–8 10^{11} m = 5 lt min

FIGURE 1–9 10^{13} m = 9 lt hrs

An area 10 billion kilometers across shows us the entire solar system in good perspective. It takes light about 10 hours to travel across the solar system. The outer planets have become visible and are receding into the distance as our journey outward continues (Fig. 1–9). This artist's conception shows a Voyager spacecraft near Saturn. Our spacecraft have now visited or passed the moon, Mercury, Venus, Mars, Jupiter, Saturn, Uranus, and their moons.

From 100 times farther away, we see little that is new. The solar system seems smaller and we see the vastness of the empty space around us. We have not yet reached the scale at which another star besides the sun is in a cube of this size (Fig. 1–10).

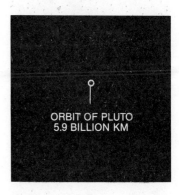

FIGURE 1–10 10^{15} m = 38 lt days

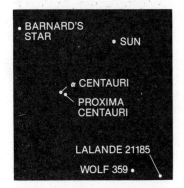

FIGURE 1–11 10^{17} m = 10 ly

As we continue to recede from the solar system, the nearest stars finally come into view. We are seeing an area 10 light years across, which contains only a few stars (Fig. 1–11), most of whose names are unfamiliar (Appendix 7). Part III of this book discusses the properties of all types of stars.

By the time we are 100 times farther away, we can see a fragment of our galaxy, the Milky Way Galaxy (Fig. 1–12). We see not only many individual stars but also many clusters of stars and many "nebulae" — regions of glowing, reflecting, or opaque gas or dust. Between the stars, there is a lot of material (most of which is invisible to our eyes) that can be studied with radio telescopes on earth or in ultraviolet or x-rays with telescopes in space.

FIGURE 1–12 10^{19} m = 10^3 ly

FIGURE 1–13 10^{21} m = 10^5 ly

In a field of view 100 times larger in diameter, we can now see an entire galaxy. The photograph (Fig. 1–13) shows a galaxy known as M74 (its number in Charles Messier's catalogue), located in the direction of the constellation Pisces, though it is far beyond the stars in that constellation. This galaxy shows arms wound in spiral form. Our galaxy also has spiral arms, though they are wound more tightly.

Next we move sufficiently far away so that we can see an area 10 million light years across (Fig. 1–14). There are 10^{25} centimeters in 10 million light years, about as many centimeters as there are grains of sand in all the beaches of the earth. Our galaxy is in a cluster of galaxies, called the Local Group, that would take up only ⅓ of our angle of vision. In this group are all types of galaxies. The photograph shows part of a cluster of galaxies in the constellation Leo.

FIGURE 1–14 10^{23} m = 10^7 ly

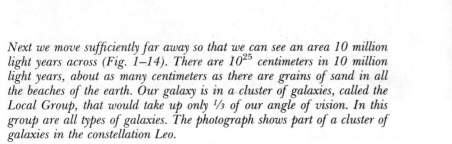

see that it would take years for us to reach the stars. Similarly, it has taken years for the light we see from stars to reach us, so we are really seeing the stars as they were years ago. In a sense, we are looking backward in time. The distance that light travels in a year is called a *light year;* note that the light year is a unit of length rather than a unit of time even though the term "year" appears in it.

If we could see a field of view 1 billion light years across, our Local Group of galaxies would appear as but one of many clusters. Before we could enlarge our field of view another 100 times we might see a supercluster—a cluster of clusters of galaxies. We would be seeing almost to the distance of the quasars. Quasars, the most distant objects known, seem to be explosive events in the cores of galaxies. Light from the most distant quasars observed may have taken 10 billion years to reach us on earth. We are thus looking back to times billions of years ago. Since we think that the universe began 13 to 20 billion years ago, we are looking back almost to the beginning of time.

We even think that we have detected radiation from the universe's earliest years. A combination of radio, ultraviolet, x-ray, and optical studies, together with theoretical work and experiments with giant atom smashers on earth, is allowing us to explore the past and predict the future of the universe.

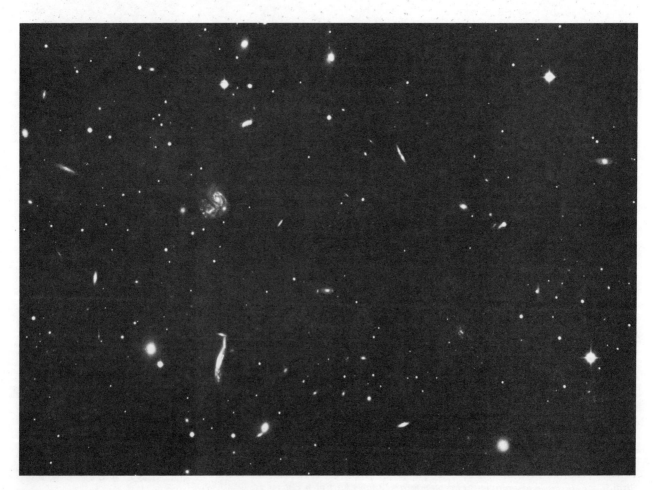

FIGURE 1–15　The cluster of galaxies in the constellation Hercules.

1.2 THE VALUE OF ASTRONOMY

Throughout history, observations of the heavens have led to discoveries that have had major impact on people's ideas about themselves and the world around them. Even the dawn of mathematics may have stemmed from ancient observations of the sky, made in order to keep track of seasons and seasonal floods in the fertile areas of the earth. Observations of the motions of the moon and the planets, which are free of such complicating terrestrial forces as friction and which are massive enough so that gravity dominates their motions, led to an understanding of gravity and of the forces that govern all motion.

We can consider the regions of space studied by astronomers as a cosmic laboratory where we can study matter or radiation, often under conditions that we cannot duplicate on earth.

Many of the discoveries of tomorrow—perhaps the control of nuclear fusion or the discovery of new sources of energy, or perhaps something so revolutionary that it cannot now be predicted—will undoubtedly be based on discoveries made through such basic research as the study of astronomical systems. Considered in this sense, astronomy is an investment in our future.

The impact of astronomy on our conception of the universe has been strong through the years. Discoveries that the earth is not in the center of the universe, or that the universe has been expanding for billions of years, affect our philosophical conceptions of ourselves and our relations to space and time.

Yet most of us study astronomy not for its technological and philosophical benefits but for its beauty, grandeur, and inherent interest. We must stretch our minds to understand the strange objects and events that take place in the far reaches of space. The effort broadens us and continually fascinates us all. Ultimately, we study astronomy because of its fascination and mystery.

KEY WORDS

radiation, spectrum, light-year, scientific notation, exponential notation, exponent

QUESTIONS

1. Why do we say that our senses have been expanded in recent years?

2. The speed of light is 3×10^5 km/s. Express this number in m/s and in cm/s.

3. List the following in order of increasing size: (a) light year, (b) distance from earth to sun, (c) size of Local Group, (d) size of football stadium, (e) size of our galaxy, (f) distance to a quasar.

4. Of the examples of scale in this chapter, which would you characterize as part of "everyday" experience? What range of scale does this encompass? How does this range compare with the total range covered in the chapter?

5. What is the largest of the scales discussed in this chapter that could reasonably be explored in person by humans with current technology?

6. (a) Write the following in scientific notation: 4642; 70,000; 34.7. (b) Write the following in scientific notation: 0.254; 0.0046; 0.10243. (c) Write out the following in an ordinary string of digits: 2.54×10^6; 2.004×10^2.

TOPIC FOR DISCUSSION

What is the value of astronomy to you? How do you rank National Science Foundation (NSF) and National Aeronautics and Space Administration (NASA) funds for research with respect to other national needs? Reanswer this question when you have completed this course.

Chapter 2

The Hubble Space Telescope being assembled for launch in late 1986. It should provide images about seven times clearer than can be obtained with ground-based telescopes.

LIGHT AND TELESCOPES: EYES AND EARS ON THE UNIVERSE

Everybody knows that astronomers use telescopes, but the telescopes that astronomers use these days are of very varied types, and most often not used directly with the eye. In this chapter, we will first discuss the telescopes that astronomers use to focus light, as they have for hundreds of years. Then we will see how astronomers now also use telescopes to study x-rays, ultraviolet, infrared, and radio waves.

2.1 LIGHT AND TELESCOPES

Almost four hundred years ago, a Dutch optician put two eyeglass lenses together, and noticed that distant objects appeared closer. The next year, in 1609, the English scientist Thomas Harriot built one of these devices and looked at the moon. But all he saw was a blotchy surface, and he didn't make anything of it. Credit for first using a telescope to make astronomical studies goes to Galileo Galilei. Hearing in 1609 that a telescope had been made in Holland, Galileo in Venice made one of his own and used it to look at the moon. Perhaps as a result of his training in interpreting light and shadow in drawings—he was surrounded by the Renaissance—Galileo realized that the light and dark patterns on the moon meant that there were craters there (Fig. 2–1). With the tiny telescopes he made—only 20 or 30 power, about as powerful as a modern pair of binoculars but showing a smaller part of the sky—he went on to revolutionize our view of the cosmos.

Whenever he could center Jupiter in the narrow field of the sky that his telescope showed, he saw that Jupiter was not just a point of light, but showed a small disk. And he spotted four points of light that moved from one side of Jupiter to another (Fig. 2–2). He eventually realized that the points of light were moons orbiting Jupiter, the first proof that not all bodies in the solar system orbited the earth. The existence of Jupiter's moons also

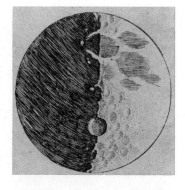

Figure 2–1 An engraving of Galileo's observations of the moon from his book *Sidereus Nuncius (The Starry Messenger)*, published in 1610. Galileo was the first to report that the moon has craters. Aristotle and Ptolemy had held that the earth was imperfect but that everything above it was perfect, so Galileo's observation contradicted them.

Figure 2–2. Some of Galileo's notes about his first observations of Jupiter's moons. Simon Marius also observed the moons at about the same time; we now use the names Marius proposed for them: Io, Europa, Ganymede, Callisto, though we call them the Galilean satellites.

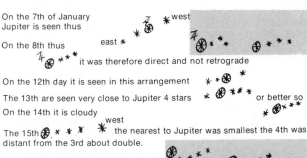

On the 7th of January Jupiter is seen thus

On the 8th thus   east   it was therefore direct and not retrograde

On the 12th day it is seen in this arrangement

The 13th are seen very close to Jupiter 4 stars   or better so

On the 14th it is cloudy

The 15th   the nearest to Jupiter was smallest the 4th was distant from the 3rd about double.

The spacing of the 3 to the west was no greater than the diameter of Jupiter and they were in a straight line.

long. 71°38' lat. 1°13'

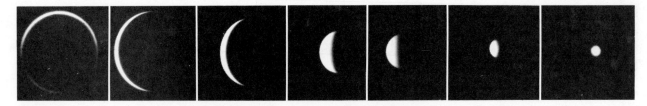

Figure 2–3   The phases of Venus. Note that Venus is a crescent only when it is in a part of its orbit that is relatively close to the earth, and so looks larger at those times.

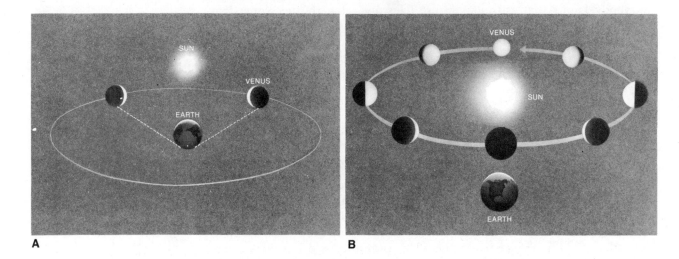

A   B

Figure 2–4   In the Ptolemaic theory *(left)*, Venus and the sun both orbit the earth. Because it is known that Venus never gets far from the sun in the sky, though, Venus could never get farther from the sun than the region restricted by the dotted lines. Thus, Venus would always appear as a crescent, though before the telescope was invented, this could not be verified. In the sun-centered theory *(right)*, Venus is sometimes on the near side of the sun, where it appears as a crescent, and it is sometimes on the far side, where we can see half or more of Venus illuminated. Thus Galileo's observations, a modern version of which appears as Figure 2–3, agree with the sun-centered theory of Copernicus.

Figure 2–5 Parallel light is diverging imperceptibly, since the stars are so far away.

showed that the ancient Greek philosophers—chiefly Aristotle and Ptolemy—who had held that the earth is at the center of all orbits were wrong. His discovery thus backed the newer theory of Copernicus, who had said in 1543 that the sun and not the earth is at the center of the universe. (We say more about the historical development of astronomy in the introduction to Part II, immediately preceding Chapter 5.) And Galileo's lunar discovery—that the moon's surface had craters—had also endorsed Copernicus's ideas, since the Greek philosophers had held that celestial bodies were all perfect.

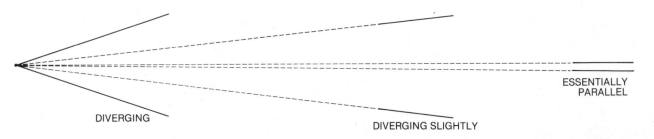

DIVERGING   DIVERGING SLIGHTLY   ESSENTIALLY PARALLEL

Galileo published these discoveries in 1610 in a book called *Sidereus Nuncius* (*The Starry Messenger*).

Galileo went on to discover that Venus went through a full set of phases, from crescent to full (Fig. 2–3), contrary to the prediction of the earth-centered theory of Ptolemy and Aristotle that only a crescent phase would be seen (Fig. 2–4). He further found that the sun had spots on it (which we now call "sunspots"), and many other exciting things.

But Galileo's telescopes had deficiencies, among them that images had tinges of color around them, caused by the way that light is bent as it passes through lenses. Toward the end of the 17th century, Isaac Newton in England had the idea of using mirrors instead of lenses to make a telescope. The light from the moon, planets, and stars is "parallel," since these objects are so far away (Fig. 2–5). A curved mirror can focus light to a single point, called the *focus* (Fig. 2–6). But for a mirror a few centimeters across, your head would block the incoming light if you tried to put your eye to the focus. Newton had the bright idea of putting a small, flat mirror just in front of the focus to reflect the focus out to the side. This *Newtonian telescope* (Fig. 2–7) is a design still in use by perhaps most amateur astronomers. We now often use mirrors that are in the shape of a *paraboloid*, since only paraboloids focus parallel light to a focus (Fig. 2–8). Many telescopes now use the *Cassegrain* design, in which a secondary mirror bounces the light back through a small hole in the middle of the primary mirror (Fig. 2–9).

Through the 19th century, telescopes using lenses—*refracting telescopes* (or refractors)—and telescopes using mirrors—*reflecting telescopes* (or reflectors)—were made larger and larger. The pinnacle of refracting telescopes was reached in the 1890's with the construction of a telescope with a lens 40 inches (1 m) across for the Yerkes Observatory in Wisconsin, now part of the University of Chicago (Fig. 2–10). It was difficult to make a lens of clear glass thick enough to support its large diameter; it had to be thick because it could be supported only around the edge. And the telescope tube had to

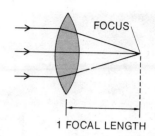

Figure 2–6 The focal length is the distance behind a lens to the point at which objects at infinity are focused. The focal length of the human eye is about 2.5 cm (1 in.).

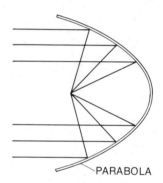

FIGURE 2–8 A paraboloid is a three-dimensional curve created by spinning a parabola on its axis. As shown here, a parabola (and a paraboloid as well) focus parallel light to a single point.

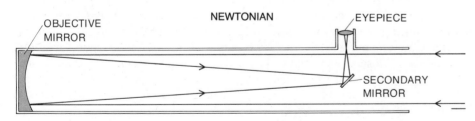

FIGURE 2–7 The path of light in a Newtonian telescope; note the diagonal mirror that brings the focus out to the side. Many amateur telescopes are of this type.

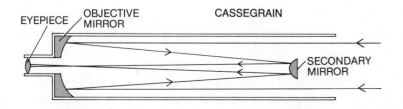

FIGURE 2–9 A Cassegrain telescope, in which a curved secondary mirror bounces light through a hole in the center of the primary mirror.

A

B

FIGURE 2–10 *(A)* The Yerkes refractor, still the largest in the world. Note Albert Einstein at right of center in this 1921 picture. *(B)* The opening of the 1-meter (40-inch) refractor of the Yerkes Observatory was the cause of much notice in the Chicago newspapers in 1893.

be tremendously long. Because of these difficulties, no larger telescope lens has ever been put into use.

The size of a telescope's main lens or mirror is particularly important because the main job of most telescopes is to collect light—to act as a "light bucket." All the light is brought to a common focus, where it is viewed or recorded. The larger the telescope's lens or mirror, the fainter the objects that can be viewed or the more quickly observations can be made. A larger telescope also provides better *resolution*—the ability to detect fine detail. For the most part, then, the fact that telescopes magnify is secondary to their ability to gather light.

From the mid-nineteenth century onward, larger and larger reflecting telescopes were made. But the mirrors, then made of shiny metal, tended to tarnish. This problem was avoided by depositing a thin coat of silver onto the mirror. More recently, a thin coating of aluminum turned out to be longer lasting, though silver with a thin transparent overcoat of tough material is now coming back into style. The 100-inch (2.5-m) reflector at the Mt. Wilson Observatory in California became the largest telescope in the world in 1917. Its use led to discoveries about distant galaxies that transformed our view of what the universe is like and what will happen to it and us in the far future.

In 1948, the 200-inch (5-m) reflecting telescope (Fig. 2–11) opened at the Palomar Observatory, also in California, and was for many years the largest in the world. (A 6-m Soviet reflector is now in operation, Color Plate 2, but its images are not as good.) New electronic imaging devices have made this and other large telescopes many times more powerful than they were when they recorded images on film.

The National Optical Astronomy Observatories, supported by the National Science Foundation, has twin 4-m reflecting telescopes at the Kitt Peak National Observatory in Arizona (Color Plate 1) and at the Cerro Tololo Inter-American Observatory in Chile. Besides these largest telescopes, over a dozen other telescopes are also at Kitt Peak (Color Plate 9), including some from universities, and a few other telescopes are in Chile. Some of the most

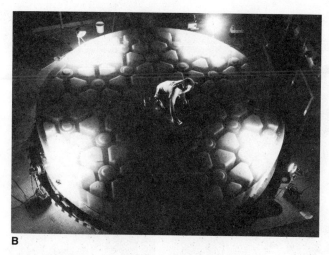

A
B

FIGURE 2–11 *(A)* The 5-m (200-inch) Hale telescope of the Palomar Observatory in southern California, owned by the California Institute of Technology. *(B)* Periodically, the Palomar telescope's 5-m mirror is cleaned and a new aluminum coating is applied. Here, before its 1982 aluminizing, we can look through the mirror's front and see the ribbed structure of its back, designed to reduce weight and stress on the mirror itself.

interesting astronomical objects are in the southern sky, so astronomers need telescopes at sites more southerly than the continental United States. For example, the nearest galaxies to our own—known as the Magellanic Clouds—are not readily observable from the continental U.S.

The observatory with the most large telescopes is now on top of the dormant volcano Mauna Kea in Hawaii, partly because its latitude is as far south as only +20° and partly because the site is so high that it is above 40 per cent of the earth's atmosphere. Further, to detect the infrared part of the spectrum, telescopes must be above as much of the water vapor in the earth's atmosphere as possible, and Mauna Kea is above 90 per cent. And in addition, the peak is above the atmospheric inversion layer that keeps the weather from rising, usually giving over 330 nights each year of clear skies with steady images. As a result, 3 of the world's 12 largest telescopes are there (Fig. 2–12). One is a joint venture of Canada, France, and the University of Hawaii; another is a United Kingdom telescope; and the third was funded by NASA.

FIGURE 2–12 The peak of Mauna Kea on the island of Hawaii in the state of Hawaii boasts three of the world's ten largest telescopes and is the prospective site of still larger telescopes.

A   B

FIGURE 2–13 *(A)* The Keck Ten-Meter Telescope, now under construction on Mauna Kea in Hawaii. It will be made of a mosaic of mirrors. *(B)* A model of the proposed National New Technology Telescope, which is to contain four 7.5-m (300-inch) individual mirrors, all directing light to the same focus.

The Mauna Kea Observatory is such an excellent site that it has been chosen for the first of the "new technology telescopes" to be built. The California Institute of Technology and the University of California are building the Keck Ten-Meter Telescope, whose mirror is twice the diameter and four times the surface area of Palomar's reflector. Since a single 10-m mirror would be prohibitively expensive even if it could be made, U.C. scientists worked out a plan to use a mirror made of many smaller segments (Fig. 2–13A). The cost will be over $85 million. Other huge telescopes are under consideration, including a 15-m telescope for the U.S. National Optical Astronomy Observatories (Fig. 2–13B), and a Japanese 7-m, both probably to go on Mauna Kea and therefore operated in collaboration with the University of Hawaii. European and Soviet large new technology telescopes are also being developed.

One successful radical departure from standard practice is the Multiple Mirror Telescope (MMT) in Arizona. The MMT uses 6 smaller mirrors that all focus their light at the same point. It gathers as much light as a 4.5-m single mirror would (Fig. 2–14), but does it much more cheaply. (It is much less expensive to make and mount several small mirrors than one large one.) The proposed U.S. national 15-m telescope will be of a similar design, with four individual 7.5-m components, each larger than the Palomar reflector!

FIGURE 2–14 Six 1.8-m mirrors are held in a single framework to make the Multiple Mirror Telescope at the Whipple Observatory in Arizona.

2.2 WIDE-FIELD TELESCOPES

Ordinary optical telescopes see a fairly narrow field of view, that is, a small part of the sky. Even the most modern show images of less than about 1° × 1°, which means it would take decades to make images of the entire sky. The German optician Bernhard Schmidt, in the 1930's, invented a way of using a thin lens ground into a complicated shape and a spherical mirror to image a wide field of sky (Fig. 2–15). The largest Schmidt telescopes, except for

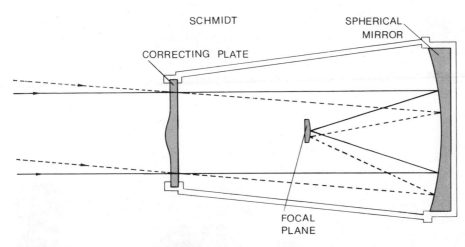

FIGURE 2–15 By having a nonspherical thin lens called a correcting plate, a Schmidt camera is able to focus a wide angle of sky onto a curved piece of film. Since the image falls at a location where you cannot put your eye, the image is always recorded on film. Accordingly, this device is often called a Schmidt camera rather than a Schmidt telescope.

one of interchangeable design, are the ones at the Palomar Observatory in California (Fig. 2–16) and the U.K. Schmidt in Australia. Both have front lenses 1.25 m (49 inches) in diameter and mirrors half again as large to allow study of objects off to the side. They can observe a field of view some 7° × 7° (Color Plates 48, 51, 55, 59, 60, 62, and 68).

The Palomar Schmidt telescope was used 35 years ago to survey the whole sky visible from southern California with film and filters that made pairs of images in red and blue light. This Palomar Observatory Sky Survey is a basic reference for astronomers. Hundreds of thousands of galaxies, quasars, nebulae, and other objects have been discovered on them. The Schmidt telescopes in Australia and Chile have since compiled the extension of this survey to the southern hemisphere. And Palomar is about to embark on a resurvey with improved films and more overlap between adjacent regions. Among other things, it will be compared with the first survey to see whether objects have changed or moved.

FIGURE 2–16 Edwin P. Hubble, whose observational work is at the basis of modern cosmology, at the guide telescope of the 1.2-meter (known as the 48-inch though the clear part of its lens is really 49 inches) Schmidt camera on Palomar Mountain.

2.3 AMATEUR TELESCOPES

It is fortunate for astronomy as a science that many people are interested in looking at the sky. Many are just casual observers, who may look through a telescope occasionally as part of a course or on an "open night," but others are quite devoted "amateur astronomers." Some amateur astronomers make their own equipment, ranging up to quite large telescopes perhaps 40 centimeters in diameter. But most amateur astronomers use one of several commercial brands of telescopes.

Most of the telescopes around are Newtonian reflectors, with mirrors 15 cm in diameter being the most popular size (Fig. 2–17). It is quite possible to shape your own mirror for such a telescope.

FIGURE 2–17 A home-built Newtonian telescope, the amateur's standard. The mirror, 15 cm (6 inches) in diameter, is at lower right. It reflects light up the tube, where it hits a diagonal mirror suspended in the center of the tube. This diagonal mirror reflects the light out to the eyepiece, which is hidden in this photograph. On the top of the telescope, we see a small "finder" telescope, whose field of view is wider than that of the main telescope to aid in locating objects.

FIGURE 2–18 *(A)* A cutaway drawing of a compound Schmidt-Cassegrain design, now widely used by amateurs because of its light-gathering power and portability. *(B)* A Schmidt-Cassegrain in use.

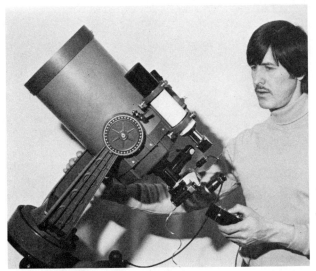

Increasingly popular in recent years have been compound telescopes that combine features of reflectors with some of Schmidt telescopes. A Schmidt-Cassegrain design (Fig. 2–18) folds the light, so that the telescope is relatively short, making it easier to transport and set up. Such telescopes can be used not only for visual observing but also for taking excellent photographs (Fig. 2–19).

FIGURE 2–19 The North America and Pelican Nebulae *(left)* and the Orion Nebula *(right)*. Both were photographed with 8-inch Schmidt cameras by undergraduates at Williams College.

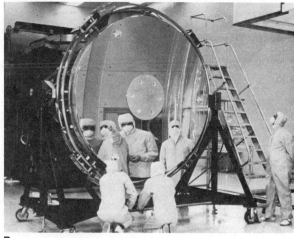

A   **B**

FIGURE 2–20 *(A)* The Hubble Space Telescope is shown in this artist's conception. It will operate from a position above the earth's atmosphere, so it will be able to probe about 7 times deeper into space than ground-based telescopes. The Space Telescope will give images with 0.1 arc sec resolution, a few times better than can be obtained from the ground. *(B)* The 2.4-m mirror of Space Telescope, after it was ground and polished and covered with a reflective coating. The workers seen reflected in the mirror are actually far off to the right; their images appear magnified. The mirror is so smooth that were it to be blown up to the size of the United States from Maine to California, the biggest bump would be only a foot high.

2.4 THE HUBBLE SPACE TELESCOPE

Though a larger mirror can, in principle, provide higher resolution—that is, allow more detail to be detected—in practice, the unsteadiness of the earth's atmosphere sets the limit. Thus the 4-m and 5-m telescopes gather more light but do not provide higher resolution than smaller telescopes. The first moderately large telescope to be launched above the earth's atmosphere will be the Hubble Space Telescope (Fig. 2–20), a NASA project now scheduled for 1988 after many delays. The HST should provide images 7 times clearer than images available from earth. Astronomers can hardly wait to see distant stars and galaxies with such high resolution. The set of instruments to be launched will be sensitive not only to visible light but also to ultraviolet radiation that doesn't pass through the earth's atmosphere.

Because of the high resolution possible, the Hubble Space Telescope will be able to concentrate the light of a star into an extremely small region. This, plus the very dark background sky at high altitude, should allow us to see fainter objects than we can now. The combination of resolution and sensitivity should lead to great advances toward solving several basic problems of astronomy. We should be able to pin down our whole notion of the size and age of the universe much more accurately, and may even be able to sight planets around other stars. But the HST will be only one 2.4-m telescope, and will probably open as many questions as it will answer. So we need still more ground-based telescopes as partners in the enterprise.

The HST will be launched by a space shuttle and will be left free in space. Astronauts can visit it every few years to fix anything that breaks; they can even bring down the whole unit every decade or so for refurbishment. A second generation of equipment may include detectors sensitive to the infrared, which is largely blocked by the earth's atmosphere. We expect the HST to last well into the 21st century.

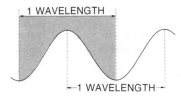

FIGURE 2–21 Waves of electric and magnetic field travelling across space are called "radiation." The wavelength is the length over which a wave repeats.

2.5 OUTSIDE THE VISIBLE SPECTRUM

We can describe a light wave by its wavelength (Fig. 2–21). But a large range of wavelengths is possible, and light makes up only a small part of this broader spectrum (Fig. 2–22). Gamma rays, x-rays, and ultraviolet have shorter wavelengths than light, and infrared and radio waves have longer wavelengths.

2.5a X-Ray and Gamma-Ray Telescopes

The shortest wavelengths would pass right through the glass or even the reflective coatings of ordinary telescopes, so special imaging devices have to be made to study them. And x-rays and gamma rays do not pass through the earth's atmosphere, so they can be observed only from satellites in space. NASA's series of three High-Energy Astronomy Observatories (HEAO's) were tremendously successful in the late 1970's (Fig. 2–23). HEAO–1 made an all-sky map of x-ray objects, and HEAO–2 (known as the Einstein Observatory) made detailed x-ray observations of individual objects (Color Plate 53) with resolution approaching that of ground-based telescopes working with ordinary light. HEAO–3 observed gamma rays and *cosmic rays*—particles of matter moving through space at tremendous speeds.

The HEAO's are not working any more, so European Space Agency, German, and Japanese x-ray satellites are now the major data gatherers. NASA is planning AXAF—Advanced X-Ray Astrophysics Facility—a project that has the highest rating from astronomers, but we will have to wait for the 1990's for it to be launched.

2.5b Ultraviolet Telescopes

We have mentioned that the Hubble Space Telescope will be sensitive to ultraviolet radiation. Ultraviolet wavelengths are longer than x-rays but still shorter than visible light. The International Ultraviolet Explorer spacecraft is currently sending back valuable ultraviolet observations; we shall discuss it in Section 20.5.

2.5c Infrared Telescopes

From high-altitude sites such as Mauna Kea, parts of the infrared can be observed from the earth's surface. Cool objects such as planets and dust around stars in formation emit most of their radiation in the infrared, so studies of planets and of how stars form have especially benefited from infrared studies.

FIGURE 2–22 The spectrum. The silhouettes represent telescopes or spacecraft used or planned for observing that part of the spectrum: the Gamma Ray Observatory, the Advanced X-Ray Astrophysics Facility, the International Ultraviolet Explorer, the dome of a ground-based telescope, the Infrared Astronomy Satellite, and a ground-based radio telescope.

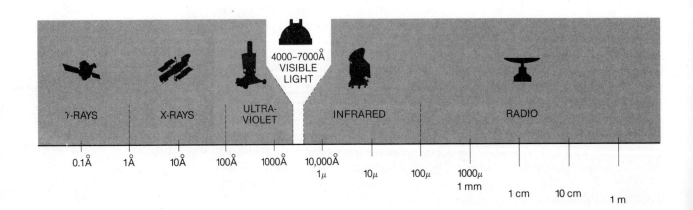

FIGURE 2–23 HEAO–2, known as the Einstein Observatory, was able to point at objects and make detailed images of them.

An international observatory, the Infrared Astronomical Satellite (IRAS), was aloft during 1983. IRAS mapped the whole sky, and discovered a half dozen comets, hundreds of asteroids, hundreds of thousands of galaxies, and many other objects. We will be studying the data for years to come (see the color essay on IRAS following page 22).

A further, more sensitive, infrared mission, the Space Infrared Telescope Facility, is on the drawing boards, and we hope for it in the early 1990's.

2.5d Radio Telescopes

Since the discovery in the 1930's that astronomical objects give off radio waves, radio astronomy has advanced greatly. Huge metal "dishes" are giant reflectors that concentrate radio waves onto antennas that enable us to detect faint signals from objects in outer space. The largest dish that can be steered to point anywhere in the sky is the 100-m radio telescope near Bonn, West Germany (Fig. 2–24), which can point anywhere in the sky. A still larger

FIGURE 2–24 The 100-m (330-foot) radio telescope near Bonn, West Germany, the largest fully steerable radio telescope in the world.

dish in Arecibo, Puerto Rico, is 1000 feet (330 m) across but only points more-or-less overhead. Still, all the planets and many other interesting objects pass through its field of view.

Radio telescopes were originally limited by their very poor resolution. The resolution of a telescope does depend on the size of the telescope's diameter, but we have to measure size relative to the wavelength of the radiation we are studying. So for a radio telescope studying waves 10 cm long, even a 100-m telescope is only 1000 wavelengths across. Even a 10-cm optical telescope studying ordinary light is 200,000 wavelengths across, so is effectively much larger and gives much finer images.

A breakthrough in providing higher resolution has been the development of arrays of radio telescopes that operate together and give the resolution of a single telescope spanning kilometers or even continents. The Very Large Array (VLA) is a set of 27 radio telescopes each 26 m in diameter (Color Plate 7). All the telescopes operate together, and powerful computers analyze the joint output to make detailed pictures of objects in space (Color Plate 52). Plans are even under way to build the Very Long Baseline Array (VLBA), which will span the whole United States and give images many times higher in resolution than even the VLA.

2.6 A NIGHT AT MAUNA KEA

An "observing run" with one of the world's largest telescopes highlights the work of many astronomers. The construction of three of the world's 12 largest telescopes at the top of Mauna Kea in Hawaii has made that mountain the site of one of the world's major observatories. Mauna Kea was chosen because of its outstanding observing conditions. Many of these conditions stem from the fact that the summit is so high—4200 m (13,800 ft) above sea level. Though there are taller mountains in the world, none has such favorable conditions for astronomy (Section 2.1).

One of the telescopes on Mauna Kea is the 3-m Infrared Telescope Facility (IRTF), sponsored by NASA and operated by the University of Hawaii (Fig. 2–25). Since it is a national facility, astronomers from all over the United States can propose observations.

During the months before your observing run, you make detailed lists of objects to observe, prepare charts of the objects' positions in the sky based on existing star maps and photographs, and plan the details of your observing procedure.

The time for your observing run comes, and you fly off to Hawaii. Your plane may be met by one of the Telescope Operators, who will be assisting you with your observations. The Telescope Operators know the telescope and its systems very well, and are responsible for the telescope, its operation, and its safety. You drive together up the mountain, as the scenery changes from tropical to relatively barren, crossing dark lava flows. First stop is Hale Pohaku, the mid-level facility at an altitude of 2750 m (9000 ft). Hale Pohaku is Hawaiian for "Stone House," in honor of an original building still on the site. The astronomers and technicians sleep and eat at Hale Pohaku.

The rules say that you must spend a full 24 hours acclimatizing to the high altitude before you can begin your own telescope run. So you overlap with the last night of the run of the people using the telescope before you. This familiarizes you with the telescope and its operation. A combination of time and altitude adjustment sends you to bed early this first night.

FIGURE 2–25 The Infrared Telescope Facility on Mauna Kea, a 3-m telescope that specializes in infrared observations.

IRAS —
The Infrared Astronomy Satellite

Beyond the rainbow of colors we can see, past the red, is the infrared. Most infrared is absorbed in the earth's atmosphere and does not reach us on earth. Further, our bodies, our telescopes, and even the sky radiate so strongly in the infrared part of the spectrum that infrared from celestial bodies is masked.

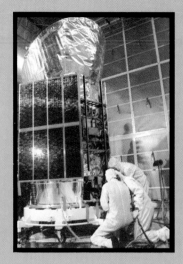

In 1983, the Infrared Astronomy Satellite (shown at right), IRAS, was launched as a joint American-Dutch-British venture. Located above the earth's atmosphere, and with its 60-cm telescope cooled to only 2 degrees above absolute zero, it could detect the faint infrared emitted by objects in space. Studying this part of the spectrum enables us to consider relatively cool bodies, those only a few tens or hundreds of degrees in temperature instead of the thousands of degrees typical of stars. The results are displayed here as false-color images, each color corresponding to a different temperature. (For the record, 12-micron radiation is blue, 25-micron radiation is yellow, 60-micron radiation is green, and 100-micron radiation is red.)

Below we see the Milky Way near the center of our own galaxy. The narrow band we see in the infrared view corresponds to dark absorbing dust we see when we look at the Milky Way in visible light. For this image, the warmest material (including warm dust in our solar system) is blue, while colder material is red. The middle-temperature yellow and green knots and blobs scattered along the band are giant clouds of interstellar gas and dust heated by nearby stars. Some are warmed by newly formed stars while others are heated by massive, hot stars thousands of times brighter than our sun.

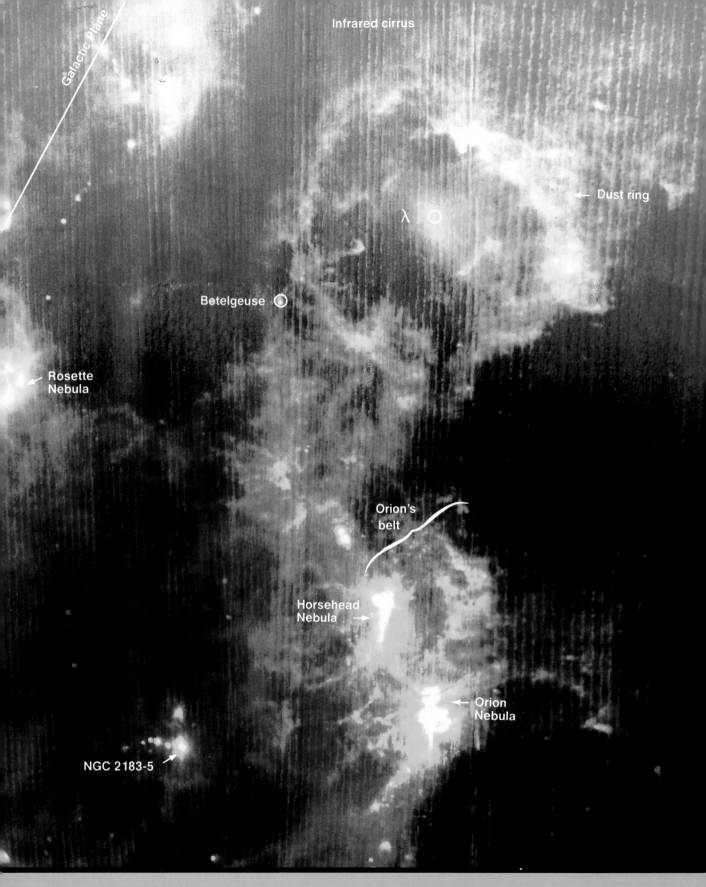

Galactic Plane

Infrared cirrus

Dust ring

λ

Betelgeuse

Rosette
Nebula

Orion's
belt

Horsehead
Nebula

Orion
Nebula

NGC 2183-5

The region surrounding the constellation Orion includes signs of different temperatures. The circular feature at top can be seen at visible wavelengths, but appears different in intensity and size here in the infrared. It corresponds to hot hydrogen gas and dust heated by the star Bellatrix. The brightest regions correspond to regions where stars are forming.

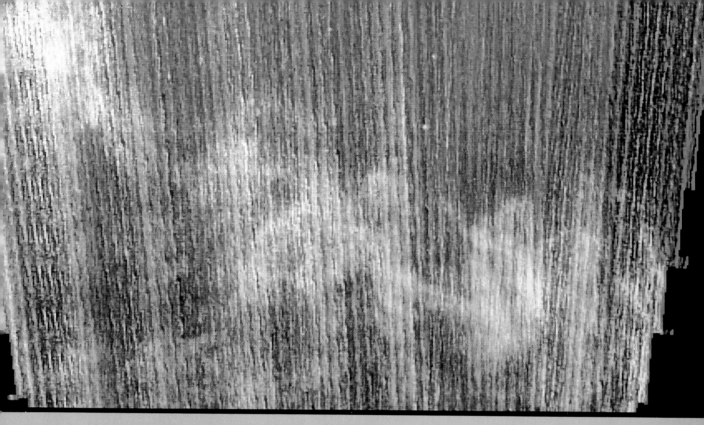

Looking past the stripes that are an artifact of the data recording in the picture above, we see wispy structures extending roughly horizontally across the picture, and located about 30° above the Milky Way as seen from earth. This "infrared cirrus" was a surprise.

The four images of the center of our galaxy, shown below, use colors to show the intensity of radiation at a given wavelength. The image at left (12-micron radiation) records mainly stars. Most dust in our galaxy is too cool to be seen here, though we can see the dust that is near the galactic center or near a hot star as at top left. The next image (25-micron) also shows the dust only in hot, dense regions. In the next image (60-micron), we begin to see the interstellar dust well, and observe streamers of dust around the center of our galaxy. The final image (100-micron) clearly shows great clouds of interstellar dust.

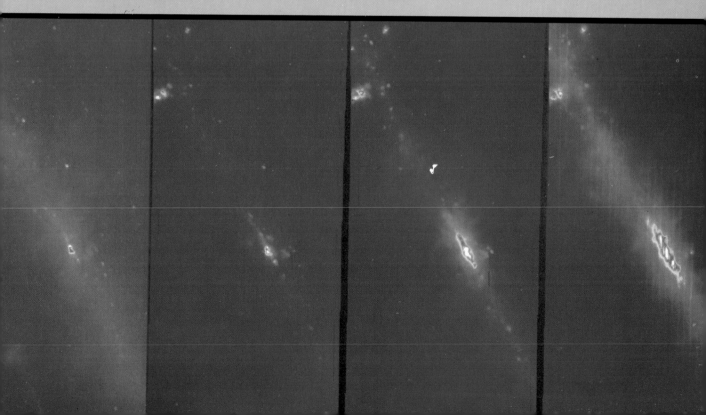

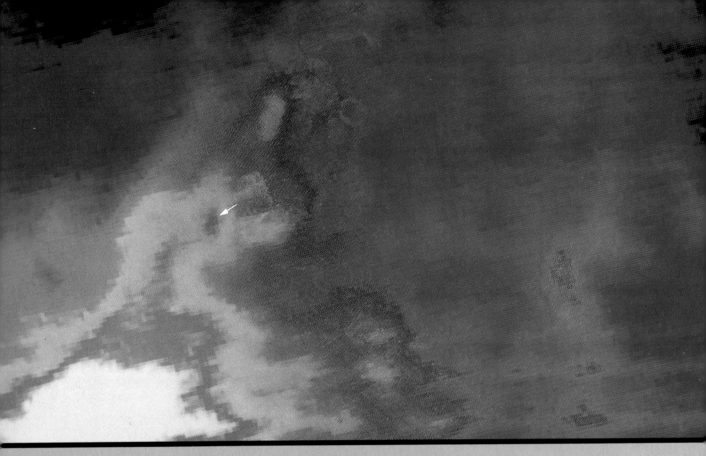

Above we see a star being born (red patch at end of arrow); it is embedded in a cloud of gas and dust known as Barnard 5, which is located only 1000 light years from us. The "protostar" is probably less than 100,000 years old, and has not reached the stage where it is shining by its own nuclear energy as do the sun and the other ordinary stars. The colors signify the intensity range in the image, taken at a wavelength of 100 microns. The white region at the bottom shows intense emission from relatively hot dust (at about our room temperature) in a region heated by a hot young star.

Below left we see Comet IRAS-Araki-Alcock, one of several comets discovered from IRAS. We see the "coma" of the comet, which is located in our solar system. The colors show intensities of 20-micron radiation from warm dust that has boiled off the solid central part of the comet and is then pushed by the pressure of light from the sun. Comets seem to be dustier than had been thought.

Below right we see an IRAS image of the Andromeda Galaxy, a spiral galaxy much like the one in which we live. Brighter areas show regions where star formation is especially active, including the galaxy's core and a ring many thousands of light years in diameter that corresponds to the inner part of the optical disk. When the liquid helium cooling IRAS's telescope ran out after 10 months (actually lasting longer than had been planned), the telescope lost its sensitivity and astronomers lost their best infrared observatory in space. The data will take years to analyze. We discuss the results at several places in this book.

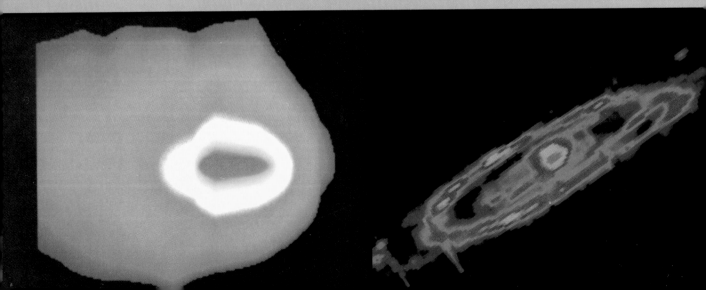

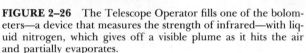

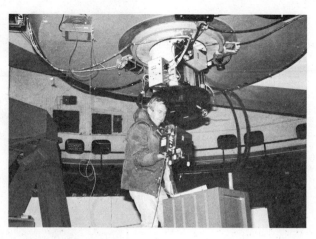

FIGURE 2–26 The Telescope Operator fills one of the bolometers—a device that measures the strength of infrared—with liquid nitrogen, which gives off a visible plume as it hits the air and partially evaporates.

FIGURE 2–27 Installing a bolometer at the Cassegrain focus. The back plate of the telescope is at the top; the 3-m mirror is on its other side.

The next night is yours. In the afternoon, you and the Telescope Operator may make a special trip up the mountain to install the systems you will be using to record data at the Cassegrain focus of the telescope (Figs. 2–26 and 2–27). Since objects at normal outdoor or room temperatures give off enough radiation in the infrared to disrupt your observations, much of the instrumentation you install is cooled. The detectors used to observe wavelengths of 10 or 20 micrometers are bolometers, devices sensitive to small temperature changes that occur as a result of incoming radiation. Liquid helium cools the bolometers to a temperature of only 2 K ($-271°C$). Surrounding the liquid helium is liquid nitrogen, at a temperature of 77 K ($-196°C$).

After dinner, you return to the summit to start your observing. You dress warmly, since the nighttime temperature approaches freezing at this altitude, even in the summertime. Since the sky is fairly dark at infrared wavelengths, you can start observing even before sunset. The telescope is operated by computer, and points at the coordinates you type in for the object you want to observe. A video screen (Fig. 2–28) displays an image of the object.

Everything is computer-controlled. You measure the brightness of the object at the wavelength you are observing by starting a computer program on a console provided for the observer.

Let's say that your main interest is a dark cloud of dust and gas in the sky, the subject of my own observing run there. You may be looking to see if there are young stars inside, hidden to visible observations but detectable in the infrared because they make the dust glow. You have to search a small area of sky to see what you can find. You have to decide on the size of the area to be searched, the length of time to spend at each point, the wavelength to use, and many other factors. There are so many decisions to make that the astronomer must be on site; leaving observing to the staff observers themselves usually doesn't work.

At some time during the night, you may go into the telescope dome to ponder the telescope at work. The telescope all but blocks your view of the

FIGURE 2–28 A Telescope Operator at her console. The television screen shows that the telescope is pointing at Saturn.

sky. Hardly anyone ever actually looks through the telescope; indeed, there is no good way to do so. Still, you may sense a deep feeling for how the telescope is looking out into space (Fig. 2–29).

You drive down the mountain in the early morning sun. Over the next few days and nights, you repeat your new routine. When you leave Mauna Kea, it is a shock to leave the pristine air above the clouds. But you take home with you data about the objects you have observed. It may take you months to study the data you gathered in a few brief days at the top of the world. But you hope that your data will allow a fuller understanding of some aspect of the universe.

FIGURE 2–29 A time exposure by moonlight showing the IRTF in the foreground and the Canada–France–Hawaii Telescope in the background.

KEY WORDS

focus, Newtonian telescope, paraboloid, Cassegrain, refracting telescopes, reflecting telescopes, resolution, cosmic rays

QUESTIONS

1. What are three discoveries that immediately followed the first use of the telescope for astronomy?

2. What advantage does a reflecting telescope have over a refracting telescope?

3. What limits a large telescope's ability to see detail?

4. List the important criteria in choosing a site for an optical observatory meant to study stars and galaxies.

5. Describe a method that allows us to make large optical telescopes more cheaply than simply scaling up designs of previous large telescopes.

6. What are the advantages of the Multiple Mirror Telescope? How does it compare with the 5-m telescope for studying faint objects?

7. What are the similarities and differences between making radio observations and using a reflector for optical observations? Compare the path of the radiation, the detection of signals, and limiting factors.

8. Why is it sometimes better to use a small telescope in orbit around the earth than it is to use a large telescope on a mountain top?

9. Why is it better for some purposes to use a medium-size telescope on a mountain instead of a telescope in space?

10. What are two reasons why the Space Telescope will be able to observe fainter objects than we can now study from the ground?

Chapter 3

Star trails over the Anglo-Australian telescope, including the south celestial pole.

OBSERVING THE STARS AND BEYOND: CLOCKWORK OF THE UNIVERSE

The sun, the moon, and the stars rise every day in the eastern half of the sky and set in the western half. If you leave your camera on a tripod with the lens open for a few minutes or hours, you will photograph the *star trails,* the trails across the sky left by the individual stars. In this chapter, we will discuss the motions of the stars and objects beyond them in the sky. In the next chapter, we will discuss the motions of objects that move with respect to the stars.

3.1 TWINKLING

If you look up at night in a place far from the city, you may see a few thousand stars with the naked eye. They will seem to change in brightness from moment to moment, that is, to *twinkle*. This twinkling comes from moving regions of air in the earth's atmosphere. The air bends starlight, just as a glass lens bends light. As the air moves, the starlight is bent by different amounts and the strength of the radiation hitting your eye varies, making the stars seem to twinkle.

Unlike stars, planets are close enough to us that they appear as tiny disks when viewed with telescopes, though we can't quite see them with the naked eye. As the air moves around, even though the planets' images move slightly, there are enough points on the image to make the average amount of light we receive keep relatively steady. So planets, on the whole, don't twinkle. But when a planet is low enough on the horizon, as Venus often is when we see it, it too can twinkle. Since the bending of light by air is different for different colors, we sometimes even see Venus or a bright star turning alternately reddish and greenish. Professional astronomers sometimes get calls that UFO's have been sighted on those occasions.

3.2 RISING AND SETTING STARS

In actuality, the earth is turning on its axis and the stars are holding steady. But we see the effect as the stars rising and setting. Extensions of the earth's axis point to places in the sky—the *celestial poles;* since the earth's axis doesn't move much, the celestial poles don't appear to move in the course of the night.

From our latitudes (the U.S. ranges from about 20° north latitude for Hawaii to about 49° north latitude for the northern continental U.S. to 65° for Alaska), we can see the north but not the south celestial pole. A star named Polaris happens to be near the north celestial pole, only about 1° away, so we call Polaris the *pole star.* If you are navigating at sea or in a forest at night, you can always go due north by heading straight toward Polaris.

FIGURE 3–1 The Little Dipper is part of the constellation Ursa Minor, the Little Bear. We see here the drawing from the star atlas of Johann Bayer, published in 1603.

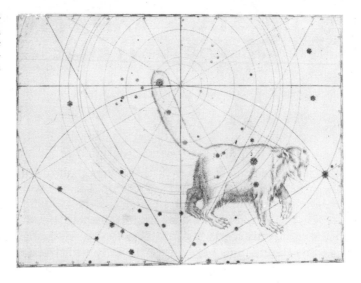

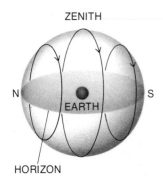

ZENITH

N EARTH S

HORIZON

FIGURE 3–2 If we were at the equator, the stars would rise straight up and set straight down.

Polaris is conveniently located at the end of the handle of the Little Dipper (Fig. 3–1), and can easily be found by following the "Pointers" at the end of the bowl of the Big Dipper (see Sky Maps). Polaris isn't especially bright, but you can find it if city lights have not brightened the sky too much.

The north celestial pole is the one simple point in our sky, since it never moves. To understand the motion of the stars, let us first consider two simple cases. If we were at the earth's equator, then the two celestial poles would be on the horizon, and stars would rise in the eastern half of the sky, go straight up and across the sky, and set in the western half (Fig. 3–2). Only a

FIGURE 3–4 Star trails over the Lick Observatory in California. We are looking at stars so far from the celestial poles that the circles of the star trails are very large, and we see only a segment that appears relatively straight.

FIGURE 3–5 Star trails around Polaris over the Kitt Peak National Observatory. Polaris is now almost 1° from the pole, so isn't quite fixed, though it makes a good approximation to a pole star.

ZENITH

EARTH

HORIZON

FIGURE 3–3 If we were at one of the earth's poles, the stars would move around the sky in circles parallel to the horizon. They would never rise or set.

FIGURE 3–6 Star trails around the south celestial pole. The Large Magellanic Cloud (LMC) and the Small Magellanic Cloud (SMC), two small companion galaxies to our own galaxy, appear in the photo though we do not see them from U.S. latitudes.

star that rose due east of us would pass directly overhead. If, on the other hand, we were at the earth's north pole, then the north celestial pole would always be directly overhead. No stars would rise and set, but they would all move in circles around the sky, parallel to the horizon (Fig. 3–3).

We live in an intermediate case, where the stars rise at an angle (Fig. 3–4). Close to the celestial pole, we can see that the stars are really circling the pole (Figs. 3–5 and 3–6). Only the pole star itself remains relatively fixed in place, although it too traces out a small circle around the north celestial pole.

As the earth spins, it wobbles slightly, like a giant top, because of the gravitational pulls of the sun and the moon. As a result of this *precession*, the axis traces out a large circle in the sky with a period of 26,000 years (Fig. 3–7). So Polaris is the pole star only for the present. The coordinates—"right ascension" (a type of longitude) and "declination" (a type of latitude)—of objects in the sky change slightly over the years. Every 50 years or so, astronomers change the "epoch" for which coordinates of objects are given. We are now in the process of switching from positions in epoch 1950.0 to epoch

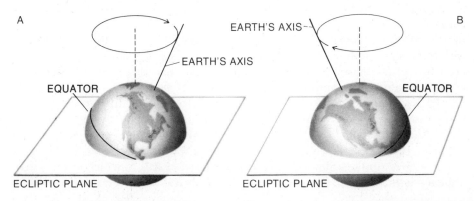

FIGURE 3–7 The wobbling of the earth's orbit leads to precession, which we see as a drifting of celestial coordinates. The period of the precession—the length of time it takes the axis to return to the same orientation—is about 26,000 years.

BOX 3.1
Photographing the Stars

It is easy to photograph the stars. If you place an ordinary 35-mm camera on a tripod, and have a dark sky far from city lights, set the lens wide open (perhaps f/1.4 or f/2). Use a cable release to open the shutter for 10 minutes or more; don't take exposures of over an hour without first testing shorter times to make sure that background skylight doesn't fog the film. You will have a picture of star trails. If the north star is centered from right to left in your field of view (it need not be centered from top to bottom), you will see the circles that the stars take as a result of the earth spinning on its axis.

With the new fast color films, including some with ASA/ISO 1000, you can record the stars in a constellation with exposures of a few seconds. Use a 50-mm or 135-mm lens and try a series: 1, 2, 4, 8, and 16 seconds. The stars will not noticeably trail on the shorter exposures. The constellation Orion, visible during the winter, is a particularly interesting constellation to photograph, since you will detect the reddish Orion Nebula in addition to the stars. Your eyes are not sensitive to such faint colors, but film will record them.

Hint: take a picture of a normal scene at the beginning of the roll, so that the photofinisher will know where to cut apart the slides or how to make the prints. Be prepared to send back negatives for printing, in spite of the photofinisher's note that they didn't come out. The photofinisher probably didn't notice the tiny specks the stars made.

2000.0 (that is, for the beginning of the year 2000). If you start observing with a telescope, right ascension and declination will quickly become second nature (Section 3.4).

3.3 APPARENT MAGNITUDE

To describe the brightness of stars in the sky, astronomers—professionals and amateurs alike—use a scale that stems from the ancient Greeks. Over two millennia ago, Hipparchus described the brightest stars in the sky as "of the first magnitude," the next brightest as "of the second magnitude," and so on. The faintest stars were "of the sixth magnitude."

We still use a similar scale, though now it is on a mathematical basis. Each difference of 5 magnitudes is a factor of 100 times in brightness. A 1st-magnitude star is exactly 100 times brighter than a 6th-magnitude star—that is, we receive exactly 100 times more energy in the form of light. Sixth magnitude is still the faintest that we can see with the naked eye, though because of urban sprawl the dark skies necessary to see such faint stars are harder to find these days than they were long ago. Objects too faint to see with the naked eye have magnitudes greater than sixth. Some stars and planets are brighter than 1st magnitude, so the scale has also been extended in the opposite direction into negative numbers.

Each difference of 1 magnitude is a factor of the fifth root of 100 (which is approximately equal to 2.512 in brightness). This is necessary to make 5 magnitudes (1 + 1 + 1 + 1 + 1, an additive process) equal to a factor of 100 (2.5 × 2.5 × 2.5 × 2.5 × 2.5, a multiplicative process).

The magnitude scale—*apparent magnitude,* since it is how bright the stars appear—is fixed by comparison with the historical scale (Fig. 3–8). If you

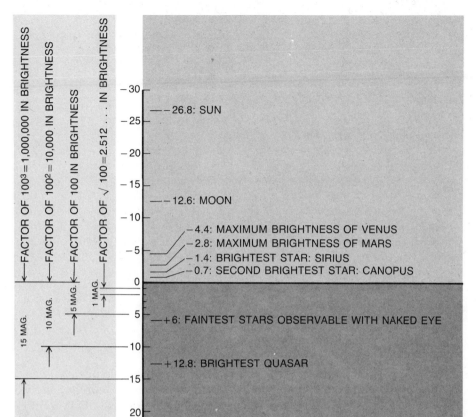

FIGURE 3-8 The scale of apparent magnitude.

read about a 13th-magnitude quasar, you should know that it is much too faint to see with the naked eye. If you read about a 20th-magnitude object, you should know that it is one of the faintest objects we can study.

3.4 CELESTIAL COORDINATES

Geographers divide the surface of the earth into a grid, so that we can describe locations. The equator is the line half-way between the poles. Lines of constant *longitude* run from pole to pole, crossing the equator perpendicularly. Lines of constant *latitude* circle the earth, parallel to the equator.

Astronomers have a similar coordinate system in the sky. (Technical definitions appear in the glossary.) They assume, for this purpose, that the celestial objects are on an imaginary sphere, the *celestial sphere,* that surrounds the earth. The *celestial equator* circles the sky on the celestial sphere, halfway between the celestial poles. It lies right above the earth's equator. Lines of constant *right ascension* run between the celestial poles, crossing the celestial equator perpendicularly. They are similar to terrestrial longitude. Lines of constant *declination* circle the celestial sphere, parallel to the celestial equator, similarly to terrestrial latitude (Fig. 3-9). The right ascension and declination of a star are essentially unchanging, just as each city on earth has a fixed longitude and latitude. (Precession actually causes the celestial coordinates to change very slowly.)

FIGURE 3–9 The celestial sphere, showing right ascension and declination.

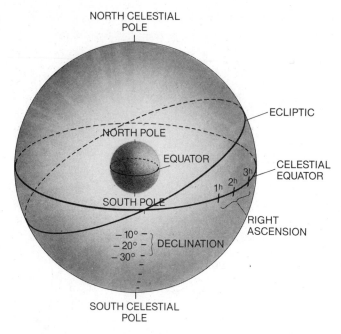

FIGURE 3–9 The celestial sphere, showing right ascension and declination.

Since a whole circle is divided into 360°, and the sky appears to turn completely once every 24 hours, the sky appears to turn 15 degrees per hour (15°/hr). Astronomers therefore measure right ascension in hours, minutes, and seconds instead of degrees. One hour of r.a. = 15°. Dividing by 60, 1 minute of r.a. = ¼° = 15 minutes of arc (since there are 60 minutes of arc in a degree). Dividing by 60 again, 1 second of r.a. = ¼ minute of arc = 15 arc seconds (since there are 60 seconds of arc in a minute of arc).

We can keep star time by noticing the right ascension of a star that is passing due south of us; we say that such a star is "crossing the meridian," where the *meridian* is a line extending due north-south and passing through the *zenith,* the point directly over our heads. This star time is called *sidereal time;* astronomers keep special clocks that run on sidereal time to show them which celestial objects are overhead. A sidereal day—a day by the stars—is about 4 minutes shorter than a solar day—a day by the sun, such as the one we usually keep track of on our watches.

3.5 USING THE STAR MAPS

The International Astronomical Union put the scheme of constellations on a definite system in 1930. The sky was officially divided into 88 constellations (see Appendix 9) with definite boundaries, and every star is now associated with one and only one constellation. But the constellations give only the directions to the stars, and not the stars' distances. Individual stars in a given constellation can be quite different distances from us.

Since the sidereal day is 4 minutes shorter than a solar day, the parts of the sky that are "up" after dark change slightly each day. By the time a season has gone by, the sky has apparently slipped a quarter of the way around at sunset as the earth has moved a quarter of the way around the sun in its yearly orbit. Some constellations are lost in the afternoon and evening glare, while others have become visible just before dawn.

Because of this seasonal difference, we have included four star maps, one of which is best for the date and time at which you are observing. Suitable combinations of date and time are marked. Hold the map while you are facing north or south, as marked on each map, and notice where your zenith is in the sky and on the map. The horizon for your latitude is also marked. Try to identify a pattern in the brightest stars that you can see. Finding the Big Dipper, and using it to locate the pole star, often helps orient yourself.

Don't let any bright planets confuse your search for the bright stars—planets usually appear to shine steadily instead of twinkling like stars.

3.5a The Autumn Sky

As it grows dark on an autumn evening, the pointers in the Big Dipper will point upward toward Polaris. Almost an equal distance on the other side of Polaris is a W-shaped constellation named Cassiopeia. Cassiopeia, in Greek mythology, was married to Cepheus, the king of Ethiopia (and the subject of the constellation that neighbors Cassiopeia to the west). Cassiopeia appears sitting on a chair, as shown in the opening illustration of Chapter 15.

Continuing across the sky away from the pointers, we next come to the constellation Andromeda, who in Greek mythology was Cassiopeia's daughter. In Andromeda, you might see a faint, hazy patch of light; this is actually the center of the nearest large galaxy to our own, and is known as the Great Galaxy in Andromeda (Color Plate 68). Though it is one of the nearest galaxies, it is much farther away than any of the individual stars that we see.

Southwest in the sky from Andromeda, but still high overhead, are four stars that appear to make a square known as the Great Square of Pegasus (Fig. 3–10). One of its corners is actually in Andromeda.

If it is really dark out (which probably means that you are far from a city and also that the moon is not full or almost full), you will see the Milky Way crossing the sky high overhead. It will appear as a hazy band across the sky, with ragged edges and dark patches and rifts in it. The Milky Way passes right through Cassiopeia.

Moving southeast from Cassiopeia, along the Milky Way, we come to the constellation Perseus; he was the Greek hero who slew the Medusa. (He flew off on Pegasus, the winged horse, who is conveniently nearby in the sky, and saw Andromeda, whom he saved.) On the edge of Perseus nearest to Cassiopeia, with a small telescope or binoculars we can see two hazy patches of stars that are really clusters of hundreds of stars called "open clusters." This "double cluster in Perseus," also known as h and χ (chi) Persei, provides two of the open clusters that are easiest to see with small

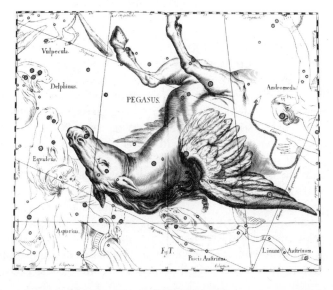

FIGURE 3–10 Pegasus, from the atlas of Hevelius, published in 1690.

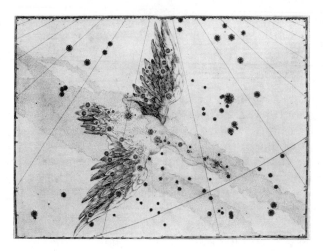

telescopes. In 1603, Johann Bayer assigned Greek letters to the brightest stars and lower-case Latin letters to less bright stars. Although h and χ Persei are clusters, they were named using Bayer's system for labelling stars. One cluster was confused with the star chi Persei—"chi of Perseus"—when the cluster was named, and the other was confused with the star *h* Persei.

In the other direction from Cassiopeia (whose W is relatively easy to find), we come to a cross of bright stars directly overhead. This "Northern Cross" is in the constellation Cygnus, the Swan. In this direction, spacecraft have detected wildly varying x-rays, and we think a black hole (Chapter 23) is located there. Also in Cygnus is a particularly dark region of the Milky Way, called the Northern Coalsack. Dust in space in that direction prevents us from seeing as many stars as we do in other directions in the Milky Way. Slightly to the west is another bright star, Vega, in the constellation Lyra (the Lyre). And farther westward, we come to the constellation Hercules, named for the mythological Greek hero who performed twelve great labors, of which the most famous was bringing back the golden apples. In Hercules is an older, larger type of star cluster called a "globular cluster." It is known as M13, the globular cluster in Hercules (see the Figure opening Chapter 17). It is visible as a fuzzy mothball in even small telescopes; larger telescopes have better resolution and so show the individual stars.

3.5b The Winter Sky

As the autumn proceeds and the winter approaches, the constellations we have discussed appear closer and closer to the western horizon for the same hour of the night. By early evening on January 1st, Cygnus is setting in the western sky, while Cassiopeia and Perseus are overhead.

To the south of the Milky Way, near Perseus, we can now see a group of six stars close together in the sky. The grouping can catch your attention as you scan the sky. It is the Pleiades (pronounced "plee'a-deez"), traditionally the Seven Sisters of Greek mythology, the daughters of Atlas. These stars are another example of an open cluster. Binoculars or a small telescope will reveal dozens of stars there; a large telescope will ordinarily show too small a region of sky for you to see the Pleiades well at all. So a bigger telescope isn't always better.

Further toward the east, rising earlier every evening, is the constellation Orion, the Hunter. Orion is perhaps the easiest constellation of all to pick

FIGURE 3–12 Orion, the Hunter, form Hevelius's atlas.

out in the sky, for three bright stars close together in a line make up its belt. Orion is warding off Taurus, the Bull, whose head is marked by a large V of stars. A reddish star, Betelgeuse (beetle-juice would not be far wrong for pronunciation, though some say "beh′tel-jooz"), marks Orion's shoulder, and symmetrically on the other side of his belt, the bright bluish star Rigel (rī′jel) marks his heel. Betelgeuse is an example of a red supergiant star; it is hundreds of millions of kilometers across, bigger itself than the earth's orbit around the sun. Orion's sword extends down from his belt. A telescope, or a photograph, reveals a beautiful reddish region known as the Great Nebula in Orion, or the Orion Nebula. Its shape can be seen in even a smallish telescope; however, only photographs reveal its color (Color Plate 56). It is a site where new stars are forming.

Rising after Orion is Sirius, the brightest star in the sky. Orion's belt points directly to it. Sirius appears blue-white, which indicates that it is very hot. Sirius is so much brighter than the other stars that it stands out to the naked eye. It is part of the constellation Canis Major, the Great Dog. (You can remember that it is near Orion by thinking of it as Orion's dog.)

Back toward the top of the sky, between the Pleiades and Orion's belt, is a group of stars that forms the V-shaped head of Taurus. This open cluster is known as the Hyades (hy′a-deez). The stars of the Hyades mark the bull's face, while the stars of the Pleiades ride on the bull's shoulder. In a Greek myth, Jupiter turned himself into a bull to carry Europa over the sea to what is now called Europe.

3.5c The Spring Sky

We can tell that spring is approaching when the Hyades and Orion get closer and closer to the western horizon each night, and finally are no longer visible when the sun sets. Now the twins (Castor and Pollux), a pair of stars, are nicely placed for viewing in the western sky at sunset. Pollux is slightly reddish, while Castor is not, and they are about the same brightness. Castor and Pollux were the twins in the Greek pantheon of gods. The constellation is called Gemini, the twins.

FIGURE 3–13   Leo, the Lion, from
Bayer's atlas.

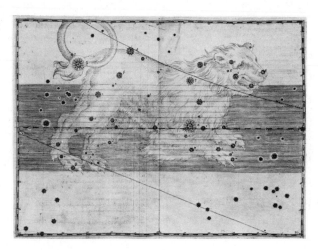

*Ursa Major was originally the
princess Callisto, an attendant
of the goddess Juno, who became jealous of her. To protect
Callisto, Jupiter turned her
into a bear. However, when
Callisto's son was about to kill
the bear one day, Jupiter
turned him into another bear
and placed both of them in the
sky.*

On Spring evenings, the Big Bear is overhead, and anything in the Big
Dipper—which is part of the Big Bear—would spill out. Leo, the Lion, is
just to the south of the zenith (follow the pointers backwards). Leo looks like
a backwards question mark, with the bright star Regulus at its base. Regulus
marks the lion's heart. The rest of Leo, to the east of Regulus, is marked by
a bright triangle of stars. Some people visualize a sickle-shaped head and a
triangular tail.

If we follow the arc made by the stars in the handle of the Big Dipper,
we come to a bright reddish star, Arcturus, a red giant. It is in the kite-
shaped constellation Boötes, the Herdsman.

Sirius sets right after sunset in the springtime; however, another bright
star, Spica, is rising in the southeast in the constellation Virgo, the Virgin. It
is farther along the arc of the Big Dipper through Arcturus. Vega, a star
that is almost as bright, is rising in the northeast. And the constellation Hercules, with its globular cluster, is rising in the east in the evening at this time
of year.

3.5d The Summer Sky

Summer, of course, is a comfortable time to watch the stars because of the
warm weather. Spica is over toward the southwest in the evening. A bright
reddish star, Antares, is in the constellation Scorpius, the Scorpion, to the
south. ("Antares" means "compared with Ares," another name for Mars, because Antares is also reddish.)

Hercules and Cygnus are high overhead, and the star Vega is prominent near the zenith. Cassiopeia is in the northeast. The center of our galaxy
is in the dense part of the Milky Way that we see in the constellation Sagittarius, the Archer, in the south.

Around August 12 every summer is a wonderful time to observe the
sky, because that is when the Perseid meteor shower occurs. One bright meteor a minute may be visible at the peak of the shower. Just lie back and
watch the sky in general—don't look in any specific direction. (An outdoor
concert is a good place to do this, if you can find a bit of grass away from
spotlights.) Although the Perseids is the most observed meteor shower,
partly because it occurs at a time of warm weather in the northern part of
the country, many other meteor showers occur during the year. The most
prominent are listed in Table 14–2.

The summer is a good time of year for observing a variable star, Delta Cephei; it appears in the constellation Cepheus, which is midway between Cassiopeia and Cygnus. Delta Cephei varies in brightness with a 5.4-day period (see Fig. 17–8). As we see in Chapter 18, studies of its variations have helped us to figure out the distances to galaxies. And this fact reminds us of the real importance of studying the sky—which is to learn **what** things are and **how** they work, and not just **where** things are. The study of the sky has led us to understand the universe, and this is the real importance and excitement of astronomy.

KEY WORDS

star trails, twinkle, celestial poles, pole star, precession, apparent magnitude, longitude, latitude, celestial sphere, celestial equator, right ascension, declination, meridian, zenith, sidereal time

QUESTIONS

1. On the picture opening the chapter, measure the angle covered by the star trails and deduce how long the exposure lasted.

2. If you look toward the horizon, are the stars you see likely to be twinkling more or less than the stars overhead? Explain.

3. Is the planet Uranus, which is in the outskirts of the solar system, likely to twinkle more or less than the nearby planet Venus?

4. Explain how it is that some stars never rise in our sky, while others never set.

5. Using the Appendices and the star maps, comment on whether Polaris is one of the twenty brightest stars in the sky.

6. Compare a 6th-magnitude star and an 11th-magnitude star in brightness. Which is brighter, and by how many times?

7. (a) Compare a 16th-magnitude quasar with an 11th-magnitude star in brightness, similarly to Question 6. (b) Now compare the 16th-magnitude quasar with the 6th magnitude star, specifying which is brighter and by how many times.

8. Since the sky revolves once a day, how many degrees does it appear to revolve in 1 hour?

9. One year contains about 365 solar days or about 366 sidereal days. Divide 24 hours by 365 to find out by how many minutes a sidereal day is shorter than a solar day.

10. Use the suitable star map to list the bright stars that are near the zenith during the summer.

PUZZLE

Every row and column contains one of the official three-letter abbreviations for a constellation (Appendix 9), and no constellation appears twice in the solution. Fill in the missing letters.

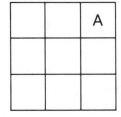
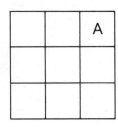
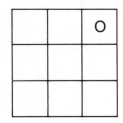

Chapter 4

The solar corona surrounds the dark disk of the moon during the 1984 total eclipse, observed from Papua New Guinea.

OBSERVING THE SUN, MOON, AND PLANETS: WANDERERS IN THE SKY

Though the stars appear to turn above the earth at a steady rate, the sun, the moon, and the planets drift among the stars. The planets were long ago noticed to be "wanderers" among the stars, ignoring the daily apparent motion of the entire sky overhead. In this chapter, we will discuss how to observe these fascinating objects.

4.1 THE PATH OF THE SUN

The path that the sun follows through the stars in the sky is known as the *ecliptic*. We can't notice this path readily because the sun is so bright that we don't see the stars when it is up and the earth's rotation causes another more rapid daily motion, but the ecliptic is marked with a dotted line on the star maps.

The earth and the other planets revolve around the sun in more-or-less a flat plane. So from earth, the paths across the stars of the other planets are all close to the ecliptic. But the earth's axis is not perpendicular to the ecliptic. It is, rather, tipped from perpendicular by 23½°.

The ecliptic is therefore tipped with respect to the celestial equator (Fig. 4–1). The two points of intersection are known as the *vernal equinox* and the *autumnal equinox*. The sun is at those points at the beginning of our northern hemisphere spring and autumn, respectively. On those days, its decli-

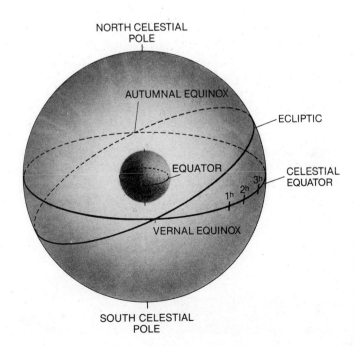

FIGURE 4–1 The ecliptic meets the celestial equator at two points, known as the equinoxes. We draw them on "the celestial sphere," a view from outside of the sphere on which the stars appear to be fixed when we look up from the earth.

FIGURE 4–2 The analemma, displayed in this unique photograph that is a multiple exposure of the sun on a single piece of film, photographed through a dense filter at ten-day intervals throughout a full year. Clouds have caused a few of the solar images to be relatively faint or missing. All the exposures were taken at 8:30 a.m. Eastern Standard Time. On three days of the year—close to the winter solstice, the summer solstice, and one of the two points at which the figure-8 crossover occurs—the shutter was left open from dawn until shortly before the exposure time for the solar image; this made the streaks. On one day in the fall, the dense filter was taken off for a brief exposure to show the foreground trees and building.

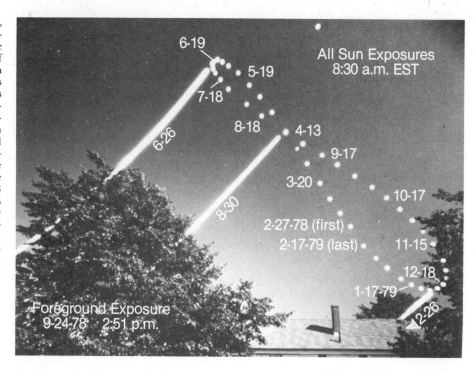

nation is zero. Three months after the vernal equinox, the sun is on the part of the ecliptic that is farthest north of the celestial equator. The sun's declination is then $+23\frac{1}{2}°$ and we say it is at the summer *solstice*. The summer is hot because the sun is above our horizon for a longer time and because it reaches a higher angle above the horizon when it is at high declinations.

The word "equinox" means "equal night," meaning that in theory the length of day and night is equal on those two occasions each year. But the equinoxes actually mark the dates at which the center of the sun crosses the celestial equator. Since the sun has a real size, and the top obviously rises before the middle, the daytime is actually a little longer than the nighttime on the day of the equinox. Also, bending of sunlight by the earth's atmosphere allows us to see the sun when it is really a little below our horizon, also lengthening the daytime. The days of equal day and night are displaced by a few days from the equinoxes.

Because of its apparent motion with respect to the stars, the sun goes through the complete range of right ascension and between $+23\frac{1}{2}°$ and

FIGURE 4–3 A montage of photographs taken hourly above the Arctic Circle for one entire day. Since the photographer was not at the north pole, the sun does not remain at a constant height, though it is always above the horizon.

$-23\frac{1}{2}°$ in declination each year. As a result, its height above the horizon varies from day to day. If we were to take a photograph of the sun at the same hour each day, over the year the sun would sometimes be relatively low and sometimes relatively high.

Also, if the earth's orbit around the sun were a circle, and if the earth's axis of rotation were perpendicular to the circle, then we would expect the sun to be at the same position across the sky (as opposed to higher or lower) each day. But the earth's orbit is not round, and the earth travels in its orbit at different speeds over the year. So the sun's speed across the sky varies over the year.

Even though our clocks are based on time from the sun's motion, we don't change the rate at which our clocks work from day to day. Rather, we tell time by a "mean sun" (using the definition of "mean" as "average"). But the real sun is not the mean sun; the real sun is almost always a bit ahead of or a bit behind the mean sun in the sky.

These two effects—the change in height in the sky and the change in horizontal position across the sky—cause the position of the sun at a given hour each day to trace out a figure 8 in the sky (Fig. 4–2). This figure 8 is called the *analemma*. The photograph is a multiple exposure (on a single piece of film!) made with a large-format camera set in a window for an entire year, bolted to the window frame so that it would not move. A very dense filter was placed over the lens so that the sun would show through as a white dot, but everything else would be filtered out.

The resulting photograph shows very plainly the changes in position of the sun in the sky over a full year. Because the sun rises much higher in the sky in the summertime than in the wintertime, summer days are longer and hotter.

If we were at or close to the north pole, we would be able to see the sun whenever it was at a declination sufficiently above the celestial equator. The phenomenon is known as the *midnight sun* (Fig. 4–3 and Color Plate 8).

4.2 MOTIONS OF THE MOON AND PLANETS

The moon goes through the full range of right ascension and its full range of declination each month as it circles the earth. The motion of the planets is much more difficult to categorize. Mercury and Venus circle the sun more quickly than does earth, and have smaller orbits than the earth. They wiggle in the sky around the sun, and are never seen in the middle of the night. They are the "evening stars" or the "morning stars," depending on which

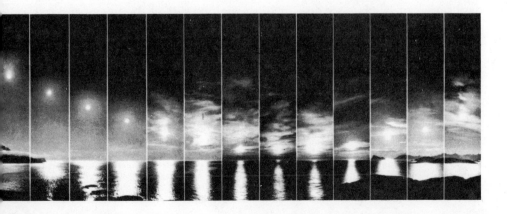

| 4 days
waxing crescent | 7 days
1st quarter | 10 days
waxing gibbous | 14 days
full |

FIGURE 4–4   The phases of the moon.

side of the sun they are on. Mars, Jupiter, and Saturn (as well as the other planets too faint to see readily) circle the sun more slowly than does earth. So they drift across the sky with respect to the stars, and don't change their right ascension or declination rapidly.

The U.S. and U.K. governments jointly put out a volume of tables each year known as *The Astronomical Almanac.* The book includes tables of the positions of the sun, moon, and planets. My *Field Guide to the Stars and Planets* includes graphs and less detailed tables.

4.3 THE PHASES OF THE MOON AND PLANETS

From the simple observation that the apparent shapes of the moon and planets change, we can draw conclusions that are important for our understanding of the mechanics of the solar system. The fact that the moon goes through a set of *phases* approximately once every month is perhaps the most familiar everyday astronomical observation (Fig. 4–4). In fact, the name "month" comes from the word "moon." The actual period of the phases, the interval between a particular phase of the moon and its next repetition, is approximately 29½ earth days (Fig. 4–5).

The explanation of the phases is quite simple: the moon is a sphere, and at all times the side that faces the sun is lighted and the side that faces away from the sun is dark. The phase of the moon that we see from the earth, as the moon revolves around us, depends on the relative orientation of the three bodies: sun, moon, and earth.

Basically, when the moon is almost exactly between the earth and the sun, the dark side of the moon faces us. We call this a "new moon." A few days earlier or later we see a sliver of the lighted side of the moon, and call this a "crescent." As the month wears on, the crescent gets bigger; we say the moon is "waxing." About 7 days after new moon, half the face of the moon that is visible to us is lighted; we sometimes call this a "half moon." Since this occurs one fourth of the way through the phases, the situation is also called a "first-quarter moon." (Instead of apologizing for the fact that

*The **phases** of moons or planets are the shapes of the sunlit areas as seen from a given vantage point.*

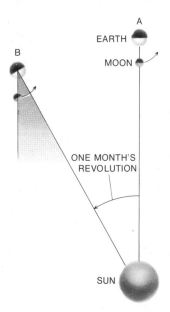

FIGURE 4–5   After the moon has completed one revolution around the earth with respect to the stars, which it does in 27⅓ days, it has moved from A to B. The moon has still not swung far enough around to again be in the same position with respect to the sun, because the earth has revolved one month's worth around the sun. It takes an extra two days for the moon to complete its revolution with respect to the sun, making its period about 29½ days. The extra angle it must cover is shaded. In the extra two days, both the earth and the moon have continued to move around the sun (toward the left in the diagram).

20 days
waning gibbous

22 days
third quarter

24 days
waning crescent

26 days
waning crescent

astronomers call the same phase both "quarter" and "half," I'll just continue with a straight face and try to pretend that there is nothing strange about it.)

When over half the moon's disk is lighted, we have a "gibbous" moon. One week after the first-quarter moon, the moon is on the opposite side of the earth from the sun, and the entire face visible to us is lighted. This is called a "full moon." Then the amount of the moon we see lighted decreases; we say the moon is "waning." One week after full moon, when we see a half moon again, we have "third-quarter moon." Then we go back to "new moon" again and repeat the cycle of phases.

Since a full moon occurs when the moon is on the opposite side of the earth from the sun, so that we see the entire lighted side, the full moon rises as the sun sets. The moon rises about one hour later each day than it did the previous day. When the moon has waned to third quarter, it rises at midnight and is high in the sky at sunrise. The new moon or thin crescents rise within a few hours of the sunrise. The first-quarter moon rises at noon and is high in the sky at sunset.

The moon is not the only object in the solar system that is seen to go through phases. Mercury and Venus both orbit inside the earth's orbit, and so sometimes we see the side that faces away from the sun and sometimes we see the side that faces toward the sun. Thus at times Mercury and Venus are seen as crescents, though it takes a telescope to observe their shapes. Spacecraft to the outer planets have looked back and seen the earth as a crescent (Fig. 4–6).

4.4 ECLIPSES

Because the moon's orbit around the earth and the earth's orbit around the sun are not precisely in the same plane (Fig. 4–7), the moon usually passes slightly above or below the earth's shadow at full moon, and the earth usually passes slightly above or below the moon's shadow at new moon. But every once in a while, up to seven times a year, the moon is at the part of its orbit that crosses the earth's orbital plane at full moon or new moon. When that happens, we have a lunar or solar eclipse. Most of the eclipses are only partial: the moon is only partly in the earth's shadow or the sun is only partly blocked by the moon. The total eclipses, which are rarer, are much more interesting to watch.

At an eclipse of the sun, the moon comes directly between the earth and the sun. You have to be in a direct line with the moon and sun to see the

FIGURE 4–6 The crescent earth seen from Pioneer Venus spacecraft; even though Venus is inside the earth's orbit, the spacecraft's trajectory took it briefly outside the earth's orbit, from which it could see the earth as a crescent.

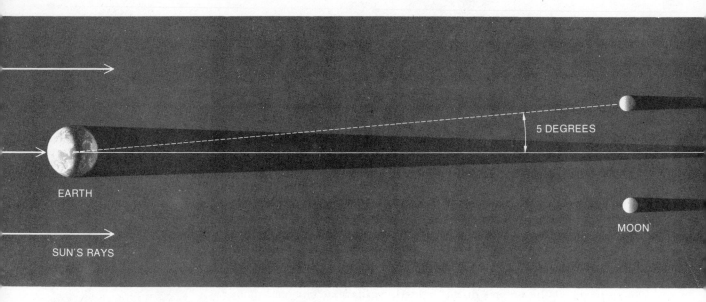

FIGURE 4–7 The moon's orbit is tilted about 5° from the earth's, so the moon usually passes above or below the earth's shadow each month.

sun totally hidden, and so must be in a narrow band only a few hundred kilometers wide. Many more people see a lunar eclipse than the total coverage at a solar eclipse. At a total lunar eclipse, the moon lies entirely in the earth's shadow and direct sunlight is entirely cut off from it (Fig. 4–8). So anywhere on the earth that the moon has risen, the eclipse is visible.

4.4a Lunar Eclipses

Totality of a lunar eclipse—when the moon is entirely within the earth's shadow—is a much more leisurely event to watch than totality of a solar eclipse. Totality of a lunar eclipse can last over an hour. During this time, the sunlight is not entirely shut off from the moon (Fig. 4–9). A small amount is refracted around the edge of the earth by our atmosphere. Most of the blue light is taken out during the sunlight's passage through our atmosphere; this explains how blue skies are made for the people part way around the globe from the point at which the sun is overhead. The remaining light is reddish, and this is the light that falls on the moon. Thus, the eclipsed moon appears reddish.

The next two total lunar eclipses visible from the United States and Canada will be on April 24, 1986 (visible from western regions only), and August 17, 1989.

4.4b Solar Eclipses

Since the outer layers of the sun are visible to the eye only at the time of a total solar eclipse, eclipses have played a major role in astronomy.

FIGURE 4–8 When the moon is between the earth and the sun, we observe an eclipse of the sun. When the moon is on the far side of the earth from the sun, we see a lunar eclipse. The distances in this diagram are not to scale.

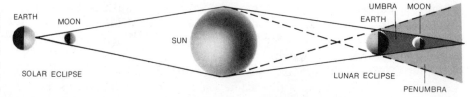

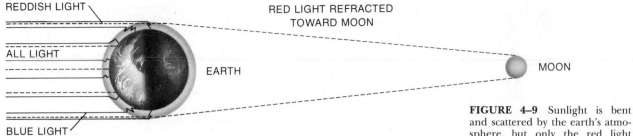

REDDISH LIGHT

RED LIGHT REFRACTED
TOWARD MOON

ALL LIGHT

EARTH

MOON

BLUE LIGHT
SCATTERED MORE
BY ATMOSPHERE

FIGURE 4–9 Sunlight is bent and scattered by the earth's atmosphere, but only the red light makes it through to illuminate the moon during a total lunar eclipse.

Solar eclipses arise because of a happy circumstance: though the moon is 400 times smaller in diameter than the solar photosphere—the bright surface of the sun that is normally visible—it is also 400 times closer to the earth. Because of this, the sun and the moon cover almost exactly the same angle in the sky—about ½° (Fig. 4–10). (Your thumb at the end of your outstretched arm covers about 2°.)

The moon's position in the sky, at certain points in its orbit around the earth, comes close to the position of the sun. This happens approximately once a month at the time of the new moon. Since the lunar orbit is inclined with respect to the earth's orbit, the moon usually passes above or below the line joining the earth and the sun. But occasionally the moon passes close enough to the earth-sun line that the moon's shadow falls upon the surface of the earth.

At a total solar eclipse, the lunar shadow barely reaches the earth's surface (Fig. 4–11). As the moon moves through space on its orbit, and as the earth rotates, this lunar shadow sweeps across the earth's surface in a band up to 300 km wide. Only observers stationed within this narrow band can see the total eclipse, which may last only seconds or may last as much as 7 minutes.

As the total phase of the eclipse—totality—begins, the bright light of the solar photosphere passing through valleys on the edge of the moon glistens like a series of bright beads, which are called *Baily's beads*. The last bead seems so bright that it looks like the diamond on a ring—the *diamond-ring effect* (Color Plate 11). About then, the *corona*—the faint outermost layer of the sun that is normally hidden by the blue sky—comes into view (Color Plate 12). For thousands of kilometers to each side of the band of totality, one sees only a partial eclipse.

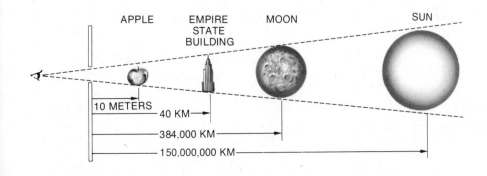

APPLE

EMPIRE
STATE
BUILDING

MOON

SUN

10 METERS
40 KM
384,000 KM
150,000,000 KM

FIGURE 4–10 The apparent angular size of objects depends not only on how large they are but also on how far away they are.

FIGURE 4–11   The circumstances of a solar eclipse, to scale. The dark central part of the shadow, from which no part of the sun can be seen, is the "umbra." A less dark part of the shadow, from which part of the sun can be seen, is the "penumbra."

FIGURE 4–12 Total eclipses through 2017, plus the American annular eclipses of 1984 and 1994. Annular eclipses are marked with dotted lines. The partial phases are visible from a much wider region than the total and annular phases visible in the paths drawn here.

The last total eclipse to cross the continental U.S. and Canada occurred in 1979. The next total eclipse in the U.S. will cross one of Alaska's Aleutian Islands in 1990, and a long total eclipse will be visible from Hawaii in 1991. The next total eclipse in the continental United States won't be until 2017. Canada won't see another total eclipse until 2024.

On the average, a total eclipse occurs somewhere in the world every year and a half. The band of totality, though, usually does not cross populated areas of the earth, and astronomers often have to travel great distances to carry out their observations. Table 4–1 and Fig. 4–12 show sites and dates of future eclipses. Astronomers took advantage of the relatively long 1980 and 1983 eclipses by sending large-scale expeditions to India and Indonesia, respectively (Fig. 4–13). The next eclipse for which favorable circumstances are expected will take place in 1988, with observing expected from the Indonesian island of Sumatra and from the Philippines.

Sometimes the moon subtends a slightly smaller angle in the sky than the sun, because the moon is on the part of its orbit that is relatively far

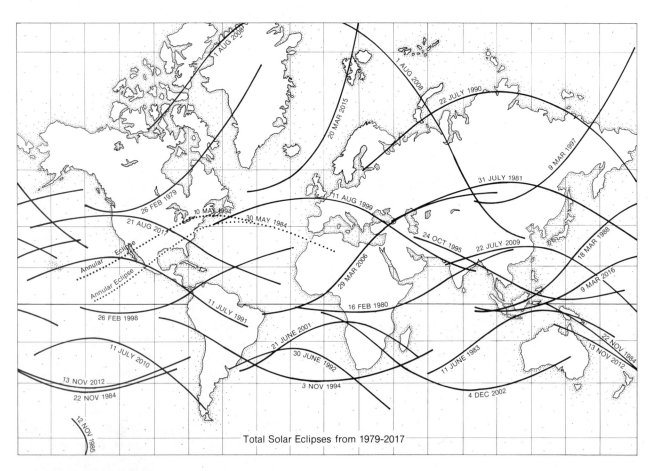

TABLE 4–1   Total Solar Eclipses

DATE	MAXIMUM DURATION	LOCATION
1988 March 18	3^m46^s	Indonesia, Philippines
1990 July 22	2^m33^s	Finland, U.S.S.R., U.S.A. (Alaska: Aleutians)
1991 July 11	6^m54^s	U.S.A. (Hawaii), Mexico, Central and South America
1992 June 30	5^m20^s	Southern Atlantic Ocean
1994 November 3	4^m24^s	South America
1995 October 24	2^m10^s	Southern Asia, Borneo

from the earth. When a well-aligned eclipse occurs in such a circumstance, the moon doesn't quite cover the sun. An annulus—a ring—of photospheric light remains visible, so we call this an *annular eclipse* (Fig. 4–14). Such an annular eclipse crossed the southeastern United States on May 30, 1984. The rest of the United States saw a partial eclipse. Another annular eclipse will cross the United States in 1994. Annular eclipses are usually much less spectacular than total eclipses.

In these days of orbiting satellites, is it worth travelling to observe solar eclipses? There is much to be said for the benefits of eclipse observing. Eclipse observations are a relatively inexpensive way, compared to space research, of observing the faint outer layers of the sun, which are not readily visible from earth. Even realizing that some eclipse experiments may not work out because of the pressure of time or because of bad weather, eclipse observations are still very cost-effective. And for some kinds of observations, those at the highest resolutions, space techniques have not yet matched ground-based eclipse capabilities. Space solar telescopes, even though they can study the outer part of the corona day to day, are not able to observe the inner and middle corona that we observe at eclipses. Coronal studies have been made much more cheaply during an eclipse than the cost of a space experiment would have been, even if it had been mounted.

FIGURE 4–13 The solar corona during the total eclipse of 1984, observed from Papua New Guinea.

4.4c A Solar Eclipse Expedition

A total solar eclipse is like an astronomical Olympics. Eclipse astronomers often travel great distances to study total eclipses when they occur, just as the world's athletes gather at long intervals. Only at a solar eclipse are the corona and other solar phenomena outside the solar limb well visible. Some eclipses are more favorable than others, for the sun may be higher in the sky, totality longer, and the weather forecasts better. The U.S. National Science Foundation sponsored national expeditions to the 1980 eclipse in India and the 1983 eclipse in Indonesia.

The 11 June 1983 eclipse path crossed the island of Java. The eclipse was long—5 minutes—and weather forecasts were excellent. Scientists from a half-dozen U.S. colleges, universities, and observatories were members of the official team. People had worked for over a year to arrange shipping and customs clearance; our equipment had been shipped out months before. My group and some others each had two tons of equipment. Our job on site for the two weeks before the eclipse was to set up, test, and align the equip-

FIGURE 4–14 Baily's beads and a bright crescent during the annular eclipse of 1984.

FIGURE 4–15 My group's equipment at the U.S. National Science Foundation site in Java, where we observed the 1983 total solar eclipse.

FIGURE 4–16 Totality viewed from our Java site. The solar corona is the bright circle at top center.

ment. Adjusting telescopes and mirrors so that they accurately track the sun is not usually done on such a short time scale. Carpenters and masons built sturdy bases for our equipment, and bamboo huts to shield it and us from the sun (Fig. 4–15).

Bit by bit, scientists checked equipment out, and tried to fix what was broken. My group actually had to have someone fly halfway round the world from Boston at the last minute with a replacement digital tape recorder. By eclipse day, all was working.

But the day before, the weather had stopped cooperating. Two weeks of sun had turned into pelting rain. When eclipse morning dawned, the rain had stopped but the clouds were heavy. We couldn't tell if we would see the eclipse or not, but the optimists among us never gave up hope.

Indeed, the clouds parted, and it was pretty clear by the time the partial phases started. The time of the eclipse can now be predicted accurately, to within seconds. As the solar crescent became smaller, the light turned eerie and shadows sharpened. Then the Baily's beads appeared on the edge of the sun. As the last one gleamed, someone called out, "Diamond ring, lens caps off." We stood in the middle of totality, the moon's shadow around us and a reddish glow on the horizon in every direction.

High in the sky was the glorious corona, looking ghostly white (Fig. 4–16). Every now and then during the five minutes of totality, we could look up from our instruments to enjoy the view. But all too soon it was over; the second diamond ring came and the sky grew rapidly brighter. We capped our equipment, and carried out calibration tests.

All experiments seemed to run well, and the data are still being studied at our various home institutions. Reports are appearing in scientific journals and at meetings of the American Astronomical Society. In these ways, the results of the eclipse are made available to the astronomical community at large.

A brief total solar eclipse was visible for 54 seconds from Papua New Guinea in 1984. The skies were clear, allowing the Baily's beads, the diamond ring, and the corona to be seen. Data about the duration of totality and about the appearances and disappearances of the Baily's beads may improve our understanding of the sizes of the sun and the moon (Fig. 4–17).

FIGURE 4–17 The diamond ring effect at the 1984 total solar eclipse, viewed from our site in Papua New Guinea.

4.5 ASTRONOMY AND ASTROLOGY

Astrology is not at all connected with astronomy, except in a historical context, so does not really deserve a place in a text on contemporary astronomy. But since so many people associate astrology with astronomy, and since astrologers claim to be using astronomical objects to make their predictions, let us use our astronomical knowledge to assess astrology's validity. Since millions of Americans believe in astrology—a number that shows no signs of decreasing—the topic is too widespread to ignore.

Astrology is an attempt to predict or explain our actions and personalities on the basis of the positions of the stars and planets now and at the instants of our births. Astrology has been around for a long time, but it has never been shown to work. Believers may cite incidents that reinforce their faith in astrology, but no successful scientific tests have ever been carried out. If something happens to you that you had expected because of an astrological prediction, you would more certainly notice that this event occurred than you would notice the thousands of other unpredicted things that happened to you that day. Yet we do enough things, have sufficiently varied thoughts, and interact with enough people that if we make many predictions in the morning, some of them are likely to be at least partially fulfilled during the day. We simply forget that the rest ever existed.

In fact, even the alignments that most astrologers use are not accurately calculated, for the precession of the earth's pole has changed the stars that are overhead at a given time of year from what they were millennia ago when astrological tables that are often still in current use were computed. At a given time of year, the sun is usually in a different sign of the zodiac from its traditional astrological one (Figs. 4–18 and 4–19). And we know that the constellations are illusions; they don't even exist as physical objects. They are merely projections of the positions of stars that may be at very different distances from us.

Studies have shown that superstition actively constricts the progress of science and technology in various countries around the world and is therefore not merely an innocent force. It is not just that some people harmlessly believe in astrology. Their lack of understanding of scientific structure may actually impede the training of people needed to solve the problems of our age, such as pollution, shortages of food, and the energy crisis. Published articles have reported that widespread superstitious beliefs have even impeded smallpox-prevention programs. Thus many scientists are not content to ignore astrology, but actively oppose its dissemination. Further, if large numbers of citizens do not understand the scientific method and the difference between science and pseudoscience, how can they intelligently vote on or respond to scientific questions that have societal implications?

A major reason why scientists in general and astronomers in particular don't believe in astrology is that they cannot conceive of a way in which it would work. The human brain is so complex that it seems most improbable that any celestial alignment can affect people, including newborns, in an overall way. The celestial forces that are known are not sufficient to set personalities nor influence day-to-day events.

Even if people do not think astrology is true but merely find it interesting, many scientists feel that so many strange and exciting explainable things are going on in the universe that we wonder why anybody should waste time with far-fetched astrological concerns that have negative consequences. After all, we will be discussing such fascinating things as quasars, pulsars, and black holes. We will consider complex molecules that have spontaneously

Most professional astronomers share sentiments something like those expressed in this section, or at least reach the same conclusion, while others may just ignore astrology, considering it unproved. Many astronomers don't even think that astrology is worth discussing, or at best, that discussions have no effect on those who "believe" and so are a waste of time. I will not take this position here.

Astrology is based on the positions of the sun, moon, and planets relative to divisions of sky near the ecliptic called "signs" and "houses," and to one another. This arrangement is called a "horoscope," and astrologers spend much time examining them. Astrologers claim to be interpreting these horoscopes to make predictions for individuals or situations. Such predictions are often vague enough so that any outcome of events will not be contradicted. Knowledge of their clients allows astrologers to mix common sense in with their statements, often resulting in reasonable sounding advice.

FIGURE 4–18 Twelve constellations through which the sun, moon, and planets pass make up the zodiac. (Drawing by Handelsman; © 1978 The New Yorker Magazine, Inc.)

FIGURE 4–19 A drawing of the signs of the zodiac made in 1496.

formed in interstellar space, and try to decide whether the universe will expand forever. We have sent a rocket into interstellar space bearing a portrait of humans, and have beamed a radio message toward a group of stars 24,000 light years away. These topics and actions are part of modern astronomy: what contemporary, often conservative, scientists are doing and thinking about the universe. How prosaic and fruitless it thus seems to spend time pondering celestial alignments and wondering whether they can affect individuals.

Moreover, astrology doesn't work. Bernie I. Silverman, a psychologist then at Michigan State University, tested specific values: Do Libras and Aquarians rank "Equality" highly? Do Sagittarians especially value "Honesty"? Do Virgos, Geminis, and Capricorns treasure the value "Intellectual"? Several astrology books agreed that these and other similar examples are values typical of those signs. Although believers often criticize the objections of skeptics on the ground that these group horoscopes are not as valuable or accurate as individualized charts, surely some general assumptions and rules hold in common.

The subjects, 1600 psychology graduate students, did not know in advance what was being tested. They gave their birthdate, and the questioners determined their astrological signs. The results: no special correlation with the values they were supposed to hold was apparent for any of the signs. Also, when asked to what extent they shared the qualities of each given sign, as many subjects ranked themselves above average as below, regardless of their astrological signs.

But if astrology is so meaningless, why does it still have so many adherents? Well, it could be the bandwagon effect, and Silverman had an ingenious test that endorsed this idea. He took twelve personality descriptions from astrology books, one for each astrological sign, and displayed them to two groups of individuals. The first group was told to which astrological signs the descriptions pertained, and was asked to write their own signs on the covers of the questionnaires. More than half the members of this group thought that the descriptions listed under their own signs were, for each individual, among the four best descriptions of themselves out of the twelve choices.

Yet, when the second group was given the twelve descriptions without mention of astrology, being told that they came from a book entitled *Twelve Ways of Life,* their choices were random. Only 30 per cent chose their own sign's description as being in the group most closely describing them. So the idea that astrology can predict personality types seems to be the result of self-delusion. When people know what they are expected to be like, they tend to identify themselves with the description. But that doesn't mean that they actually satisfy the description that astrology predicts for them better than any other description.

A team of psychologists from California State University at Long Beach arranged for a magician to perform three psychic-like stunts in front of psychology classes. Even when they emphasized to the students that the performer was a magician performing tricks, 50 per cent of the class still believed the magician to be psychic.

From an astronomer's view, astrology is meaningless, unnecessary, and impossible to explain if we accept the broad set of physical laws we have conceived over the years to explain what happens on the earth and in the sky. Astrology snipes at the roots of all pure science. Moreover, astrology patently doesn't work. If people want to believe in it as a religion, or have a personal astrologer act as a psychologist, let them not try to cloak their beliefs with a scientific astronomical gloss. The only reason people may believe that they have seen astrology work is that it is a self-fulfilling means of prophecy, conceived of long ago when we knew less about the exciting things that are going on in the universe. Let's all learn from the stars, but let's learn the truth.

Science is more than just a set of facts, since a methodology of investigation and standards of proof are involved, but science is more than just a methodology since many facts have been well established. In this course, you are supposed not only to learn certain facts about the universe but also to appreciate the way that theories and facts come to be accepted. (See also Section 14.6.) If, by the end of this course, you still believe in astrology, then you have not understood some of the basic lessons.

KEY WORDS

ecliptic, vernal equinox, autumnal equinox, solstice, analemma, midnight sun, phases, Baily's beads, diamond-ring effect, corona, annular eclipse

QUESTIONS

1. Describe why Christmas comes in the summer in Australia.

2. Are the lengths of day and night equal at the vernal equinox? Explain.

3. Explain why Figure 4–2 shows a figure 8.

4. Does the sun pass due south of you at noon on your watch each day? Explain.

5. Explain why the sun changes its altitude in the sky over the year for an observer at the north pole, while the stars do not.

6. What is the difference between a new moon and an eclipse?

7. How can the moon block the sun to make a solar eclipse, given that the moon is so much smaller than the sun?

8. Use the map of eclipses to see how far you will be from the band in which an annular eclipse will be visible in 1994. The rest of the United States will see a partial eclipse.

PART II The Solar System

The orbits of the planets, with the exception of Pluto's, are ellipses that are not too squashed and are close to being in the same plane. The orbits are shown for relative scale against the silhouette of the United States.

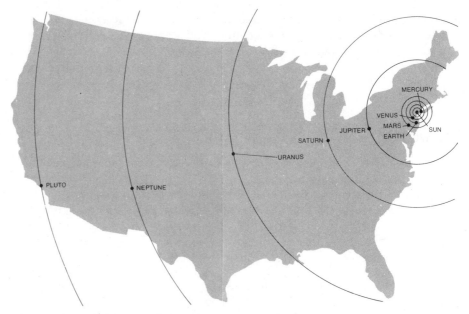

The Earth and the rest of the solar system may be important to us, but they are only minor companions to the stars. In "Captain Stormfield's Visit to Heaven," by Mark Twain, the Captain races with a comet and gets off course. He comes into heaven by a wrong gate, and finds that nobody there has heard of "the world" ("**the** world, there's billions of them!" says a gate-keeper). Finally, someone goes up in a balloon to try to detect "the world" on a huge map. The balloonist rises into clouds and after a day or two of searching comes back to report that he has found it: an unimportant planet named "the Wart."

Nowadays we realize that the Earth is but a minor player on the stage of the Universe. But for thousands of years, people thought that the Earth was the center and that everything else revolved around it. This view of the Universe, especially as formulated by the Greek philosophers Aristotle (350 B.C.) and Ptolemy (A.D. 140), dominated thinking through the Middle Ages.

One main problem was to describe why the planets, as they drifted across the sky from night to night with respect to the stars, sometimes moved more rapidly than the stars and sometimes more slowly. This backwards motion of a planet with respect to the stars was known as *retrograde motion*. Aristotle and Ptolemy explained it as the effect of the planet's motion around a small circle—an *epicycle*—whose center moved on a larger circle. The predictions of their model did not match the sky perfectly, but this did not trouble people for a long time.

In the 1500's, the Polish cleric Nicolaus Copernicus worked out a model in which the Sun is at the center of the solar system. He published his ideas in 1543 in a book from which we mark the origin of modern astronomy (Color Plate 12). In the Copernican theory, the planets orbit the Sun at different speeds. As the Earth passes a planet, the planet appears to be moving backward for a while, just as you might think a neighboring car on a highway is moving backward as you pass it.

The Copernican theory, with the Sun at the center, gained powerful support when the Italian scientist Galileo first turned a telescope on the sky in 1609 and 1610. Galileo discovered that the Moon had craters on it and that the Sun had spots on it. This differed from the ideas of Aristotle and Ptolemy that the celestial bodies were perfect. Galileo also discovered that Jupiter had moons revolving around it, indicating that the Earth wasn't the only body with something revolving around it, also in disagreement with Aristotle and Ptolemy. And Galileo discovered that Venus went through a full set of phases and changed in size as it did so. He could explain this with the Copernican theory, since Venus would sometimes be on the near side of the Sun and sometimes on the far side, but the Earth-centered theory had no explanation.

At about the same time, Johannes Kepler discovered three laws that governed the motion of the planets and that allowed accurate predictions to be made about their positions in the sky. His laws were based on the Copernican theory. The first law is that the planets travel in ellipses, with the Sun at one of the pair of crucial points inside each ellipse. The positions of these two points—the "foci"—along with a size factor determine the ellipse's shape. Kepler's second law governs how fast the planets travel in their orbits, going faster when they are closer to the Sun. Kepler's third law links the length of time the planet takes to complete an orbit with the size of the orbit.

Kepler's laws were based merely on observations, but by 1687, the English scientist Isaac Newton had worked them out on the basis of theory. Newton's ideas of motion and of gravity provide the basis for modern physics and astronomy. Newton derived Kepler's third law in a way that showed how to determine how much mass a body has by studying the smaller bodies that revolve around it. Thus studying the orbit of our Moon tells us the mass of the Earth, and studying the orbits of Jupiter's moons tells us Jupiter's mass. Studying the orbit of any of the planets tells us the mass of the Sun. We still use Kepler's and Newton's laws to navigate spacecraft around the solar system and to study far-off planets and stars.

In this Part, we will see how astronomers since Copernicus and Galileo have found out about the bodies in our solar system. We will see what we have learned with observations from telescopes on Earth and from equipment aboard spacecraft. And we will learn about future plans for space observations.

Chapter 5

Earth photographed from a satellite in synchronous orbit. A grid of latitude and longitude and outlines of states and continents have been added to make it easier to follow weather patterns. Computer calculations using the data gathered by satellites and equations similar to those that govern stars are making it possible to improve weather predictions. The symbol for the Earth appears at the top of the facing page.

Color Plate 1 (top, left): The 4-m Mayall optical telescope on Kitt Peak, near Tucson, Arizona. (Kitt Peak National Observatory photo)

Color Plate 2 (top, right): The soviet 6-m telescope, the largest in the world. Note the figures standing on top of the dome. (Courtesy of J.M. Kopylov, Special Astrophysical Observatory)

Color Plate 3 (bottom, left): The Greenbelt, Maryland, control room of the International Ultraviolet Explorer Spacecraft. The image displayed on the screen shows a stellar spectrum in the ultraviolet. (Photo by the author)

Color Plate 4 (bottom, right): The 67-m radio telescope at the Australian National Radio Observatory at Parkes, N.S.W. (Photo by the author)

Plate 5 (bottom): Fraunhofer's original spectrum from 1814. Only the D and H lines retain their notation from that time. The C line is Hα

Color Plate 6 (top): HEAO-1 (High-Energy Astronomy Observatory 1), launched in 1977 to study x-rays and gamma rays. (NASA and TRW Systems Group photo)

Color Plate 7 (bottom): The VLA (Very Large Array), the aperture synthesis radio telescope near Socorro, New Mexico. It is composed of 27 dishes, each 26 m in diameter, arranged in the shape of a "Y" over a flat area 27 km in diameter. The third arm of the "Y" extends to the right. (National Radio Astronomy Observatory photo)

Color Plate 8 (top): Multiple exposures at 15-minute intervals taken in Alaska. The upper section shows the phenomenon of the midnight sun (which never dips below the horizon) on the day of the summer solstice, and the lower section shows the sun on the shortest day of the year. (Photo by Mario Grassi)

Color Plate 9 (bottom): Lightning flashes illuminate the Kitt Peak National Observatory in this 45-second exposure. (Photo ©1972 Gary Ladd)

Color Plate 10 (top): This cluster of telescopes on top of Mauna Kea on the island of Hawaii contains 3 of the world's largest dozen telescopes. They will soon be joined by still larger telescopes. (Institute for Astronomy, University of Hawaii, photo by J. M. Pasachoff)

Color Plate 11 (bottom): The diamond-ring effect at the beginning of totality at the June 30, 1973, eclipse. These photographs were taken from Loiengalani, Kenya, by the author's expedition.

THE EARTH: OUR PLANET

When astronauts first reached the Moon, they looked back and saw the Earth floating in space (Color Plate 18). The realization that Earth is an oasis in space helped inspire our present concern for our environment. Until fairly recently, we studied the Earth in geology courses and the other planets in astronomy courses, but now the lines are very blurred. Not only have we learned more about the interior, surface, and atmosphere of the Earth but we have also seen the planets in enough detail to be able to make meaningful comparisons with Earth. The study of *comparative planetology* is helping us to understand weather, earthquakes, and other topics. This expanded knowledge will help us improve life on our own planet.

"The Moon" is often capitalized to distinguish it from moons of other planets; in general writing, it is usually written with a small "m." In Part II, we shall capitalize "Earth" to put it on a par with the other planets and the Moon. For consistency, we shall also capitalize "Sun" and "Universe."

5.1 THE EARTH'S INTERIOR

The study of the Earth's interior and surface is called *geology*. Geologists study how the Earth vibrates as a result of large shocks, such as earthquakes. Much of our knowledge of the structure of the Earth's interior comes from *seismology*, the study of these vibrations. The vibrations travel through different types of material at different speeds. From seismology and other studies, geologists have been able to develop a picture of the Earth's interior (Fig. 5–1).

The Earth's innermost region, the *core*, consists primarily of iron and nickel. The central part of the core may be solid, but the outer part is probably a very dense liquid. Outside the core is the *mantle*, and on top of the mantle is the thin outer layer called the *crust*. The upper mantle and crust are rigid, while the lower mantle is partially melted.

How did such a layered structure develop? The Earth probably formed from a cloud of gas and dust, along with the Sun and the other planets, and

FIGURE 5–1 The structure of the Earth and stages in its evolution. *(A)* Tens of millions of years after its formation, the heat from radioactive elements added to the effects of gravitational compression and the impact of debris produced melting and differentiation. Heavy materials sank inward and light materials floated outward. *(B)* The heaviest materials form the core, and the lightest materials form the crust. *(C)* The crust has broken into rigid plates that carry the continents and move very slowly away from areas of sea-floor spreading.

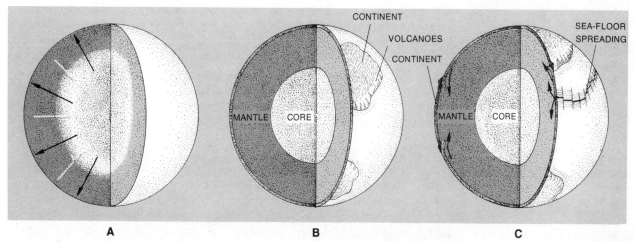

BOX 5.1
Density

Densities

water	1 gm/cm^3
aluminum	3 gm/cm^3
rock	3 gm/cm^3
iron	8 gm/cm^3
lead	11 gm/cm^3

The density of an object is its mass divided by its volume. The scale of mass in the metric system was set up so that 1 cubic centimeter of water would have a mass of 1 gram. Though this varies slightly with the water's temperature and pressure, the density of water is always about 1 gm/cm^3. Density is always expressed in units of mass per volume (with the "per" usually written as a slash). In "customary units," the density of water is about 62 pounds/cubic foot (though, technically, pounds are a measure of weight—gravitational force—rather than of mass).

To measure the density of any amount of matter, astronomers must independently measure its mass and its size. They are often able to identify what materials are present in an object on the basis of density. For example, since the density of iron is about 8 gm/cm^3, the measurement that Saturn's density is less than 1 gm/cm^3 shows that Saturn is not made largely of iron. On the other hand, the measurement that the Earth's density is about 5 gm/cm^3 shows that the Earth has a substantial content of dense elements of which iron is a likely candidate.

Astronomical objects cover the extremes of density. In the space between the stars, the density is tiny fractions of a gram in each cubic centimeter. And in neutron stars, the density exceeds billions of tons per cubic centimeter.

so was surrounded by a lot of debris in the form of dust and rocks. The young Earth was probably subject to constant bombardment from this debris. This bombardment heated the surface to the point where it began to melt, producing lava.

Much of the original heat for the Earth's interior came from gravitational energy released as particles came together to form the Earth; such energy is released from gravity between objects when the objects move closer together, but we will not discuss any details here. But the major source of the present-day heat for the interior is the natural radioactivity within the Earth. Certain forms of atoms are unstable, that is, they spontaneously change into more stable forms. In the process, they give off energetic particles that collide with the atoms in the rock and give some of their energy to these atoms. The rock heats up.

After about a billion years, the Earth's interior had become so hot that the iron melted and sank to the center since it was denser, forming the core. Eventually other materials also melted. As the Earth cooled, various materials, because of their different densities and freezing points (the temperature at which they change from liquid to solid), solidified at different distances from the center. This process is responsible for the present layered structure of the Earth.

The rotation of the Earth's metallic core helps generate a magnetic field on Earth. The magnetic field has a north magnetic pole and a south magnetic pole that are not quite where the regular north and south geographic poles are—the Earth's north magnetic pole is near Hudson Bay, Canada.

FIGURE 5–2 Thermal activity beneath the Earth's surface results in geysers. The thermal area shown here is in Rotorua, New Zealand. Geothermal steam can be used to generate electricity; The Geysers, a geothermal area in California, provides much of San Francisco's electricity.

5.2 CONTINENTAL DRIFT

Some geologically active areas (Fig. 5–2) exist in which heat flows from beneath the surface at a rate much higher than average. The outflowing *geothermal energy* in some of these regions is being tapped as an energy source.

The Earth's rigid outer layer is segmented into *plates,* each thousands of kilometers in extent but only about 50 km thick. Because of the internal heating, the top layers float on an underlying hot layer where the rock is soft, though it is not hot enough to melt completely. This hot material beneath the rigid plates of the surface churns very slowly. The churning carries the plates around. This theory, called *plate tectonics,* explains the observed *continental drift*—the drifting over eons of the continents from their original positions. ("Tectonics" comes from the Greek word meaning "to build.")

Although the notion of continental drift once seemed unreasonable, it is now generally accepted. The continents were once connected as two supercontinents. These may have, in turn, separated from a single supercontinent called Pangaea, which means "all lands." Over two hundred million years or so, the continents have moved apart as plates have separated. We can see from their shapes how they originally fit together (Fig. 5–3). We even find similar fossils and rock types along two opposite coastlines, which were once adjacent but are now widely separated. In the future, we expect California to separate from the rest of the United States, Australia to be linked to Asia, and the Italian "boot" to disappear.

The boundaries between the plates (Fig. 5–4) are geologically active areas. Therefore, these boundaries are traced out by the regions where

FIGURE 5–3 The NASA satellite LAGEOS (Laser Geodynamic Satellite), launched into a circular Earth orbit in 1976, provides information about the Earth's rotation and the crust's movements. It is expected to survive in orbit for 8 million years. In case the satellite is discovered in the distant future, it bears a series of views of the continents in their locations 270 million years in the past *(top)*, at present *(middle)*, and 270 million years in the future *(bottom)*, according to our knowledge of continental drift. (The dates are given in the binary system.) Data on continental drift are gathered by reflecting laser beams from telescopes on Earth off the 426 retroreflectors that cover its surface. (Each retroreflector is a cube whose interior reflects incident light back in the direction from which it came.) Measuring the time until the beams return can be interpreted to show the accurate position of the ground station.

FIGURE 5–4 The San Andreas fault in California.

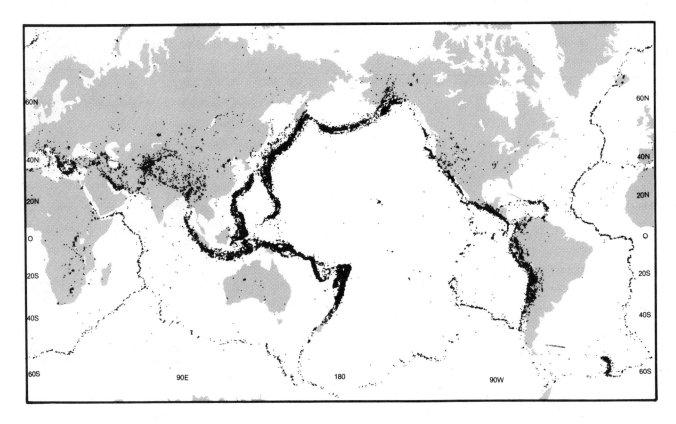

FIGURE 5–5 This plot of earthquakes from 1963 through 1973 greater than 4.5 on the Richter scale shows that earthquakes occur preferentially at plate boundaries.

earthquakes (Fig. 5–5) and most of the volcanoes (Fig. 5–6) occur. The boundaries where two plates are moving apart mark regions where molten material is being pushed up from the hotter interior to the surface, such as ridges in the middle of the oceans. The motion of the plates is also responsible for the formation of the great mountain ranges. When two plates come together, one may be forced under the other and the other raised. The "ring

FIGURE 5–6 *(A)* Mt. St. Helens, shown here during its May 18, 1980, eruption, lies on a plate boundary. *(B)* After its devastating eruption. *(C)* Mauna Loa on the island of Hawaii during its 1984 eruption. Lava flows from the lower right center downhill toward the left. Light from the glowing lava is reflected toward us in the columns at right center and the horizontal clouds.

A

B

C

of fire" volcanoes around the Pacific Ocean (including Mt. St. Helens in Washington) were formed in this way. New evidence indicates that the North American plate may itself be splitting apart: a geological rift running through Colorado and New Mexico is widening about 1 millimeter per year.

5.3 TIDES

It has long been accepted that tides are most directly associated with the Moon. This is because the tides—like the Moon's passage by your meridian (the imaginary line in the sky passing from north to south through the point overhead)—occur about an hour later each day. Tides result from the fact that the force of gravity exerted by the Moon (or any other body) gets weaker as you get farther away from it. Tides depend on the **difference** between the gravitational attraction of a massive body at different points on another body.

To explain the tides in Earth's oceans, consider, for simplicity, that the Earth is completely covered with water. We might first say that the water closest to the Moon is attracted toward the Moon with the greatest force and so is the location of high tide as the Earth rotates. If this were all there were to the case, high tides would occur about once a day. However, two high tides occur daily, separated by roughly 12½ hours (Fig. 5–7).

To see how we get two high tides a day, consider three points, A, B, and C, where B represents the solid Earth, A is the ocean nearest the Moon, and C is the ocean farthest from the Moon (Fig. 5–8). Since the Moon's gravity weakens with distance, it is greater at A than at B, and greater at point B than at C. If the Earth and Moon were not in orbit about each other, all these points would fall toward the Moon, moving apart as they fell.

Thus the high tide on the side of the Earth that is near the Moon is a result of the water being pulled away from the Earth. The high tide on the opposite side of the Earth results from the Earth being pulled away from the water. In between the locations of the high tides, the water has rushed elsewhere so we have low tides. Since the Moon is moving in its orbit around

FIGURE 5–7 Low and high tides in Nova Scotia on the Bay of Fundy, site of the world's highest tides. Though the explanation of tides given is valid, details of tides at individual locations on the shore also depend on such factors as the depth of the ocean floor or the shape of the channel; the latter is especially important here. A proposal to build a dam to generate hydroelectric power at the end of the Bay of Fundy is proving controversial. By changing the length of the bay, it will set up a resonance in the sloshing of the water, and would change the height of the tides as far south as Cape Cod.

FIGURE 5–8 A schematic representation of the tidal effects caused by the Moon. The arrows represent for each point the acceleration that results from the gravitational pull of the Moon (exaggerated in the drawing). The water at point A has greater acceleration toward the Moon than does point B; since the Earth is solid, the whole Earth moves with point B. (Tides in the solid Earth exist, but are much smaller than tides in the oceans.) Similarly, the solid Earth is pulled away from the water at point C.

Tidal forces are important in many astrophysical situations, including disks of matter orbiting neutron stars and black holes (Chapters 22 and 23), and rings around planets (Chapters 10 through 12).

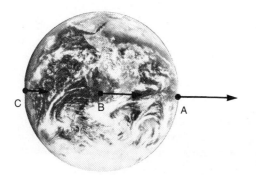

FIGURE 5–9 The layers of the Earth's atmosphere showing the location of the part of the stratosphere where ozone is formed. Once formed, the ozone is transported to lower stratospheric layers. Ozone is important in shielding us from solar ultraviolet rays. Aerosols with chemical propellants that destroy ozone have been restricted in the United States.

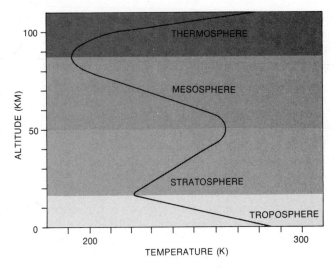

the Earth, a point on the Earth's surface has to rotate longer than 12 hours to return to a spot nearest to the Moon. Thus the tides repeat about every 12½ hours.

The Sun's effect on the Earth's tides is only about half as much as the Moon's. Though the Sun exerts a greater gravitational force on the Earth than does the Moon, the Sun is so far away that its force does not change very much from one side of the Earth to the other. And it is only the change in force that counts for tides.

5.4 THE EARTH'S ATMOSPHERE

We name layers of our atmosphere (Fig. 5–9) according to the composition and the physical processes that determine the temperature. The Earth's weather is confined to the very thin *troposphere*. A major source of heat for the troposphere is infrared radiation emitted from the ground, so the temperature of the troposphere decreases with altitude.

Observations from satellites have greatly enhanced our knowledge of our atmosphere. (See the figure opening this chapter.) Scientists carry out calculations using the most powerful supercomputers to interpret the global data and to predict how the atmosphere will behave. The equations are essentially the same as those for the internal temperature and structure of stars, except that the sources of energy are different.

The rotation of the Earth also has a very important effect in determining how the winds blow. Comparison of the circulation of winds on the Earth (which rotates in 1 Earth day), on slowly rotating Venus (which rotates in 243 Earth days), and on rapidly rotating Jupiter and Saturn (each of which rotates in about 10 Earth hours), helps us understand the weather on Earth.

5.5 THE VAN ALLEN BELTS

In January 1958, the first American space satellite carried aloft, among other things, a device to search for particles carrying electric charge that might be orbiting the Earth. This device, under the direction of James A.

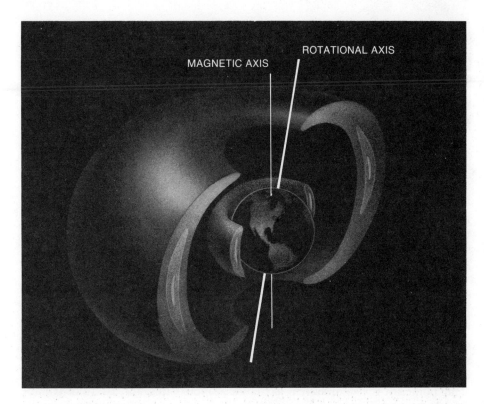

MAGNETIC AXIS

ROTATIONAL AXIS

FIGURE 5-10 The doughnut-shaped Van Allen belts. Though often called "radiation belts," they are actually regions of charged particles trapped by the Earth's magnetic field.

Van Allen of the University of Iowa, detected a region filled with charged particles having high energies. We now know that two such regions—the *Van Allen belts*—surround the Earth, like a small and a large doughnut (Fig. 5–10).

The particles in the Van Allen belts are trapped by the Earth's magnetic field. Charged particles preferentially move in the direction of magnetic-field lines, and not across the field lines.

Charged particles, often from solar magnetic storms, are guided by the Earth's magnetic field toward the Earth's poles. When they interact with air molecules they cause our atmosphere to glow, which we see as the beautiful northern and southern lights—the *aurora borealis* or *aurora australis*, respectively (Fig. 5–11 and Color Plate 26).

5.6 THE FORMATION OF THE SOLAR SYSTEM

Many scientists studying the Earth and the planets are particularly interested in an ultimate question: How did the Earth and the rest of the solar system form? We can accurately date the formation by studying the oldest objects we can find in the solar system and allowing a little more time. For example, astronauts found rocks on the Moon older than 4.4 billion years. We think the solar system formed about 4.6 billion years ago.

Our best current idea is that some 4.6 billion years ago, a huge cloud of gas and dust in space started collapsing. Just why the collapse started isn't known.

You have undoubtedly noticed that ice skaters spin faster when they pull their arms in. Similarly, the solar system began to spin faster as it col-

FIGURE 5–11 The aurora forms an oval around the north pole, in this view from the Dynamics Explorer spacecraft, launched in 1981. The observations were made in ultraviolet light emitted by glowing oxygen; glowing oxygen also causes the typical green auroral light. The auroral oval traces the "feet" of the outer Van Allen belt and extends across the terminator between the sunlighted and night sides of the Earth. The aurora is caused by an interaction of the solar wind with the Earth's magnetic field. Charged particles hit atoms in the earth's atmosphere, giving them energy that they then radiate.

FIGURE 5–12 The leading model for the formation of the solar system has a cloud of gas and dust collapsing. Between stages *C* and *D*, the protosun and a large number of small bodies called planetesimals are formed. The planetesimals clumped together to form protoplanets *(D)*; they, in turn, contract to become planets. Some of the planetesimals may have become moons, asteroids, comets, or meteoroids.

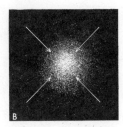
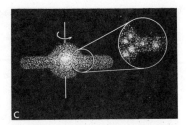

lapsed. (The original spin it had may have just been randomly belonging to the gas and dust.) Objects that are spinning around tend to fly off, and this force eventually became strong enough to counteract the force of gravity pulling inward. Thus the solar system stopped collapsing in one plane. Perpendicular to this plane, there was no spin to stop the collapse, so the solar system wound up as a disk.

In the disk of gas and dust, we think that the material began to clump. Smaller clumps joined together to make larger ones, and eventually *planetesimals*, bodies a few hundred kilometers across, were formed. Eventually, gravity pulled many planetesimals together to make *protoplanets* (pre-planets) orbiting a *protosun* (pre-Sun). Subsequently, the protoplanets contracted and cooled to make the planets we have today, and the protosun contracted and started shining, becoming the Sun (Fig. 5–12). Some of the planetesimals may still be orbiting the Sun; that is why we are so interested in studying small bodies of the solar system like comets, meteorites, and asteroids (Chapter 13).

One interesting aspect of this model of solar-system formation is that nothing about it implies that the solar system is unique. There may well be other systems of planets around other stars. We are increasingly finding indirect evidence that such planets may exist. As we shall discuss in Section 14.2, a disk has even been seen around a nearby star, which may indicate that a solar system is now forming there.

KEY WORDS

comparative planetology, geology, seismology, core, mantle, crust, geothermal energy, plates, plate tectonics, continental drift, troposphere, Van Allen belts, aurora borealis, aurora australis, planetesimals, protoplanets, protosun

QUESTIONS

1. Consult an atlas and compare the sizes of the Grand Canyon in Arizona and the Rift Valley in Africa. How do they compare in size with the giant canyon on Mars, which is about as long as the distance from New York to California?

2. Plan a set of experiments or observations that you, as a Martian scientist, would have an unmanned spacecraft carry out on Earth to find out if life existed here. What data would your spacecraft radio back if it landed in a corn field? In the Sahara? In the Antarctic? In Times Square?

3. How did the layers of the Earth arise? Where did the heat come from?

4. What carries the continental plates around over the Earth's surface?

5. (a) Explain the origins of tides. (b) If the Moon were twice as far away from the Earth as it actually is, how would tides be affected?

6. Draw a diagram showing the positions of the Earth, Moon, and Sun at a time when there is the least differ-

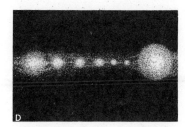
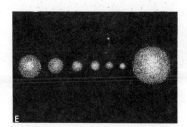

ence between high and low tides.

7. What is the source of most of the radiation that heats the troposphere?

8. Look at a globe and make a list or sketches of which pieces of the various continents probably lined up with each other before the continents drifted apart.

9. Space-shuttle astronauts orbit 175 km above the Earth. Communications satellites are 5½ Earth radii above the surface. Compare their locations with the Van Allen belts, which can cause false readings on instruments.

10. What role do planetesimals play in the origin of the planets?

Chapter 6

Scientist-astronaut Harrison Schmitt collecting small rocks and rock chips with a lunar rake during the Apollo 17 mission to the Moon.

THE MOON: OUR NATURAL SATELLITE

The Earth's nearest celestial neighbor—the Moon—is only 380,000 km (238,000 miles) away from us on the average, close enough that it appears sufficiently large and bright to dominate our nighttime sky. The Moon's stark beauty has called our attention since the beginning of history. Now we can study the Moon not only as an individual object but also as an example of a small planet or a large planetary satellite, since spacecraft observations have told us that there may be little difference between small planets and large moons.

6.1 THE MOON'S APPEARANCE

Even binoculars—Galileo's telescope was less powerful than modern binoculars—reveal that the Moon's surface is pockmarked with craters. Other areas, called *maria* (pronounced mar'ē-a; singular, *mare*, pronounced mar'ā), are relatively smooth, and indeed the name comes from the Latin word for sea (Fig. 6–1). But there are no ships sailing on the lunar seas and no water in them; the Moon is a dry, airless, barren place.

The gravity at the Moon's surface is only ⅙ that of the Earth. You would weigh only 20 or 30 pounds there if you stepped on a scale! Gravity is so weak that any atmosphere and any water that may once have been present would long since have escaped into space.

The Moon rotates on its axis at the same rate as it revolves around the Earth, always keeping the same face in our direction. The Earth's gravity has locked the Moon in this pattern, pulling on a bulge in the distribution of the lunar mass to prevent the Moon from rotating freely. As a result of this interlock, we always see essentially the same side of the Moon from our vantage point on Earth.

When the Moon is full, it is bright enough to cast shadows or even to read by. But full moon is a bad time to try to observe lunar surface structure, for any shadows we see are short. When the Moon is a crescent or even a half moon, however, the part of the Moon facing us is covered with long shadows. The lunar features then stand out in bold relief (Fig 6–2).

Shadows are longest near the *terminator,* the line separating day from night. The Moon revolves around the Earth every 27⅓ days with respect to the stars. But during that time, the Earth has moved part way around the Sun, so it takes a little more time for the Moon to complete a revolution with respect to the Sun. The cycle of phases that we see from Earth repeats with this 29½-day period.

In some sense, before the period of exploration by the Apollo program, we knew more about almost any star than we did about the Moon. As a solid body, the Moon reflects the spectrum of sunlight rather than emitting its own spectrum, so we were hard pressed to determine even the composition or the physical properties of the Moon's surface.

FIGURE 6–1 This Earth-based photograph shows Mare Imbrium at the upper left, the Apennine Mountains at the lower right, and many craters, some of which have central peaks (which arise from a rebound as the crater forms). The largest crater shown is Archimedes.

FIGURE 6–2 The full moon.
Note the dark maria and the
lighter, heavily cratered highlands. The positions of the 6
American Apollo (A) and 3 Soviet
Luna (L) missions from which material was returned to Earth for
analysis are marked. This and
other ground-based photographs
in this chapter are oriented with
north up, as we see the moon with
our naked eyes or through binoculars; a telescope normally inverts
the image.

6.2 LUNAR EXPLORATION

The space age began on October 4, 1957, when the U.S.S.R. launched its
first Sputnik (the Russian word for "travelling companion") into orbit. The
shock of this event galvanized the American space program, and within
months American spacecraft were also in Earth orbit.

In 1959, the Soviet Union sent its Luna 3 spacecraft around the Moon;
Luna 3 radioed back the first murky photographs of the Moon's far side.
Now that we have high-resolution maps, it is easy to forget how big an ad-
vance that was.

In 1961, President John F. Kennedy proclaimed that it would be a U.S.
national goal to put a man on the Moon by 1970 and bring him safely back
to Earth. The American lunar program, under the direction of the National
Aeronautics and Space Administration (NASA), proceeded in gentle stages.
The ability to carry out manned space flight was developed first with single-
astronaut suborbital and orbital capsules, called Project Mercury, and then
with two-astronaut orbital spacecraft, called Project Gemini. Simultaneously,
a series of unmanned spacecraft was sent to the Moon.

The manned and unmanned trains of development merged with Apollo 8, in which three astronauts circled the Moon on Christmas Eve, 1968, and returned to Earth. The next year, Apollo 11 brought humans to land on the Moon for the first time. It went into orbit around the Moon after a three-day journey from Earth, and a small spacecraft called the Lunar Module separated from the larger Command Module. On July 20, 1969, Neil Armstrong and Buzz Aldrin left Michael Collins orbiting in the Command Module and landed on the Moon (Fig. 6–3). In the preceding days there had been much discussion of what Armstrong's historic first words should be, and millions listened as he said "One small step for man, one giant leap for mankind." (He meant to say "for a man." In 1984, the space-shuttle astronaut who became the first human to fly freely in orbit, untethered, joked "That may have been one small step for Neil, but it's a heck of a big leap for me.")

The Lunar Module carried many experiments, including devices to test the soil, a camera to take stereo photos of lunar soil, a sheet of aluminum with which to capture particles from the solar wind, and a seismometer. Later Lunar Modules carried additional experiments, some even including a vehicle (Fig. 6–4). Six Apollo missions in all, ending with Apollo 17 in 1972, carried people to the Moon (Color Plates 14–17).

The Soviet Union sent three unmanned spacecraft to land on the lunar surface, collect lunar soil, and return it to Earth.

6.3 THE RESULTS FROM APOLLO

The kilometers of film exposed by the astronauts, the 382 kg of rock brought back to Earth (Fig. 6–5), the lunar seismograph data recorded on tape, and other data have been studied by hundreds of scientists from all over the Earth. The data have led to new views of several basic questions, and have raised many new questions about the Moon and the solar system.

6.3a The Lunar Surface

The rocks that were encountered on the Moon are types that are familiar to terrestrial geologists. All the rocks are the type that were formed by the cooling of lava.

FIGURE 6–3 Neil Armstrong, the first person to set foot on the Moon, took this photograph of his fellow astronaut Buzz Aldrin climbing down from the Lunar Module of Apollo 11 on July 20, 1969. The site is called Tranquillity Base, as it is in the Sea of Tranquillity (Mare Tranquillitatis).

FIGURE 6–4 (A) Eugene Cernan riding on the Lunar Rover during the Apollo 17 mission. The mountain in the background is the east end of the South Massif. (B) Drawing by Alan Dunn; © 1971 The New Yorker Magazine, Inc.

A

B

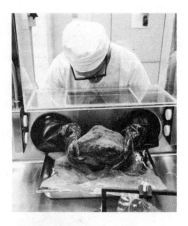

FIGURE 6–5 A moon rock from Apollo 14 being handled in the Lunar Receiving Laboratory at the Johnson Space Center in Houston.

Meteorites hit the Moon with such high velocities that huge amounts of energy are released at the impact. The effect is that of an explosion, as though TNT or an H-bomb had exploded.

Radioactive isotopes are those that decay spontaneously; that is, they change into other isotopes even when left alone. Stable isotopes remain unchanged. For certain pairs of isotopes— one radioactive and one stable—we know the proportion of the two when the rock was formed. Since we know the rate at which the radioactive one is decaying, we can calculate how long it has been decaying from a measurement of what fraction is left.

In the maria, the rocks are mainly *basalts* (Fig. 6–6*A*). In both the maria and the highlands, some of the rocks are *breccias* (Fig. 6–6*B*), mixtures of fragments of several different types of rock that have been compacted and welded together.

The Moon and the Earth seem to be similar chemically, though significant differences in overall composition do exist. Some elements that are rare on Earth—such as uranium, thorium, and the rare-earth elements—are found in greater abundances. (Will we be mining on the Moon one day?) Since none of the lunar rocks contain any trace of water bound inside their minerals, clearly water never existed on the Moon. This eliminates the possibility that life evolved there; water is essential for life as we know it.

One way of dating the surface of a moon or planet is to count the number of craters in a given area, a method that was used before Apollo. Surely those locations with the greatest number of craters must be the oldest (Fig. 6–7). Relatively smooth areas—like maria—must have been covered over with volcanic material at some relatively recent time (which is still billions of years ago).

A few craters on the Moon have thrown out obvious rays of lighter-colored matter (Fig. 6–8). The rays are material ejected when the crater was formed. Since these rays extend over other craters, the craters with rays must have formed later. The youngest rayed craters may be very young indeed—perhaps only a few hundred million years. The rays darken with time, so rays that may have once existed near other craters are now indistinguishable from the rest of the surface.

Crater counts and the superposition of one crater on another give only relative ages. We could find the absolute ages only when rocks were physically returned to Earth. Scientists worked out the dates by comparing the current ratio of radioactive forms of atoms to nonradioactive forms present in the rocks with the ratio that they would have had when they were formed. The oldest rocks that were found on the Moon were formed 4.42 billion years ago. The youngest rocks were formed 3.1 billion years ago.

The observations can be explained on the basis of the following general picture: The Moon formed 4.6 billion years ago. We know that the top 100 km or so of the surface was molten about 200 million years later. Then the surface cooled. From 4.2 to 3.9 billion years ago, bombardment by interplanetary rocks caused most of the craters we see today. About 3.8 billion years ago, the interior of the Moon heated up sufficiently (from radioactive elements inside) that vulcanism began. Lava flowed onto the lunar surface and

FIGURE 6–6 *(A)* A basalt brought back to Earth by the Apollo 15 astronauts. It weighs 1.5 kg. Note the many spherical cavities that arise in volcanic rock because of gas trapped at the time of the rock's formation. *(B)* A breccia, dark gray and white in color, returned to Earth by Apollo 15.

Under a microscope. one can easily see the contrast between lunar and terrestrial rocks. The lunar rocks contain no water, so we know that they never underwent the reactions that terrestrial rocks undergo. The lunar rocks also show that no oxygen was present when they were formed.

A

B

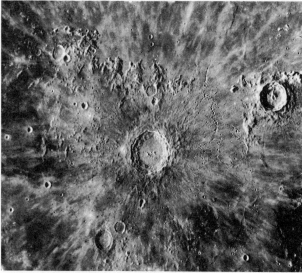

FIGURE 6–7   This view from the orbiting Apollo 15 Command Module shows a smaller crater, Krieger B, superimposed on a larger crater, Krieger. Obviously, the smaller crater is younger than the larger one. Several rilles (clefts along the lunar surface that can be hundreds of kilometers in length) and ridges are also visible. Sometimes the areas in the centers of craters, the floors, are smooth; other craters have central peaks.

FIGURE 6–8   The crater Copernicus, seen in this ground-based photograph, has rays of light material emanating from it. This light material was thrown out radially when the meteorite impacted and formed the crater.

filled the largest basins that resulted from the earlier bombardment, thus forming the maria (Fig. 6–9). By 3.1 billion years ago, the era of vulcanism was over. The Moon has been geologically pretty quiet since then.

Up to this time, the Earth and the Moon shared similar histories. But active lunar history stopped about 3 billion years ago, while the Earth continued to be geologically active. Almost all the rocks on the Earth are younger than 3 billion years of age; erosion and the remolding of the continents as they move slowly over the Earth's surface, according to the theory of plate tectonics (Section 5.2), have taken their toll. The oldest single rock ever dis-

FIGURE 6–9   Artist's views of the formation of the lunar surface compared with a recent photographic map. *(A)* The Moon before the formation of the present mare surface material about 4 billion years ago. The concentric rings of the Imbrium basin, now Mare Imbrium, show prominently at upper left. From current lunar features, the artist has removed most mare material and late craters, and freshened certain early craters. *(B)* The Moon soon after the formation of most of the mare material approximately 3.3 billion years ago. Young craters such as Tycho and Copernicus are absent. The Moon looked much like it does now, though surface details differ. *(C)* A modern lunar map from the U.S. Geological Survey. (Drawings by Donald E. Davis under the guidance of Don E. Wilhelms of the U.S. Geological Survey)

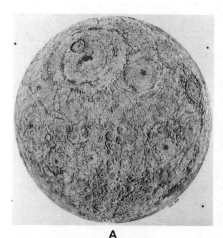

A   B   C

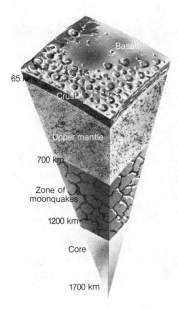

FIGURE 6–10 The Moon's interior. The depth of basalts is greater under maria, which are largely on the side of the Moon nearest the Earth. Almost all the 10,000 moonquakes observed originated in a zone halfway down toward the center of the Moon, a distance ten times deeper than most terrestrial earthquakes. This fact can be used to interpret conditions in the lunar interior. If too much of the interior of the Moon were molten, the zone of moonquakes probably would have sunk instead of remaining suspended there. The deep moonquakes came from about 80 locations, and were triggered at each location twice a month by tidal forces resulting from the variation in the Earth-Moon distance.

We are obtaining additional evidence about the capture model by studying the moons of Jupiter and Saturn. The outermost moon of Saturn, for example, is apparently a captured asteroid.

covered on Earth has an age of 4.1 or 4.2 billion years, and few rocks are older than 3 billion years. So we must look to extraterrestrial bodies—the Moon or meteorites—that have not suffered the effects of plate tectonics or erosion (which occurs in the presence of water or an atmosphere) to study the first billion years of the solar system.

6.3b The Lunar Interior

Before the Moon landings, it was widely thought that the Moon was a simple body, with the same composition throughout. But we now know it to be differentiated (Fig. 6–10), like the planets. Most scientists believe that the Moon's core is molten, but the evidence is not conclusive.

Tracking the orbits of the Apollo Command Modules and other satellites that orbited the Moon told us about the lunar interior. If the Moon were a perfect, uniform sphere, the spacecraft orbits would have been perfect ellipses. But they weren't. One of the major surprises of the lunar missions was the discovery in this way of *mascons*, regions of mass concentrations near and under most maria. The mascons may be lava that is denser than the surrounding matter, providing a stronger gravitational force on satellites passing overhead.

6.3c The Origin of the Moon

Among the models for the origin of the Moon that have been considered in the decade surrounding the Apollo missions are the following:

1. *Fission:* the Moon was separated from the material that formed the Earth; the Earth spun up and the Moon somehow spun off;
2. *Capture:* the Moon was formed far from the Earth in another part of the solar system, and was later captured by the Earth's gravity; and
3. *Condensation:* the Moon was formed near to and simultaneously with the Earth in the solar system.

But recent work in the mid-1980's has all but ruled out the first two of these and has made the third less likely than it seemed. The two models currently under the most study have not been adequately investigated, but research is continuing on them. They are:

4. *Ejection of a gaseous ring:* A Mars-size planetesimal hit the proto-Earth, ejecting matter in gaseous (and perhaps some in liquid or solid) form. This matter would have ordinarily fallen back onto the Earth but, since it was gaseous, pressure differences could exist. These presumably caused some of the matter to start moving rapidly enough to go into orbit, and some asymmetry established an orbiting direction; the other matter fell back. The orbiting material eventually coalesced into the Moon.
5. *Interaction of Earth-orbiting and Sun-orbiting planetesimals:* Collisions of planetesimals orbiting the Earth and planetesimals orbiting the Sun led to the breakup of the material and the eventual formation of the Moon. In the collisions, rock stuff and iron got treated differently.

It had been hoped that landing on the Moon would enable us to clear up the problem of the lunar origin. But though the lunar programs have led to modifications and updating of the original three models described above, none of these models has been entirely ruled out. And over time researchers provide still additional models. Even a supposedly simple question such as "is the chemical composition of the Moon the same as that of the Earth's mantle" has no accepted answer. At a 1984 conference on the origin of the Moon, opposed factions saying "yes" and "no" could not agree, nor did they change the other side's position.

6.4 FUTURE LUNAR STUDIES

One surprise came recently when it was realized that a handful of meteorites found in Antarctica probably came from the Moon (Fig. 6–11). They may have been ejected from the Moon when craters formed. So we are still getting new moon rocks to study.

It is hard to believe that the era of manned lunar exploration not only has begun but also has already ended. At present we have no plans to send more people to the Moon. The United States does not even have any unmanned missions definitely planned, though some are in a prospective group of spacecraft suggested to NASA. The Soviet Union is planning an unmanned spacecraft to orbit the Moon from pole to pole, mapping the entire surface. A similar American mission had been proposed in the 1970's, and is now being considered again. The European Space Agency and Japan have also expressed interest in such a Lunar Polar Orbiter.

The orbiting manned space stations now planned for the 1990's by the U.S. and the U.S.S.R. may provide the capability of again sending people out beyond Earth orbit. Perhaps by the turn of the 21st century, manned lunar exploration will resume. Twenty or thirty years from now, we may each be able to visit the Moon as researchers or even as tourists.

FIGURE 6–11 This rock is one of many meteorites found in the Earth's continent of Antarctica. Under a microscope and in mineralogical and isotopic analyses, it seems like a sample from the lunar highlands and quite unlike any terrestrial rock or any other meteorite.

KEY WORDS

maria, mare, terminator, basalts, breccias, mascons, fission, capture, condensation, ejection of a gaseous ring, interaction of Earth-orbiting and Sun-orbiting planetesimals

QUESTIONS

1. To what location on Earth does the terminator on the Moon correspond?
2. What does cratering tell you about the age of the surface of the Moon, compared to that of the Earth's surface?
3. From looking at the photograph of the Moon, identify a rayed crater by name.

4. Why are we more likely to learn about the early history of the Earth by studying the rocks from the Moon than those on the Earth?
5. What are mascons?
6. Describe three of the proposed theories to explain the origin of the Moon.
7. How can we get lunar material on Earth to study?

TOPICS FOR DISCUSSION

1. Consider the scientific value of the proposed manned space station for Earth orbit, and the relation of its funding to other space projects. Does generally increasing budgets mean more money for all projects—the analogy that a rising tide raises all boats is often cited—or would

the space station take money away from exploration?
2. Discuss the scientific, political, and financial arguments for resuming (a) unmanned and (b) manned exploration of the Moon.

OBSERVING PROJECT

1. Use binoculars to identify on the Moon all the features shown in the overall photograph.
2. Observe the Moon with your naked eye on as many nights as possible for a month, noting its rising time or

position in the sky at a given time of night. Interpret your observations in terms of the Earth-Moon-Sun angles that cause the phases.

Chapter 7

Mercury, photographed at a distance of 200,000 km from Mariner 10, revealed a cratered surface. The symbol for Mercury appears at the top of the facing page.

MERCURY:
A LIFELESS WORLD

Mercury is the innermost planet. Its average distance from the Sun is ⁴/₁₀ of the Earth's average distance. Since the Earth's average distance from the Sun is known as an *Astronomical Unit* (*A.U.*), Mercury is 0.4 A.U. from the Sun. Except for distant Pluto, its orbit around the Sun is the most elliptical.

Since we on the Earth are outside Mercury's orbit looking in at it, Mercury always appears close to the Sun in the sky. At times it rises just before sunrise, and at times it sets just after sunset, but it is never up when the sky is really dark. As a result, whenever Mercury is visible, its light travels a long path through the Earth's atmosphere, which leads to blurred images. Even the best photographs taken from the Earth show Mercury as only a fuzzy ball with faint, indistinct markings (Fig. 7–1).

On rare occasions, Mercury goes into transit across the Sun; that is, we see it as a black dot crossing the Sun. The next transit is on November 13, 1986, but will take place in the night-time for the U.S., so it won't be visible from our side of the Earth. Subsequent transits will be in 1993, 2003, and 2006.

7.1 THE ROTATION OF MERCURY

From studies of ground-based drawings and photographs, astronomers did as well as they could to describe Mercury's surface. A few features could barely be distinguished, and the astronomers watched to see how long those features took to rotate around the planet. From these observations they decided that Mercury rotated in the same amount of time that it took to revolve around the Sun. Thus they thought that one side always faced the Sun and the other side always faced away from the Sun. This led to the fascinating conclusion that Mercury could be both the hottest planet and the coldest planet in the solar system.

But when the first measurements were made of Mercury's radio radiation, the planet turned out to be giving off more energy than had been expected. This meant that it was hotter than expected. The dark side of Mercury was too hot for a surface that was always in the shade. (The light we see is merely sunlight reflected by Mercury's surface and doesn't tell us the surface's temperature. The radio waves are actually being emitted by the surface.)

Later, we became able to transmit radio signals from Earth to Mercury and detect the echo. This technique is called *radar*, the acronym for **ra**dio **d**etection **a**nd **r**anging. Radar can show how fast something is moving, just as a police radar measures the speed of cars. The results about Mercury were a surprise: scientists had been wrong about the period of Mercury's rotation. It actually rotates every 59 days.

Mercury's 59-day period of rotation with respect to the stars is exactly ⅔ of the 88-day period that it takes to revolve around the Sun, so the planet rotates three times for each two times it revolves. Mercury's rotation and revolution combine to give a value for the rotation of Mercury relative to the Sun (that is, a Mercurian solar day) that is neither 59 nor 88 days long (Figure 7–2). If we lived on Mercury we would measure each day and each night to be 88 Earth days long. We would alternately be fried and frozen for 88 Earth days at a time.

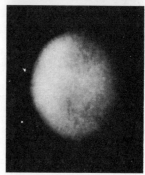

FIGURE 7–1 From the Earth, we cannot see much surface detail on Mercury. These views, among the best ever taken from Earth, were made by the New Mexico State University Observatory.

FIGURE 7–2 Follow the arrow that starts facing rightward toward the Sun in the image of Mercury at the left of the figure (A), as Mercury revolves along the dotted line. Mercury, and thus the arrow, rotates once with respect to the stars in 59 days (E), when Mercury has moved only ⅔ of the way around the Sun. (Our view is as though we were watching from a distant star.) Note that after one full revolution of Mercury around the Sun, the arrow faces away from the Sun (G). It takes another full revolution, a second 88 days, for the arrow to again face the Sun. Thus the rotation period with respect to the Sun is twice 88, or 176, days.

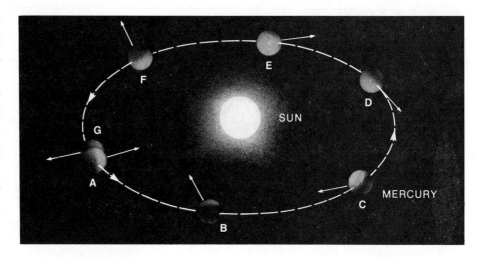

In the kelvin temperature scale, 1 kelvin (symbol K) is equal to 1°C. But the zero point of the kelvin scale is 273°C lower than the zero point of the Celsius scale. To find the temperature in kelvins, add 273 to the temperature in °C. Appendix 10 discusses the kelvin, Celsius, and Fahrenheit scales.

The **albedo** *(from the Latin for whiteness) is the ratio of light reflected from a body to light received by it. The concept of albedo is widely used in the study of solid bodies in our solar system.*

FIGURE 7–3 *Albedo* is the fraction of the radiation hitting an object that is reflected. A surface of low albedo looks dark.

Since each point on Mercury faces the Sun at some time, the heat doesn't build up forever at the place under the Sun nor does the coldest point cool down as much as it would if it never received sunlight. The hottest temperature is about 700 K (800°F). The minimum temperature is about 100 K (−280°F).

No harm was done by the scientists' original misconception of Mercury's rotational period, but the story teaches all of us a lesson: we should not be too sure of so-called facts. Don't believe everything you read here, either.

7.2 MERCURY FROM THE GROUND

Even though the details of the surface of Mercury can't be seen very well from the Earth, other properties of the planet can be better studied. For example, we can measure Mercury's *albedo,* the fraction of the sunlight hitting Mercury that is reflected (Fig. 7–3). We can measure the albedo because we know how much sunlight hits Mercury (we know the brightness of the Sun and the distance of Mercury from the Sun). Then we can easily calculate at any given time how much light Mercury reflects, from both (1) how bright Mercury looks to us and (2) its distance from the Earth. Once we have a measure of the albedo, we can compare it with the albedo of materials on the Earth and on the Moon and thus learn something of what the surface of Mercury is like.

Let us consider some examples of albedo. An ideal mirror reflects all the light that hits it; its albedo is thus 100 per cent. (The very best real mirrors have albedos of as much as 96 per cent.) A black cloth reflects essentially none of the light that hits it; its albedo (in the visible part of the spectrum, anyway) is almost 0 per cent. Mercury's overall albedo is only about 6

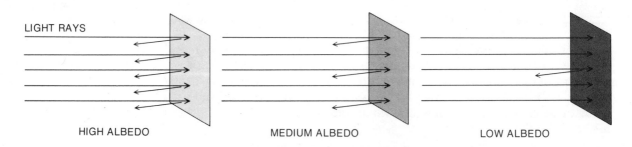

LIGHT RAYS

HIGH ALBEDO   MEDIUM ALBEDO   LOW ALBEDO

per cent. Its surface, therefore, must be made of a dark—that is, poorly reflecting—material. The albedo of the Moon is similarly low. In fact, Mercury (or the Moon) appears bright to us only because it is contrasted against a relatively dark sky; if it were silhouetted against a bedsheet, it would look relatively dark, as if it had been washed in Brand X instead of Tide.

Mercury's density (its mass divided by its volume) is roughly the same as the Earth's (Appendix 3). So Mercury's core, like those of Venus and the Earth, must be heavy; it too must be made of iron.

7.3 MARINER 10

In 1974, we learned most of what we know about Mercury in a brief time. We flew right by. The tenth in the series of Mariner spacecraft launched by the United States went to Mercury. First it passed by Venus and then had its orbit changed by Venus' gravity to direct it to Mercury.

When Mariner 10 flew by Mercury the first time it took 1800 photographs. The most striking overall impression is that Mercury is heavily cratered. At first glance, it looks like the Moon! But there are several basic differences between the features on the surface of Mercury and those on the lunar surface. For example, Mercury's craters seem flatter than those on the Moon, and have thinner rims (Fig. 7–4). Mercury's higher surface gravity may have caused the rims to slump more.

Some craters have rays of higher albedo emanating from them (Fig. 7–5), just as some lunar craters do. The ray material represents relatively recent crater formation (that is, within the last hundred million years). The ray material must have been tossed out in the impact that formed the crater.

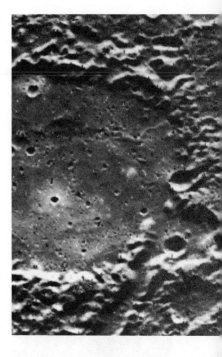

FIGURE 7–4 Mercury, photographed from a distance of 35,000 km as Mariner 10 approached the planet for the first time, shows a heavily cratered surface with many low hills. The valley at the bottom is 7 km wide and over 100 km long. The large flat-floored crater is about 80 km in diameter.

A   **B**

FIGURE 7–5 *(A)* Many rayed craters can be seen on this photograph of Mercury, taken six hours after Mariner 10's closest approach on its first pass. The north pole is at the top and the equator extends from left to right about ⅔ of the way down from the top. *(B)* A field of rays radiating from a crater off to the top left, photographed on the second pass of Mariner 10. The crater at top is 100 km in diameter.

BOX 7.1
Naming the Features of Mercury

Seeing features on the surface of Mercury led to a need for names. The scarps were named for historical ships of discovery and exploration, such as Endeavour (Captain Cook's ship), Santa Maria (Columbus's ship), and Victoria (the first ship to sail around the world, which it did in 1519–1522 under Magellan and his successors). Some plains were given the name of Mercury in different languages, such as Tir (in ancient Persian), Odin (an ancient Norse god), and Suisei (Japanese). Craters were named for non-scientific authors, composers, and artists, in order to complement the lunar naming system, which honors scientists.

FIGURE 7–6 A scarp *(arrow)* more than 300 km long extends from top to bottom in this second-pass picture.

FIGURE 7–7 This first-pass view of Mercury's northern limb shows a prominent scarp *(arrow)* extending from the limb near the middle of the photograph. The photograph shows an area 580 km from side to side.

Lines of cliffs hundreds of miles long are visible on Mercury; on Mercury, as on Earth, such lines of cliffs are called *scarps*. The scarps are particularly apparent in the region of Mercury's south pole (Figs. 7–6 and 7–7). Unlike fault lines on the Earth, such as the San Andreas fault in California, on Mercury there are no signs of geologic tensions like rifts or fissures nearby. These scarps are global in scale, not just isolated.

The scarps may actually be wrinkles in Mercury's crust. Mercury's core, judging by the fact that Mercury's average density is about the same as the Earth's, is probably iron and takes up perhaps 50 per cent of the volume or 70 per cent of the mass. Perhaps the core was once molten, and shrank by 1 or 2 km as it cooled. This shrinking would have caused the crust to buckle, creating the scarps in the quantity that we now observe.

One part of the Mercurian landscape seems particularly different from the rest (Fig. 7–8). It seems to be grooved, with relatively smooth areas between the grooves. It is called the "weird terrain." No other areas like this are known on Mercury and only a couple have been found on the Moon. The weird terrain is 180° around Mercury from the Caloris Basin, the site of a major meteorite impact. Shock waves from that impact may have been focused halfway around the planet.

Data from Mariner's infrared radiometer indicate that the surface of Mercury is covered with fine dust, as is the surface of the Moon, to a depth of at least several centimeters. Astronauts sent to Mercury, whenever they go, will leave footprints behind.

The biggest surprise of the mission was the detection of a magnetic field in space near Mercury. The field is weak; the value calculated for Mercury's surface on the basis of the spacecraft measurements shows that it is only about 1 per cent of the Earth's surface field. It had been thought that magnetic fields were generated by the rapid rotation of molten iron cores in planets, but Mercury is so small that its core would have quickly solidified. So the magnetic field is not now being generated. Perhaps the magnetic field

:as been frozen into Mercury since the time when its core was molten. We don't know.

Scientists and engineers were able to find an orbit around the Sun that brought Mariner 10 back to Mercury several times over. Every six months the spacecraft and Mercury returned to the same place at the same time. On its second visit, in September 1974, Mariner 10 studied the region around Mercury's south pole for the first time. On its third visit, in March 1975, it had the closest encounter ever—only 300 km above the surface (Fig. 7–9). Thus it was able to photograph part of the surface with such especially high resolution. Then the spacecraft ran out of gas for the small jets that control its pointing, so even though it still passes close to Mercury every few months, it can no longer take clear photographs or send them back to Earth.

No moons of Mercury have ever been detected, although Mariner 10 did an especially careful job of searching.

FIGURE 7–8 This mosaic of pictures from the first pass of Mariner 10 shows a "weird" terrain—hills and ridges cutting across many of the craters and the intercrater areas.

FIGURE 7–9 A high-resolution view of the fractured and ridged plains of the Caloris Basin, photographed from a distance of 21,000 km on Mariner 10's third pass, 34 minutes after the spacecraft's closest approach.

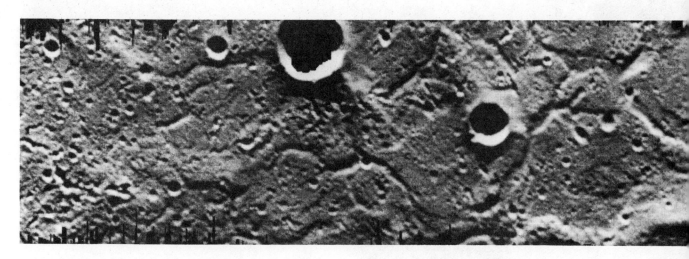

KEY WORDS
Astronomical Unit (A.U.), radar, transit, albedo, scarps

QUESTIONS

1. Assume that on a given day, Mercury sets after the Sun. Draw a diagram, or a few diagrams, to show that the height of Mercury above the horizon depends on the angle that the Sun's path in the sky makes with the horizon as the Sun sets. Discuss how this depends on the latitude or longitude of the observer.

2. If Mercury did always keep the same side toward the Sun, does that mean that the night side would always face the same stars? Draw a diagram to illustrate your answer.

3. Explain why a day on Mercury is 176 Earth days long.

4. What did radar tell us about Mercury? How?

5. If ice has an albedo of 70–80 per cent, and basalt has an albedo of 5–20 per cent, what can you say about the surface of Mercury based on its measured albedo?

6. If you increased the albedo of Mercury, would its temperature increase or decrease? Explain.

7. List those properties of Mercury that could best be measured by spacecraft observations.

8. How would you distinguish an old crater from a new one?

9. List three major findings of Mariner 10.

Chapter 8

Venus, photographed from the Pioneer Venus Orbiter spacecraft in 1979. The picture shows the cloud tops; we see a turbulent atmosphere. The small mottled features near the center appear to be cells of convection—the method of transferring energy we see in a boiling pot of water—caused by the heating of the atmosphere by solar radiation. The symbol for Venus appears at the top of the facing page.

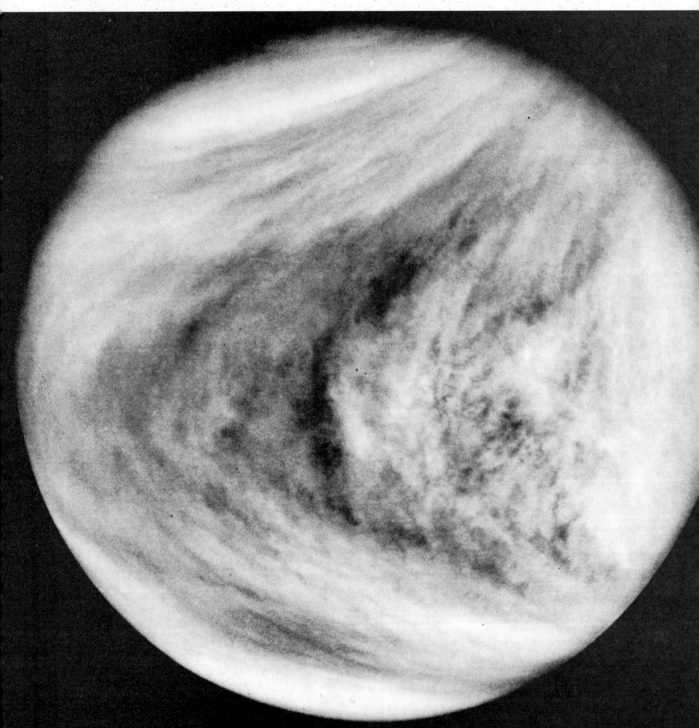

VENUS: SO DIFFERENT FROM EARTH

Venus and the Earth are sister planets: their sizes, masses and densities are about the same. But they are as different from each other as the wicked sisters were from Cinderella. The Earth is lush; it has oceans and rainstorms of water, an atmosphere containing oxygen, and creatures swimming in the sea, flying in the air, and walking on the ground. On the other hand, Venus is a hot, foreboding planet with temperatures constantly over 750 K (900°F), a planet on which life seems unlikely to develop. Why is Venus like that? How did these harsh conditions come about? Can it happen to us here on Earth?

Venus orbits the Sun at a distance of 0.7 A.U. Although it comes closer to us than any other planet—as close as 45 million kilometers—we did not know much about it until recently because it is always shrouded in heavy clouds (Fig. 8–1).

8.1 THE ATMOSPHERE OF VENUS

Studies from the Earth show that the clouds on Venus are primarily composed of droplets of sulfuric acid, H_2SO_4, with water droplets mixed in. Sulfuric acid may seem like a peculiar constituent of a cloud, but the Earth, too, has a significant layer of sulfuric acid droplets in its stratosphere, a higher layer of the atmosphere. However, the water in the lower layers of the Earth's atmosphere, circulating because of weather, washes the sulfur compounds out of these layers. Venus has sulfur compounds in the lower layers of its atmosphere in addition to those in its clouds.

Sulfuric acid takes up water very efficiently, so there is little water vapor above Venus' clouds. Painstaking work conducted at high-altitude sites on the Earth revealed the presence of the small amount of venusian water vapor above Venus' clouds. This observation was difficult, because the signs of water vapor from Venus were masked by the water vapor in the Earth's own atmosphere.

Observations from Earth show a high concentration of carbon dioxide in the atmosphere of Venus. In fact, carbon dioxide makes up over 90 per cent of Venus' atmosphere (Fig. 8–2). The Earth's atmosphere, in comparison, is mainly nitrogen, with a fair amount of oxygen as well. Carbon dioxide makes up less than 1 per cent of the terrestrial atmosphere.

Because of the large amount of carbon dioxide in Venus' atmosphere, its surface pressure is 90 times higher than the pressure of Earth's atmosphere. Carbon dioxide on Earth dissolved in sea water and eventually formed our terrestrial rocks, often with the help of life forms. (Limestone, for example, has formed from marine life under the Earth's oceans.) If this carbon dioxide were released from the Earth's rocks, along with other carbon dioxide trapped in sea water, our atmosphere would become as dense and have as high a pressure as that of Venus. Venus, slightly closer to the

FIGURE 8–1 A crescent Venus, observed with the Palomar Observatory 5-m telescope in blue light. We see only a layer of clouds.

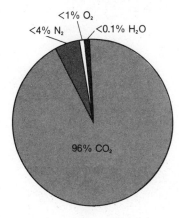

FIGURE 8–2 The composition of Venus' atmosphere.

Sun than Earth and thus hotter, had no oceans in which the carbon dioxide could dissolve nor life to help take up the carbon. Thus the carbon dioxide remains in Venus' atmosphere.

8.2 THE ROTATION OF VENUS

In 1961, radar astronomy penetrated Venus' clouds, allowing us to determine accurately how fast Venus rotates. Venus rotates in 243 days with respect to the stars in the direction opposite from the other planets. Venus revolves around the Sun in 225 Earth days. Venus' periods of rotation and revolution combine, in a way similar to that in which Mercury's sidereal day and year combine (Section 7.1), so that a solar day on Venus corresponds to 117 Earth days; that is, the planet's rotation brings the Sun back to the same position in the sky every 117 days.

The slow rotation of Venus' solid surface contrasts with the rapid rotation of its clouds. The tops of the clouds rotate in the same sense as the surface rotates but about 60 times more rapidly, once every 4 days.

8.3 THE TEMPERATURE OF VENUS

We can determine the temperature of Venus' surface by studying its radio emission, since radio waves emitted by the surface penetrate the clouds. The surface is very hot, about 750 K (900°F).

In addition to measuring the temperature on Venus, scientists theoretically calculate what the temperature would be if Venus' atmosphere allowed all radiation hitting it to pass through it. This value—less than 375 K (215°F)—is much lower than the measured values. The high temperatures derived from radio measurements indicate that Venus traps much of the solar energy that hits it.

The process by which this happens on Venus is similar to the process that is generally—though incorrectly—thought to occur in greenhouses here on the Earth. The process is thus called the *greenhouse effect* (Fig. 8–3): Sunlight passes through the venusian atmosphere in the form of radiation in the

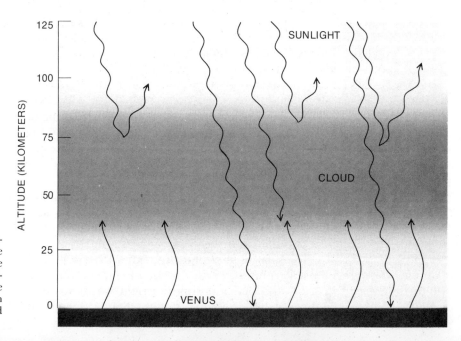

FIGURE 8–3 Sunlight can penetrate Venus' clouds, so the surface is illuminated with radiation in the visible part of the spectrum. Venus' own radiation is mostly in the infrared. This infrared radiation is trapped, a phenomenon called the greenhouse effect.

visible part of the spectrum. The sunlight is absorbed by the surface of Venus, which heats up. At the temperatures that result, the radiation that the surface gives off is mostly in the infrared. But the carbon dioxide and other constituents of Venus' atmosphere are together opaque to infrared radiation, so the energy is trapped. Thus Venus heats up far above the temperature it would reach if the atmosphere were transparent. The surface radiates more and more energy as the planet heats up. Finally, a balance is struck between the rate at which energy flows in from the Sun and the rate at which it trickles out (as infrared) through the atmosphere (Fig. 8–4). The situation is so extreme on Venus that we say a "runaway greenhouse effect" is taking place there. Understanding such processes involving the transfer of energy is but one of the practical results of the study of astronomy.

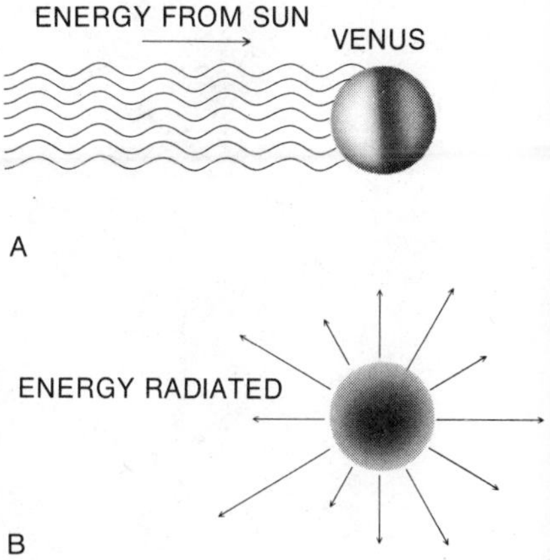

Greenhouses on Earth don't work quite this way. In actual greenhouses, closed glass on Earth prevents the mixing of air inside, heated by conduction from the warmed ground, with cooler outside air. The trapping of solar energy by the "greenhouse effect"—the inability of infrared radiation, once formed, to get out—is a less important process in an actual greenhouse or in a car when it is left in the sun. Try not to be bothered by the fact that the greenhouse in your backyard is heated not by the "greenhouse effect" but by another way of passively trapping solar energy.

If the Earth's atmosphere were to gain enough carbon dioxide, the oceans could boil, and the carbon in rocks could be released and enter the atmosphere as carbon dioxide. This would increase the greenhouse effect, which could also become a runaway. We are presently increasing the amount of carbon dioxide in the Earth's atmosphere by burning fossil fuels; Venus has shown us how important it is to be careful about the consequences of our energy use so that we do not disturb the balance in our atmosphere.

8.4 EARTHBOUND RADAR MAPPING

We can study the surface of Venus by using radar to penetrate Venus' clouds. Radars using huge Earth-based radio telescopes, such as the giant 1000-ft (305-m) dish at Arecibo, Puerto Rico, have mapped a small amount of Venus' surface with a resolution of up to about 20 km. (By that resolution, we mean that only features larger than about 20 km in size can be detected.) Regions that reflect a large percentage of the radar beam back at Earth show up as bright, and other regions as dark. The Earth-based radar maps of Venus (Fig. 8–5) show large-scale surface features, some very rough and others relatively smooth, similar to the variation we find of surfaces on the Moon. Many round areas that are probably craters have been found, ranging up to 1000 km across. Most have probably been formed by meteoritic impact.

From Venus' size and from the fact that its mean density is similar to that of the Earth, we conclude that its interior is also probably similar to that of the Earth. This means that we expect to find volcanoes and mountains on Venus, and that venusquakes probably occur too. A bright area near a large plain has been called Maxwell; we will see later on that Maxwell turns out to be a huge mountain, perhaps a volcano.

A huge peak, Beta, whose base is over 750 km in radius and which has a depression 40 km in radius at its summit, has also been detected by radar. It too is probably a giant volcano. It has long tongues of rough material extending as far as 480 km from it in an irregular fashion, like lava flows.

A long trough at Venus' equator, 1500 km in length, seems to resemble the Rift Valley in East Africa, the Earth's largest canyon.

ENERGY FROM SUN   VENUS

A

ENERGY RADIATED

B

FIGURE 8–4 Venus receives energy from only one direction, and heats up and radiates energy (mostly in the infrared) in all directions. From balancing the energy input and output, astronomers can calculate what Venus' temperature would be if Venus had no atmosphere. This type of calculation—balancing quantities theoretically to make a prediction to be compared with observations—is typical of those made by astronomers.

In Section 8.5, we will see how spacecraft radar observations have mapped Venus' surface more completely.

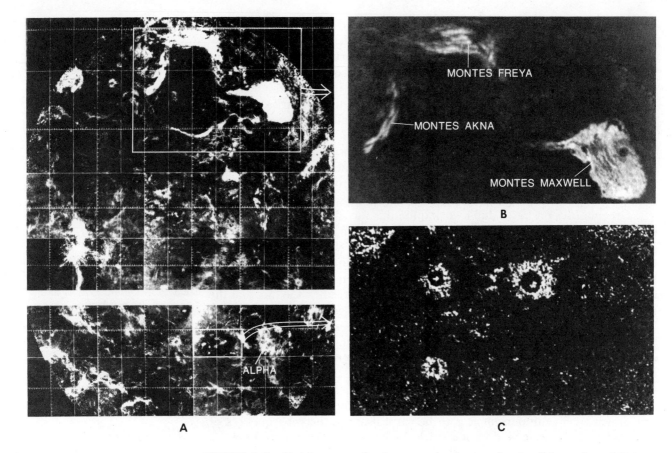

FIGURE 8–5 *(A)* A montage of radar maps showing one-fourth of the surface of Venus. Whiter areas show regions where more power bounces back to us from Venus, which usually corresponds to rougher terrain than darker areas. *(B)* The low reflectivity region Planum Lakshmi (a plain) and the high reflectivity region Montes Maxwell (mountains) make up most of a continental-sized region, Terra Ishtar. This image has a resolution of 10 km. *(C)* The craters are 30 to 65 km in diameter. They are relatively smooth and are surrounded by rough areas of ejected material. The craters are probably the result of impacts on the surface, but a volcanic origin cannot be ruled out as yet. This image, released in 1983, has a resolution of 3 km. (Courtesy of D. B. Campbell, Arecibo Observatory)

8.5 SPACECRAFT OBSERVATIONS

Venus was an early target of both American and Soviet space missions. During the 1960's, American spacecraft flew by Venus, and Soviet spacecraft dropped through its atmosphere. In 1970 and 1972, Soviet spacecraft survived the high temperature and pressure on the surface of Venus long enough to confirm the Earth-based results of high temperatures, pressures, and carbon dioxide content.

The United States spacecraft Mariner 10 took thousands of photographs of Venus in 1974 as it passed by en route to Mercury. Structure in the clouds shows only when viewed in ultraviolet light (Fig. 8–6). We could observe much finer details than from Earth. The clouds appeared as long, delicate streaks, like terrestrial cirrus clouds.

No magnetic field was detected. This may indicate either that Venus does not have a liquid core or that the core does not have a high conductivity.

FIGURE 8–6 A series of photographs of the circulation of the clouds of Venus, photographed at 7-hour intervals from Mariner 10 at ultraviolet wavelengths slightly shorter than the human eye can see. The contrast has been electronically enhanced so that small actual differences in contrast are made apparent. The size of the feature indicated by the arrows is about 1000 km.

Such studies of Venus have great practical value. The principles that govern weather on Venus are the same that govern weather on Earth. The better we understand the interaction of solar heating, planetary rotation, and chemical composition in setting up an atmospheric circulation, the better we will understand our Earth's atmosphere. We then may be better able to predict the weather and discover jet routes that would aid air travel, for example. The potential financial return from this knowledge is enormous: it would be many times the investment we have made in planetary exploration.

Since 1975, a series of Soviet spacecraft have landed on Venus and sent back photographs. They also found that the soil resembles basalt in chemical composition and density, in common with the Earth, the Moon, and Mars.

The United States and the Soviet Union both sent spacecraft to Venus in 1978. NASA's Pioneer Venus 1 was the first spacecraft to go into orbit around Venus, allowing it to make observations over a lengthy time period. Its dozen experiments included cameras to study Venus' weather by photographing the planet regularly in ultraviolet light (Fig. 8–7).

The spacecraft also carried a small radar to study the topography of Venus' surface. The resolution was not quite as good as that of the best Earth-based radar studies of Venus, but the orbiting radar mapped a much wider area. It mapped over 90 per cent of the surface with a resolution typically better than 100 km. It was an early version of a new method of radar, called *synthetic-aperture radar*. In this method, the radar images are recorded over a period of time while the spacecraft moves. The views from these different aspects are assembled during the data reduction, giving resolution equivalent to a radar of much larger aperture; the larger aperture has been "synthesized."

FIGURE 8–7 Four photographs of Venus taken over a one-month span. They show relatively dark equatorial bands with many small features, probably from convection, superimposed. Polar bands of clouds are relatively bright.

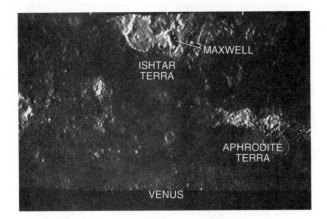

FIGURE 8–8   Venus' surface, based on the Pioneer Venus Orbiter's radar map, compared with Earth's surface at the same resolution (50–100 km). Two continents exist on Venus: Aphrodite Terra, which is comparable in scale with Africa, and Ishtar Terra, which is comparable in scale with the continental United States or Australia. The volcano Maxwell, 10.8 km above venusian "sea level" (the average radius of Venus), is on the right of Ishtar Terra and is the highest point on Venus. Sixty per cent of Venus' surface is covered with a huge rolling plain. Only about 16 per cent of Venus' surface is covered with lowlands, compared with over two-thirds of Earth's surface covered wtih oceans. Mid-ocean ridges, a sign of spreading plates, are not visible on Venus though this comparison shows that they would be detectable on Earth at this resolution.

Aphrodite, the Greek goddess of love, was the equivalent of the Roman goddess Venus. Ishtar was the Babylonian goddess of love and war, daughter of the moon and sister of the sun. Major features on Venus are being named after mythical goddesses, minor features after other mythical female figures, and still smaller circular features after famous women.

FIGURE 8–9   Pioneer Venus 1, the orbiter, is in the background. Pioneer Venus 2, which contained the basic cylinder called the "bus," is in the foreground. A large probe and three smaller probes (of which one is hidden) were mounted on the bus.

From the radar map (Color Plate 25 and Fig. 8–8), we now know that 60 per cent of Venus' surface is covered by a rolling plain, flat to within plus or minus 1 km. Only about 16 per cent of Venus' surface lies below this plain, a much smaller fraction than the two-thirds of the Earth covered by ocean floor.

A large feature, the size of small Earth continents, extends several km above the mean elevation. This northern continent, Terra Ishtar, is about the size of the continental United States. A giant mountain on it known as Maxwell is 11 km high, 2 km taller than Earth's Mt. Everest stands above terrestrial sea level.

A smaller elevated feature, Beta Regio, on the basis of Pioneer Venus and terrestrial radar results and on the results of the Soviet landers, appears to be a pair of volcanoes. Another highland feature, Alpha Regio, has rough terrain that may resemble the western United States. A giant canyon, 1500 km long, is 5 km deep and 400 km wide. It is the largest in the solar system and may be left over from an earlier stage of Venus' geological evolution.

Venus, unlike Earth, is apparently made of only one continental plate. We observe nothing on Venus equivalent to Earth's mid-ocean ridges, at which new crust is carried up. Venus may well have such a thick crust that any plate tectonics that existed in the distant past was choked off.

The orbiter has provided evidence that volcanoes may be active on Venus. The abundance of sulfur dioxide it found on arrival was far higher than the upper limits of previous observations, and then dropped by a factor of 10 in the next five years. Over the same five-year period, the haze dropped drastically, as it had about 20 years earlier. The effect could come from eruptions at least ten times greater than Mexico's El Chichon in 1982.

FIGURE 8–10   The view from Venera 13, showing a variety of sizes and textures of rocks. The spacecraft landed in the foothills of a mountainous region south of Venus' equator and below the region Beta. The spacecraft survived for 127 minutes.
  The camera first looked off to the side, then scanned downward as though looking at its feet, and then scanned up to the other side. As a result, opposite horizons are visible as slanted boundaries at upper left and right.

Pioneer Venus 2 arrived at Venus at about the same time as Pioneer Venus 1, and carried several spacecraft together on a basic "bus" (Fig. 8–9). The bus and its four probes all penetrated Venus' atmosphere and descended to the surface. The Pioneer probes found that high-speed winds at the upper levels are coupled to other high-speed winds at lower altitudes. The lowest part of Venus' atmosphere, however, is relatively stagnant.

Both American and Soviet missions, from the radio signals they detected, reported lightning at a frequent rate. The Soviet landers detected up to 25 pulses of energy per second between 5 and 11 km above the surface, later confirmed by the Pioneer Venus Orbiter. The lightning is concentrated over Beta Regio and the eastern part of Aphrodite Terra. This is strong evidence that those regions are sites of active volcanoes. On Earth, it is common for the dust and ash ejected by volcanoes to rub together, generating static electricity that produces lightning. But the lightning may be produced in other ways.

Two Soviet spacecraft arrived at Venus in March 1982. Photographs they took revealed a variety of sizes and shapes of rocks (Figs. 8–10 and 8–11). The color photographs showed that Venus' sky is orange. The landing

Now that the Pioneer Venus results have enabled us to understand the greenhouse effect on Venus, we can much better understand the effect of adding carbon dioxide from burning fossil fuels to Earth's climate. We have already increased the amount of carbon dioxide in Earth's atmosphere by 15 per cent; some scientists have predicted that it may even double in the next 50 years, which could result in a worldwide temperature rise of a few °C and eventually even in a runaway greenhouse effect. But such long-range predictions are very uncertain.

FIGURE 8–11   The view from Venera 14, showing flat structures without the smaller rocks of the other site. The flat structures may be large flat rocks or may be fine material held together. The site is 1000 km east of the Venera 13 site and at a lower elevation. The chemical composition resembles that of basalts found on the Earth's ocean floor and the Moon's maria. Venera 13, on the other hand, found basalts more typical of the Earth's continental crust and the Moon's highlands.

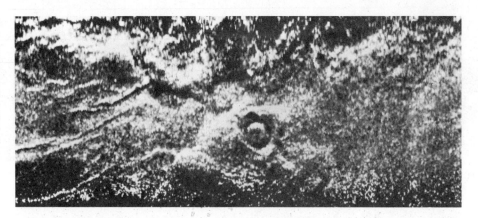

FIGURE 8–12 A radar image apparently showing a young volcanic crater about 40 km wide, transmitted from the Venera 15 or 16 spacecraft in 1983. The region is Metis Regio, a highland area about 500 km across located just west of Ishtar. The linear features at left may be scarps and faults, perhaps formed or modified by vulcanism. The resolution is 1 or 2 km.

sites were chosen in consultation with American scientists on the basis of Pioneer Venus data, the first example of Soviet-American cooperation on planetary exploration.

The Soviet Veneras 15 and 16 have been orbiting Venus since October 1983. They are carrying infrared equipment to study the atmosphere, and a radar for determining topography (Fig. 8–12).

Our recent space results, coupled with our ground-based knowledge, show us that Venus is even more different from the Earth than had previously been imagined. Among the differences are Venus' slow rotation, its one-plate surface, the absence of a satellite, the extreme weakness or absence of a magnetic field, the lack of water in its atmosphere, and its high surface temperature.

The most recent mission to Venus was a pair of joint Soviet-French probes carried in 1985 by Soviet rockets en route to Halley's Comet. The mission was called "Vega": "Ve" for Venus and "ga" for the Russian-language equivalent of Halley (there is no "h" in the Russian alphabet).

Each Vega spacecraft deployed both a lander and a balloon. The lan-

FIGURE 8–13 The Venus Radar Mapper will use a synthetic-aperture radar to obtain images of the surface that will appear like optical photographs, showing enough detail to be useful for geological studies. A 1988 launch is hoped for.

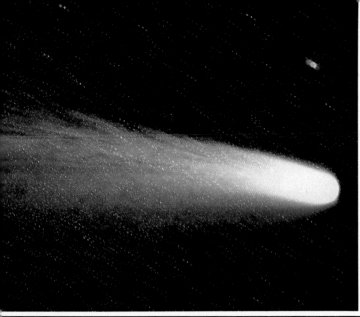
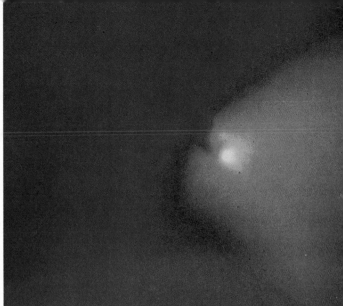

Color Plate 12 (top left): Halley's Comet on March 12, 1986. The print is combined from three single-color originals. Since the telescope was tracking the comet rather than the stars, star images appear once in each color. (credit: David Malin, ©1986 Anglo-Australian Telescope Board)

Color Plate 13 (top right): Halley's Comet, seen from the Giotto spacecraft, at a distance of 18,000 km. Its nucleus is the dark potato-shaped object at upper left, 15 km long. Two bright gas jets point rightward toward the sun.

Color Plate 14 (center): Apollo 17 view of the Taurus-Littrow Valley on the Moon. This huge, fragmented boulder had rolled a kilometer down the side of the North Massif to here. Scientist-astronaut Harrison Schmitt is at the left. The Lunar Rover is on the right. (NASA photo)

Color Plate 15 (bottom, left): The blast-off of Apollo 17, a night launch. (Photo by the author)

Color Plate 16 (bottom, center): An Apollo 17 astronaut with an instrument package. (NASA photo)

Color Plate 17 (bottom, right): The crescent Earth seen from Apollo 11 in orbit around the Moon. (NASA photo)

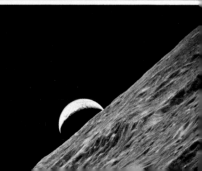

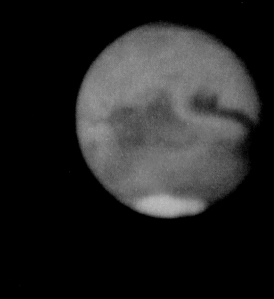

Color Plate 18 (top, left): Earth photographed from Apollo 11. Africa and the Middle East are clearly visible. (NASA photo)

Color Plate 19 (top, right): Mars, photographed from Earth as part of the International Planetary Patrol.

Color Plate 20 (bottom, left): Mars, photographed from the Viking 1 spacecraft as it approached in June 1976. Olympus Mons, the large volcano, is toward the top right of the picture. The Tharsis Mountains, a row of three other volcanoes, are also visible. To the left of these volcanoes, the irregular white area may be surface frost or ground fog. The large impact basin, Argyre, is the circular feature at the bottom of the disk. (NASA photo)

Color Plate 21 (bottom, right): The other side of Mars, photographed from Viking 1. The giant canyon in the upper hemisphere has a length larger than the coast-to-coast diameter of the United States.

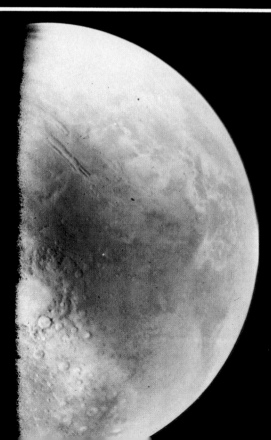

Color Plate 22 (top): Viking 1's sampler scoop is in the foreground of this view from the Martian surface. Angular rocks of various types can be seen. Large blocks one to two meters across can be seen on the horizon, which is about 100 meters from the spacecraft. The horizon may be the rim of a crater. (NASA photo)

Color Plate 23 (bottom, left): Mars' Utopia Planitia as seen from Viking 2, with the Lander's boom in the foreground. Many of the rocks are porous and sponge-like, similar to some of the Earth's volcanic rocks. (NASA photo; color-corrected version courtesy of Friedrich O. Huck)

Color Plate 24 (bottom, right): Mars' Utopia Planitia as seen from Viking 2. This winter scene reveals frost under some of the rocks. (NASA photo; color-corrected version courtesy of Friedrich O. Huck)

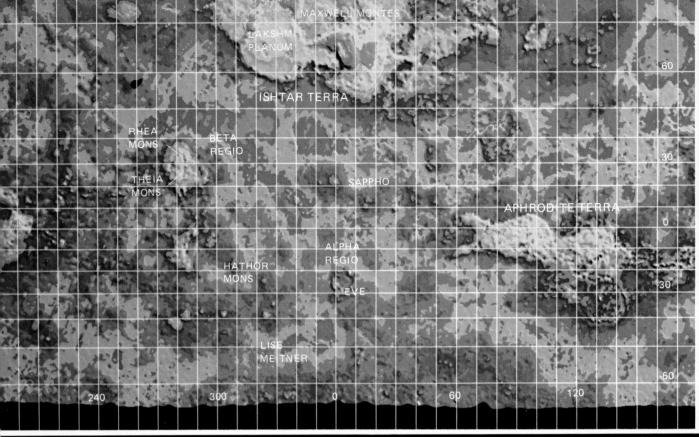

Color Plate 25 (top): A radar map of Venus from Pioneer Venus Orbiter. Most of Venus is covered by a rolling plain, shown in green and blue. The highlands (yellow and brown contours) sit atop the plain, like continents. Aphrodite Terra is half as big as Africa. Ishtar Terra is the size of Australia, but looks relatively large in this Mercator projection. The lowlands, though resembling Earth's ocean basins, cover only 16 percent of the planet. (Experiment and data — MIT; maps — U.S. Geological Survey; NASA/Ames spacecraft)

Color Plate 26 (bottom): An aurora borealis photographed in the Goldstream Valley, Alaska. (Gustav Lamprecht photo)

Color Plate 27 (top): Jupiter, its Great Red Spot, and 2 of its moons, in this Voyager 1 photograph. The satellite Io can be seen against Jupiter's disk. The satellite Europa is visible off the limb at the right.

Color Plate 28 (bottom): Jupiter's Great Red Spot, photographed from Voyager 2. Gas in both the Spot and the adjacent white oval is circulating in the same direction.

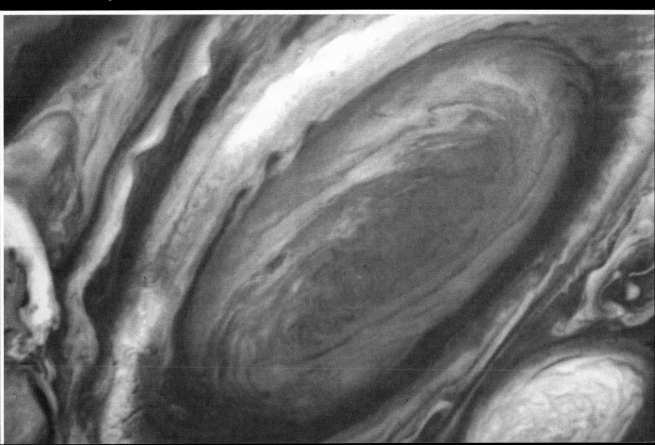

Color Plate 29 (top): Jupiter, photographed from Voyager 1. Io is visible above the Great Red Spot. Europa is also visible.

Color Plate 30 (bottom): (A) Io, photographed at a range of 826,000 km from Voyager 1. (B) Volcanoes erupting on Io, photographed from Voyager 2. Two volcanic eruption plumes that are about 100 km high and strongly scatter blue light appear on the limb.

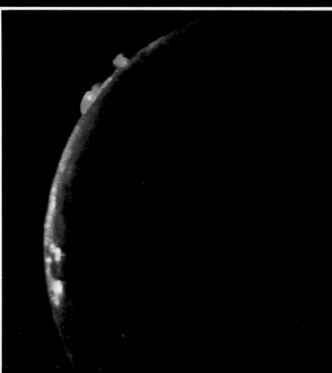

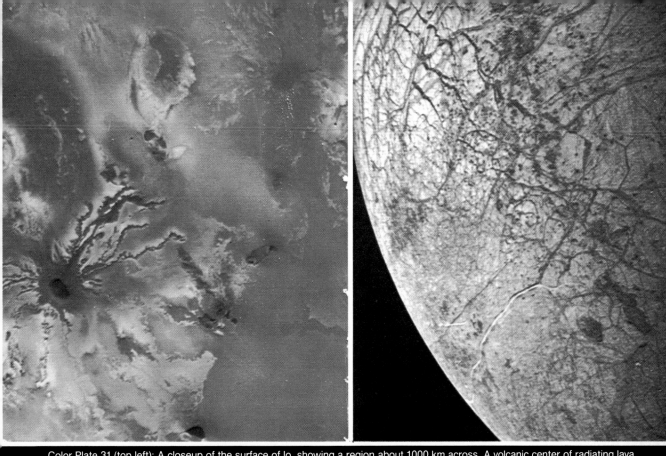

Color Plate 31 (top left): A closeup of the surface of Io, showing a region about 1000 km across. A volcanic center of radiating lava flows is visible at left center.

Color Plate 32 (top right): Europa, photographed from Voyager 2, showing its very smooth and fractured crust.

Color Plate 33 (bottom left): Ganymede, photographed from Voyager 2. The dark, cratered, circular feature is about 3200 km in diameter.

Color Plate 34 (bottom right): Callisto, photographed from Voyager 1. The bull's-eye, named Valhalla, is a large impact basin. The outer ring is about 2600 km across.

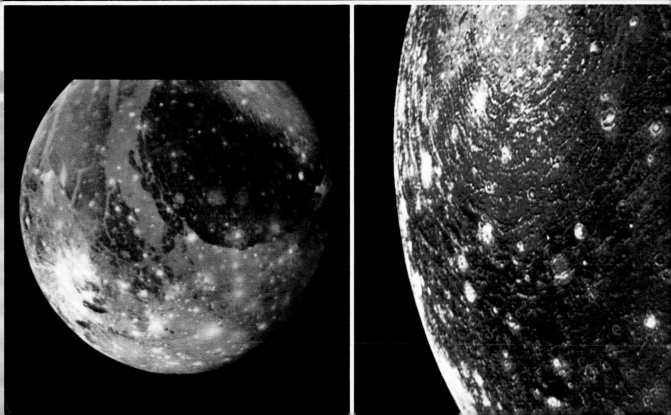

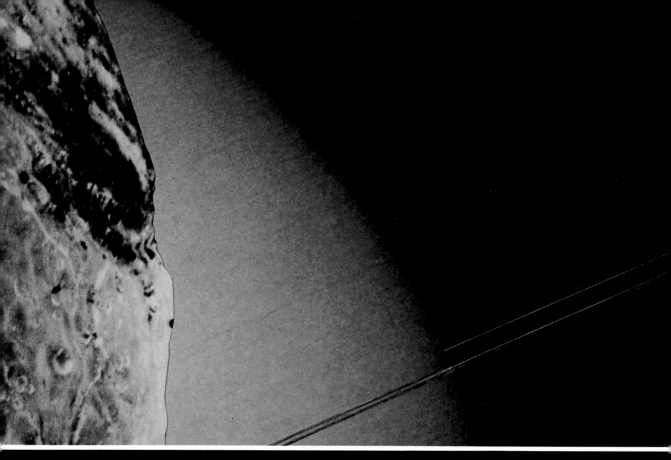

Color Plate 35 (top): Rings drawn by an artist on a Voyager 2 image of Uranus. A Voyager 2 image of Miranda, showing one of its giant canyons, is superimposed.

Color Plate 36: A montage of Saturn and some of its moons, photographed from the Voyager 1 and 2 spacecraft. Clockwise from upper right, the moons are Titan (reddish), Iapetus (with one dark side), Tethys, Mimas (a small moon with a giant crater), Enceladus (lower left), Dione, and Rhea. (JPL/NASA)

ders analyzed the clouds on their way down, and were equipped to sample the surface material and study the relative amounts of elements present. They detected lightning, a sign of volcanic activity. Each helium-filled balloon sent data for 46 hours about the densest cloud layer, 55 km in altitude; from Earth, scientists monitored how they moved in the wind. Each balloon traveled 12,000 km across Venus in strong winds blowing 250 km/hr.

NASA plans to launch a radar mapper called Magellan in 1988 (Fig. 8–13). Its synthetic-aperture radar should map most of Venus with ¼- to ½-km resolution, compared with the 20-km best resolution of Pioneer Venus and the 1-km best resolution of Venera 15 and 16. It will cover Venus' equatiorial regions while Venera 15 and 16 covered the north polar region. Maybe, with resolution on the order of 1 km, we can detect volcanic changes on Venus' surface during the 1-Venus-year lifetime of the Venera or Venus Radar Mapper missions.

KEY WORDS

greenhouse effect, synthetic-aperture radar

QUESTIONS

1. Make a table displaying the major similarities and differences between the Earth and Venus.

2. Why does Venus have more carbon dioxide in its atmosphere than does the Earth?

3. Why do we think that there have been significant external effects on the rotation of Venus?

4. If observers from another planet tried to gauge the rotation of the Earth by watching the clouds, what would they find?

5. Suppose a planet had an atmosphere that was opaque in the visible but transparent in the infrared. Describe how the effect of this type of atmosphere on the planet's temperature differs from the greenhouse effect.

6. Why do radar observations of Venus provide more data about the surface structure than a flyby with close-up cameras?

7. If one removed all the CO_2 from the atmosphere of Venus, the pressure of the remaining constituents would be how many times the pressure of the Earth's atmosphere?

8. How might the Earth's atmosphere change if it was moved to Venus' position in the solar system? If the Earth had been formed there?

9. Why do we say that Venus is the Earth's "sister planet"?

10. Do radar observations of Venus study the surface or the clouds? Explain.

11. Describe the major radar results from the Venus Pioneer Orbiter.

12. What is the advantage of synthetic-aperture radar over ordinary radar?

13. What signs of vulcanism are there on Venus?

Chapter 9

A view from Viking Orbiter 2 as it approached the dawn side of Mars, in early August 1976. At the top, with water-ice cloud plumes on its western flank, is Ascreaus Mons, one of the giant Martian volcanoes. In the middle is the great rift canyon called Valles Marineris, and near the bottom is the large, frosty crater basin called Argyre. The south pole is at the bottom. The symbol for Mars appears at the top of the facing page.

MARS: OUR HOPE FOR LIFE

Mars has long been the planet of greatest interest to scientists and nonscientists alike. Its interesting appearance as a reddish object in the night sky (Color Plate 19) coupled with some past scientific studies have made Mars the prime object of speculation as to whether or not extraterrestrial life exists there.

In 1877, the Italian astronomer Giovanni Schiaparelli published the results of a long series of telescopic observations he had made of Mars. He reported that he had seen "canali" on the surface. When this Italian word for "channels" was improperly translated into "canals," which seemed to connote that they were dug by intelligent life, public interest in Mars increased.

Over the next decades, there were endless debates over just what had been seen. We now know that the channels or canals Schiaparelli and other observers reported are not present on Mars—the positions of the "canali" do not even always overlap the spots and markings that are actually on the martian surface (Fig. 9–1). But hope of finding life elsewhere in the solar system springs eternal, and the latest studies have indicated the presence of considerable quantities of liquid water in Mars' past, a fact that leads many astronomers to hope that life could have formed during those periods. Still, as we shall describe, the Viking spacecraft that landed on Mars found no signs of life.

9.1 CHARACTERISTICS OF MARS

Mars is a small planet, 6800 km across, which is only about half the diameter and one-eighth the volume of Earth or Venus, although somewhat larger than Mercury. Mars' atmosphere is thin—at the surface its pressure is only 1 per cent of the surface pressure of Earth's atmosphere—but it might be sufficient for certain kinds of life.

Unlike the orbits of Mercury or Venus, the orbit of Mars is outside the Earth's, so we can observe Mars (Fig. 9–2) in the late night sky.

Mars revolves around the Sun in 23 Earth months. The axis of its rotation is tipped at a 25° angle from the plane of its orbit, nearly the same as the Earth's 23½° tilt. Because the tilt of the axis causes the seasons, we know that Mars goes through a year with four seasons just as the Earth does.

We have watched the effects of the seasons on Mars over the last century. In the martian winter, in a given hemisphere, there is a polar cap. As the martian spring comes to the northern hemisphere, the north polar cap shrinks and material at more temperate zones darkens. The surface of Mars is always mainly reddish, with darker gray areas that appear blue-green for physiological reasons of color contrast (see Color Plate 19). In the spring, the darker regions spread. Half a martian year later, the same thing happens in the southern hemisphere.

These changes were once thought to be biological: martian vegetation could be blooming or spreading in the spring. But the current theory is that

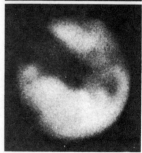

FIGURE 9–1 A drawing of Mars and a photograph, both made at a close approach. We now realize that the "canals" seen in the past do not usually correspond to actual surface features.

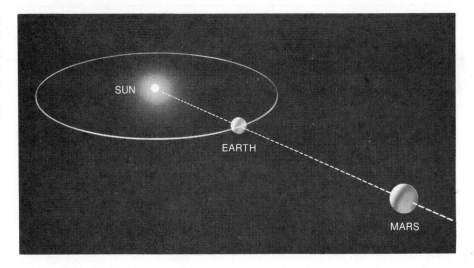

each year at the end of southern-hemisphere springtime, a global dust storm
starts, covering Mars' entire surface with light dust. Then the winds reach
velocities as high as hundreds of kilometers per hour and blow fine, light-
colored dust off some of the slopes. This exposes the dark areas underneath.
Gradually, as Mars passes through its seasons over the next year, the location
where the dust is stripped away changes, mimicking the color change we
would expect from vegetation. Astronomers still debate why the dust is reddish; the presence of iron oxide (rust) would explain the color in general,
but might not satisfy the detailed measurements of the dust.

From Mars' mass and radius we can easily calculate that it has an average density about that of the Moon, substantially less than the density of
Mercury, Venus, and Earth. The difference indicates that Mars' overall composition must be fundamentally different from that of these other planets.
Mars probably has a smaller core and a thicker crust than does Earth.

9.2 EARLY SPACE OBSERVATIONS

Mars has been the target of series of spacecraft launched by both the United
States and the Soviet Union. The earliest of these spacecraft were flybys,
which photographed and otherwise studied the martian surface and atmosphere during the few hours they were in good position as they passed by.
The first flyby sent back photos showing only a cratered surface. This
seemed to indicate that Mars resembles the Moon more than it does the
Earth. Later spacecraft confirmed the cratering, and also showed some signs
of erosion.

In 1971, the United States sent out a spacecraft, Mariner 9, not just to
fly by Mars for only a few hours, but actually to orbit the planet and send
back data for a year or more. But when it reached Mars, our first reaction
was as much disappointment as elation. While Mariner 9 was en route, a
tremendous dust storm had come up, almost completely obscuring the entire
surface of the planet. Only the south polar cap and 4 dark spots were visible.
The storm began to settle after a few weeks, with the polar caps best visible
through the thinning dust. Finally, three months after the spacecraft arrived, the surface of Mars was completely visible, and Mariner 9 could proceed with its mapping mission. Over the next year, it mapped the entire
surface at a resolution of 1 km.

The data showed four major types of surface on Mars: volcanic regions, canyon areas, expanses of craters, and terraced areas near the poles. A chief surprise of the Mariner 9 mission was the discovery of extensive areas of vulcanism. The four dark spots on the martian surface proved to be volcanoes. The largest of the volcanoes—which corresponds in position to the surface marking long known as Nix Olympica, "the snow of Olympus"—is named Olympus Mons, "Mount Olympus." It is a huge volcano—600 kilometers at its base and about 25 kilometers high (Fig. 9–3). It is crowned with a crater 65 km wide; Manhattan island could be easily dropped inside. (The tallest volcano on Earth is Mauna Kea, in the Hawaiian islands, if we measure its height from its base deep below the ocean. Mauna Kea is taller than Everest, though still only 9 km high.) The three other large volcanoes are on a 200-km-long ridge known as Tharsis.

Another surprise on Mars was the discovery of systems of canyons. One tremendous canyon—about 5000 kilometers long—is as big as the United States and comparable in size to the Rift Valley in Africa, the longest geological fault on Earth.

Perhaps the most amazing discovery on Mars was the presence of sinuous channels. These are on a smaller scale than the "canali" that Schiaparelli had seen, and are entirely different phenomena. Some of the channels show tributaries (Fig. 9–4), and the bottoms of some of the channels show the same characteristic features as stream beds on Earth. Even though water cannot exist on the surface of Mars under today's conditions, it is difficult to think of ways to explain the channels satisfactorily other than to say that they were cut by running water in the past.

This indication that water most likely flowed on Mars is particularly interesting because biologists feel that water is necessary for the formation and evolution of life. The presence of water on Mars, therefore, even in the past, may indicate that life could have formed and may even have survived. Where has the water all gone? Most of the water would probably be in a permafrost layer—permanently frozen subsoil—beneath middle latitudes and polar regions.

Some of the water is bound in the polar caps (Fig. 9–5). Up to the time of Mariner 9, we thought that the polar caps were mostly frozen carbon

FIGURE 9–3 Olympus Mons, from Mariner 9. The image shows a region 800 km across.

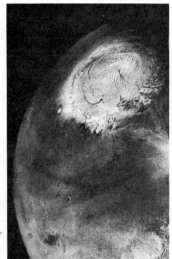

FIGURE 9–5 This mosaic of Mariner 9 views shows the north polar cap, which is shrinking as a result of martian springtime. The volcano Olympus Mons is visible at the bottom.

FIGURE 9–4 Stream channels with tributaries indicate to most scientists that water flowed on Mars in the past.

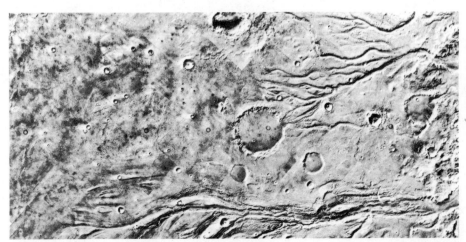

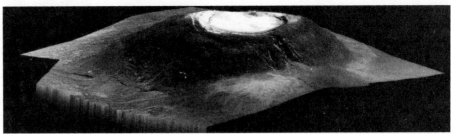

FIGURE 9–7   The martian volcano Arsia Mons, the southernmost of the line of 3 volcanoes in Tharsis. The image was plotted based on Viking Orbiter data.

dioxide—"dry ice"; the presence of water was controversial. We now know that the large polar caps that extend to latitude 50° during the winter are carbon dioxide. But when a cap shrinks during its hemisphere's summer, a residual polar cap of water ice remains.

We found that the martian atmosphere is composed of 90 per cent carbon dioxide with small amounts of carbon monoxide, oxygen, and water. The surface pressure is less than 1 per cent of that near the Earth's surface. The atmosphere is too thin to affect the surface temperature, in contrast to the huge effects that the atmospheres of Venus and the Earth have on climate.

9.3 VIKING

In the summer of 1976, two U.S. spacecraft named Viking reached Mars. Each Viking contained two parts: an orbiter and a lander. The orbiter served two roles: it mapped and analyzed the martian surface using its cameras and other instruments and relayed the lander's radio signals to Earth. The lander served two purposes as well: it studied the rocks and weather near the surface of Mars and sampled the surface in order to decide whether there was life on Mars!

FIGURE 9–6   A high-resolution view of the caldera crater on top of Olympus Mons, photographed by the Viking 1 orbiter.

FIGURE 9–8   A mosaic of 102 photographs of Mars from Orbiter 1 in 1980, near the end of the spacecraft's lifetime. Valles Marineris, as long as the United States is wide, stretches across the center. The sharp line near them, marked with an arrow, is an unusual feature that is either a weather front or an atmospheric shock wave. The lower arrow shows the shadows of a group of clouds that are at an elevation of 28 km (91,000 ft). The largest cloud is nearly 32 km long.

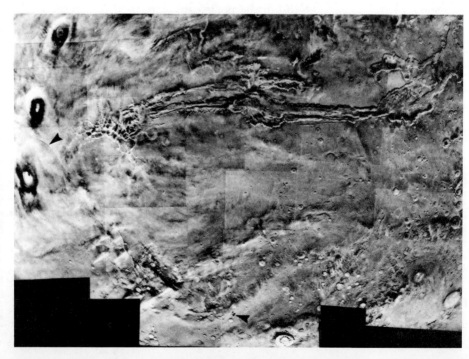

Viking 1 reached Mars in June of 1976 after a flight that led to spectacular large-scale views of the martian surface (Color Plates 20 and 21). The
orbiter observed the volcanoes (Figs. 9–6 and 9–7) and the huge valley (Fig.
9–8) at higher resolution than did Mariner 9.

These detailed views of Mars allow us to interpret better the similarities
and differences that this planet—with its huge canyons and gigantic volcanoes—has with respect to the Earth. For example, Mars has exceedingly
large, gently sloping volcanoes but no signs of the long mountain ranges or
deep mid-ocean ridges that on Earth tell us that plate tectonics has been and
is taking place. Many of the large volcanoes on Mars are "shield volcanoes"—
a type that has gently sloping sides formed by the rapid spread of lava. On
Earth, we also have steep-sided volcanoes, which occur where the continental
plates are overlapping, as in Mt. St. Helens, the Aleutian Islands, or on
Mount Fujiyama. No steep-sided volcanoes exist on Mars.

Perhaps the volcanic features on Mars can get so huge because continental drift is absent there. If molten rock flowing upward causes volcanoes
to form, then on Mars the features just get bigger and bigger for hundreds
of millions of years, since the volcanoes stay over the sources and do not
drift away. (Each of the Hawaiian islands was formed over a single "hot
spot," but drifted away from it.)

On July 20, 1976, exactly seven years after the first manned landing on
the Moon, Viking 1's lander descended safely onto a plain called Chryse.
The views showed rocks of several kinds (Color Plates 22–24), covered with
yellowish-brown material that is probably an iron oxide (rust) compound.
Sand dunes were also visible (Fig. 9–9). The sky on Mars turns out to be
yellowish-brown, almost pinkish (Color Plates 22–24); the color is formed as
sunlight is scattered by dust suspended in the air as a result of one of Mars'
frequent dust storms.

A series of experiments aboard the lander was designed to search for
signs of life. A long arm was deployed (Color Plate 22), and a shovel at its
end dug up a bit of the martian surface. The soil was dumped into three
experiments that searched for such signs of life as respiration and metabo-
lism. The results were astonishing at first: the experiments sent back signals
that seemed similar to those that would be caused on Earth by biological
rather than by mere chemical processes. But later results were less spectacular, and non-biological explanations seem more likely. It is probable that
some strange chemical process mimicked life in these experiments.

One important experiment gave much more negative results for the
chance that there is life on Mars. It analyzed the soil and looked for traces
of organic compounds. On Earth, many organic compounds left over from
dead forms of life remain in the soil; the life forms themselves are only a
tiny fraction of the organic material. Yet these experiments found no trace

FIGURE 9–9 This 100° view of
the martian surface was taken
from the Viking 1 Lander, looking
northeast at left and southeast at
right. It shows a dune field with
features similar to many seen in
the deserts of Earth. From the
shape of the peaks, it seems that
the dunes move from upper left to
lower right. The large boulder at
the left is about 8 meters from the
lander and is 1 × 3 meters in size.
The boom that supports Viking's
weather station cuts through the
center of the picture. ("Chance of
precipitation," the local newscaster
would say, "0 per cent.")

FIGURE 9–10   The first photograph that Viking 2 took on the surface of Mars. A wide variety of rocks lie on a surface of fine-grained material. Most of the rocks are 10–20 cm across. One of the lander's footpads is at lower right.

FIGURE 9–11   This oblique view from the Viking 1 Orbiter shows Argyre, the smooth plain at left center. It is surrounded by heavily cratered terrain. The martian atmosphere was unusually clear when this photograph was taken, and craters can be seen nearly to the horizon. The horizon is bright mainly because of a thin haze. Detached layers of haze can be seen to extend from 25 to 40 km above the horizon and may be crystals of carbon dioxide.

of organic material. On Mars, who knows? Perhaps life forms evolved that efficiently used up their predecessors. Still, the absence of organic material from the martian soil is a strong argument against the presence of life on Mars.

Viking 2, when the Lander descended, found a smaller variety of types of rocks, with more large rocks and more pitted rocks (Fig. 9–10 and Color Plate 23). Such a distribution would result if the surface there had been ejected and flowed from a large crater 200 km away. The atmosphere at the second site, Utopia Planitia, contains three or four times more water vapor than had been observed at the first landing site 7500 km away. This observation makes sense, since Utopia Planitia is closer to the pole. Viking 2's experiments in the search for life sent back data similar to Viking 1's.

Even if the life signs detected by Viking come from chemical rather than biological processes, as seems likely, we have still learned of fascinating new chemistry going on. When life arose on Earth, it probably took up chemical processes that had previously existed. Similarly, if life began on Mars in the past or will begin there in the future (assuming our visits didn't contaminate Mars and ruin the chances for indigenous life), we might expect the life forms to use chemical processes that already existed. So even if we haven't detected life itself, we may well have learned important things about its origin.

Observations from the orbiters showed Mars' atmosphere (Fig. 9–11). The lengthy period of observation led to the discovery of weather patterns on Mars. With its rotation period similar to that of Earth, many features of Mars' weather are similar to our own (Fig. 9–12). Surface temperatures measured from the landers ranged from a low of 150 K (−190°F) at the northern site of Lander 1 to over 300 K (80°F) at Lander 2. Temperature varied each day by 35–50°C (60–90°F). Studies of Mars' weather have already helped us better understand windstorms in Africa that affect weather as far away as North America.

Further, studies of the effect of martian dust storms on the planet have led to the idea that the explosion of a nuclear bomb on Earth would lead to

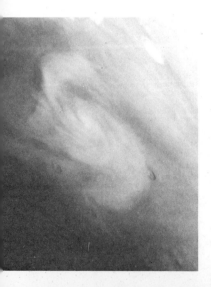

FIGURE 9–12   This Martian storm on August 9, 1978, resembles satellite pictures of storms on Earth, showing the counterclockwise circulation typical at northern latitudes. The temperature is too warm for the clouds to be carbon dioxide; they therefore must be made of water ice. The frost-filled crater at top right is 92 km in diameter. The white patches at top are outlying regions of the north polar remnant cap.

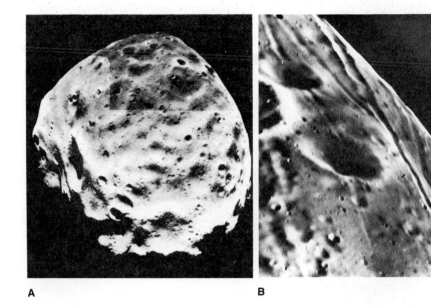

A   B

FIGURE 9–13 *(A)* A Viking mosaic of Phobos. Its two largest craters have been named after Asaph Hall, who discovered the moons of Mars a century ago, and after Angelina Stickney, his wife, who encouraged the search. *(B)* A close-up view of Phobos from Viking Orbiter 2 shows an area 3 × 3.5 km which, surprisingly, has grooves and chains of small craters. The grooves were probably associated with the formation of the crater Stickney. Similar chains of small craters on Earth's Moon, on Mars, and on Mercury were formed by secondary cratering from a larger impact.

In Greek mythology, Phobos (Fear) and Deimos (Terror) were companions of Ares, the equivalent of the Roman war god, Mars.

a "nuclear winter." Dust thrown into the air would shield the Earth's surface from sunlight for a lengthy period, with dire consequences for life on Earth. The possibility is so important that it is now being studied at high levels in countries around the globe. Improvement of computer models for the circulation of atmospheres will contribute to the investigations. These models, and observations of Mars, also help us to understand the smaller amounts of matter we are putting into the atmosphere from factories and power plants and by the burning of forests.

Mars has two moons, Phobos and Deimos. They are mere chunks of rock, only 27 and 15 kilometers across, respectively, too small for gravity to have pulled them into round shapes. Mariner 9 and the Vikings made close-up photographs and studies (Figs. 9–13 and 9–14).

What's the next step? A pair of Soviet spacecraft will be launched to reach Mars in 1988. The first spacecraft's orbit will be so similar in size and shape to that of Phobos that it will essentially hover only 50 meters above Phobos' surface! A part of the spacecraft will even land on Phobos and analyze its surface. If this spacecraft is successful, the second spacecraft would do the same for Deimos.

NASA is preparing a Mars orbiter for launch in 1990, to map surface chemistry and to study seasonal weather patterns. The European Space Agency is considering Kepler, an orbiter, among other space projects; the one chosen would be launched in about 1992. Manned space flights to Mars would be much more expensive and so are even farther off. People are now talking seriously about sending astronauts to Mars in the 2020's.

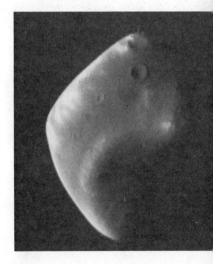

FIGURE 9–14 A Viking view of Deimos. The lighted portion measures about 12 × 18 km. The surface is heavily cratered and thus presumably very old.

QUESTIONS

1. Outline the features of Mars that made scientists think that it was a good place to search for life.

2. Compare the tallest volcanoes on Earth and Mars relative to the diameters of the planets.

3. What evidence exists that there is, or has been, water on Mars?

4. Why is Mars' sky pinkish?

5. Describe the composition of Mars' polar caps.

6. Compare the temperature ranges on Mercury, Venus, Earth, and Mars.

7. List the evidence from Viking for and against the existence of life on Mars.

8. Aside from the biology experiments, list three types of observations made from the Viking landers.

9. Why aren't Phobos and Deimos regular, round objects like the Earth's Moon?,

Chapter 10

A mosaic of Jupiter and its four Galilean satellites made from the Voyager 1 spacecraft. Callisto is at lower right, Ganymede is at lower left, relatively bright Europa is shown against Jupiter, and Io appears in the background at left. The symbol for Jupiter appears at the top of the facing page.

JUPITER: GIANT OF OUR SOLAR SYSTEM

Jupiter, Saturn, Uranus, and Neptune are *giant planets*. They are much bigger and more massive, and less dense, than the inner planets. The internal structure of these giant planets is entirely different from that of the four inner planets.

Jupiter, the largest and most massive planet, dominates the Sun's planetary system. It alone contains two-thirds of the mass in the solar system outside of the Sun, 318 times as much mass as the Earth. Jupiter has at least 16 moons of its own and so is a miniature planetary system in itself. It is often seen as a bright object in our night sky, and observations with even a small telescope reveal bands of clouds across its surface and show four of its moons.

Jupiter is more than 11 times greater in diameter than the Earth. From its volume and mass, we calculate its density to be 1.3 grams/cm^3, not much greater than the 1 gram/cm^3 density of water. This tells us that any core of heavy elements (such as iron) makes up only a small fraction of Jupiter's mass. Jupiter, rather, is mainly composed of the light elements hydrogen and helium. Jupiter's chemical composition is closer to that of the Sun and stars than it is to that of the Earth (Fig. 10–1). Jupiter has no crust. At deeper and deeper levels, its gas just gets denser and denser, eventually liquefying.

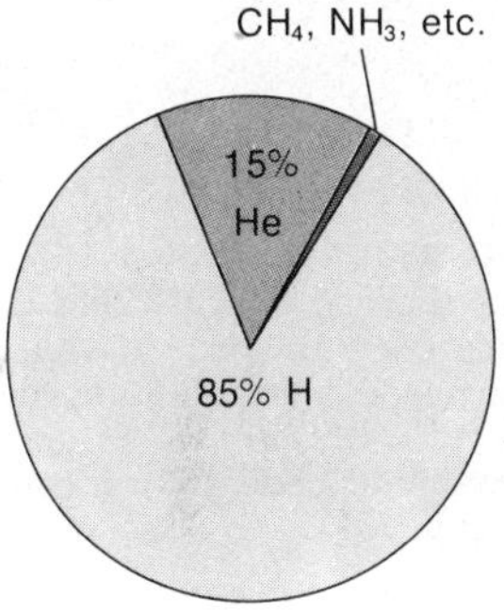

Different latitudes on Jupiter's surface rotate at slightly different speeds. The different speeds determine where the bands of clouds are. The clouds are in constant turmoil; the shapes and distribution of bands can change within days.

The most prominent feature of Jupiter's surface is a large reddish oval known as the *Great Red Spot* (Color Plates 27 and 28). It is many times larger than the Earth, and drifts about slowly with respect to the clouds as the planet rotates. The Great Red Spot has been visible for at least 150 years, and maybe 300 years. Sometimes it is relatively prominent and colorful, and at other times the color may even disappear for a few years. Other, smaller spots are also present.

Jupiter emits radio waves, which indicates that Jupiter has a strong magnetic field and strong *radiation belts* (actually, belts filled with magnetic fields in which particles are trapped, large-scale versions of the Van Allen belts of Earth). High-energy particles passing through space interact with Jupiter's magnetic field to produce the radio emission.

FIGURE 10–1 The composition of Jupiter.

10.1 JUPITER'S MOONS: EARLY VIEWS

Four of the innermost satellites were discovered by Galileo in 1610 when he first looked at Jupiter with his small telescope. These four moons are called the *Galilean satellites* (Fig. 10–2). One of these moons, Ganymede, at 5276 km in diameter, is the largest satellite in the solar system and is larger than the planet Mercury.

<table>
<tr><td>

BOX 10.1
Jupiter's Satellites in
Mythology

</td><td>

All the moons except Amalthea are named after lovers of Zeus, the Greek equivalent of Jupiter. Amalthea, a goat-nymph, was Zeus' nurse, and out of gratitude he made her into the constellation Capricorn. Zeus changed Io into a heifer to hide her from his jealous wife Hera; in honor of Io, the crescent moon has horns.

Ganymede was a Trojan youth carried off by an eagle to be Jupiter's cup bearer (the constellation Aquarius). Callisto was changed into a bear as punishment for her affair with Zeus. She was then slain by mistake, and rescued by Zeus by being transformed into the Great Bear in the sky. Jealous Hera persuaded the sea god to forbid Callisto ever to bathe in the sea, which is why Ursa Major never sinks below the horizon.

</td></tr>
</table>

FIGURE 10–2 Jupiter with the four Galilean satellites. These satellites were named Io, Europa, Ganymede, and Callisto by Simon Marius, a German astronomer who independently discovered them in 1611.

The Galilean satellites have played a very important role in the history of astronomy. The fact that these particular satellites were noticed to be going around another planet, like a solar system in miniature, supported Copernicus' Sun-centered model of our solar system. Not everything revolved around the Earth!

At least a dozen additional moons have been discovered, all smaller than the Galilean satellites. Some have been found by flyby spacecraft; others are still being found with telescopes on Earth.

10.2 SPACECRAFT OBSERVATIONS

Pioneer 10 and Pioneer 11 gave us our first close-up views of the colossal planet in 1973 and 1974. A second revolution in our understanding of Jupiter occurred in 1979, when Voyager 1 and Voyager 2 also flew by.

Each spacecraft carried many types of instruments to measure properties of Jupiter, its satellites, and the space around them. The observations made with the imaging equipment were of the most popular interest. The resolution of the Voyager images was five times better than the best images we can get from Earth. In only 48 seconds, the Voyagers could send a full digitally-coded picture back to Earth, where detailed computer work improved the images. This was remarkable for a vehicle so far away that its signals travelled 52 minutes at the speed of light to reach us. The energy in the signal would have to be collected for billions of years to light a Christmas tree bulb for just one second!

10.2a The Great Red Spot

The Great Red Spot shows very clearly in many of the images (Fig. 10–3 and Color Plate 28). It is a gaseous island many times larger across than the Earth. The Spot may be the vortex of a violent, long-lasting storm, similar to large storms on Earth.

Time-lapse observations from Voyager can be used to study the Spot's rotation. We also see how it interacts with surrounding clouds and smaller spots (Fig. 10–4).

Why has the Great Red Spot lasted this long? Perhaps heat energy flows into the storm from below it, maintaining its energy supply. The storm also contains more mass than hurricanes on Earth, which makes it more stable. Further, unlike Earth, Jupiter has no continents or other structure to break up the storm. Also, we do not know how much energy the Spot gains from

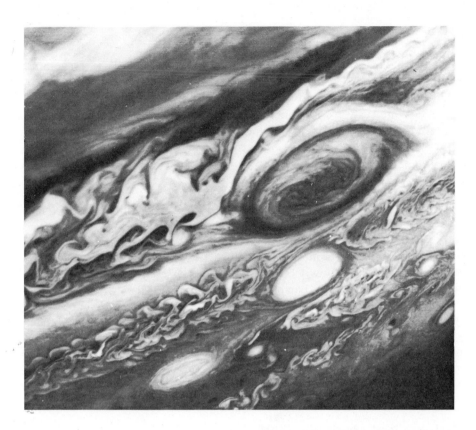

FIGURE 10–3 A mosaic showing the Great Red Spot, made from Voyager 2 at a range of 2 million km from Jupiter. The white oval south of the Great Red Spot is similar in structure. Both rotate in the anticyclonic—counterclockwise—sense.

the circulation of Jupiter's upper atmosphere and eddies (rotating regions) in it. Until we can sample lower levels of Jupiter's atmosphere, we will not be able to decide definitively. Studying the eddies in Jupiter's atmosphere helps us interpret features on Earth. For example, one theory of Jupiter's spots holds that they are similar to circulating rings that break off from the Gulf Stream in the Atlantic Ocean.

10.2b Jupiter's Atmosphere

Heat emanating from Jupiter's interior churns the atmosphere. (In the Earth's atmosphere, on the other hand, most of the energy comes from the outside—from the Sun.) The bright bands (Color Plates 27 and 29) on Jupiter, called *zones,* are rising currents of gas. The dark bands, the *belts,* are

FIGURE 10–4 A time-lapse sequence of the Great Red Spot, showing the flow of gas. The pictures are taken every other rotation of Jupiter, making the interval 22 Earth hours. Note how a white cloud enters the Spot's circulation and begins to be swept around. The Great Red Spot rotates with a period of about 6 Earth hours.

falling gas. The tops of these dark belts are somewhat lower (about 20 km) than the tops of the zones and so are about 10 K warmer.

Voyager measurements of wind velocities showed that each hemisphere of Jupiter has a half-dozen currents blowing eastward or westward. The Earth, in contrast, has only one westward current at low latitudes (the trade winds) and one eastward current at middle latitudes (the jet stream).

Extensive lightning storms, including giant-sized lightning strikes called "superbolts," were discovered from the Voyagers, as were giant aurorae.

10.2c Jupiter's Interior

Most of Jupiter's interior is in liquid form. Jupiter's central temperature may be between 13,000 and 35,000 K. The central pressure is 100 million times the pressure of the Earth's atmosphere measured at our sea level because of Jupiter's great mass pressing in. Because of this high pressure, Jupiter's interior is probably composed of ultracompressed hydrogen surrounding a rocky core consisting of 20 Earth masses of iron and silicates.

Jupiter radiates 1.6 times as much heat as it receives from the Sun. It must have an internal energy source—perhaps the energy remaining from its collapse from a primordial gas cloud 20 million km across. Jupiter is undoubtedly still contracting. It lacks the mass necessary by a factor of about 75, however, to have heated up enough to become a star, generating energy by nuclear processes.

10.2d Jupiter's Magnetic Field

The Pioneer missions showed that Jupiter's tremendous magnetic field is even more intense than many scientists had expected, a result confirmed by the Voyagers (Fig. 10–5). At the height of Jupiter's clouds the magnetic field is 10 times that of the Earth, which itself has a strong field.

The inner field is shaped like a doughnut, containing several shells like giant versions of the Earth's Van Allen belts. The satellites Amalthea, Io,

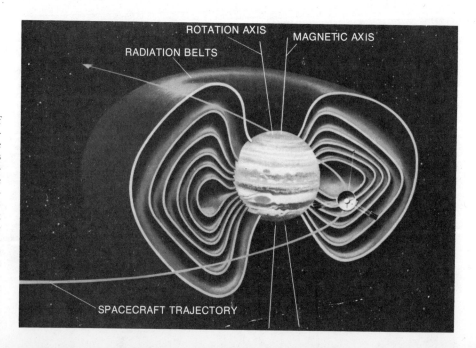

FIGURE 10–5 An artist's view of Jupiter's magnetosphere, the region of space occupied by the planet's magnetic field. It rotates at hundreds of thousands of kilometers per hour around the planet, like a big wheel with Jupiter at the hub. The inner magnetosphere is shaped like a doughnut with the planet in the hole. The highly unstable outer magnetosphere is shaped as though the outer part of the doughnut has been squashed. The outer magnetosphere is "spongy" in that it pulses in the solar wind like a huge jellyfish. It often shrinks to one-third of its largest size.

ROTATION AXIS

MAGNETIC AXIS

RADIATION BELTS

SPACECRAFT TRAJECTORY

Europa, and Ganymede travel through this region. The middle region, with charged particles being whirled around rapidly by the rotation of Jupiter's magnetic field, does not have a terrestrial counterpart. The outer region of Jupiter's magnetic field interacts with the particles flowing outward from the Sun—the solar wind—and forms a shock wave, as does the bow of a ship plowing through the ocean. When the solar wind is strong, Jupiter's outer magnetic field (shaped like a flattened pancake) is pushed in.

10.3 JUPITER'S MOONS CLOSE UP

Through Voyager close-ups, the satellites of Jupiter have become known to us as worlds with personalities of their own. The four Galilean satellites, in particular, were formerly known only as dots of light. Since these satellites range between 0.9 and 1.5 times the size of our own Moon, they are substantial enough to have interesting surfaces and histories (Color Plates 30–34 and the picture opening this Chapter).

Io provided the biggest surprises. Scientists knew that Io gave off particles as it went around Jupiter, but Voyager 1 discovered that these particles resulted from active volcanoes on the satellite (Fig. 10–6). Eight volcanoes were seen actually erupting, many more than erupt on the Earth at any one time. When Voyager 2 went by a few months later, most of the same volcanoes were still erupting.

Io's surface (Figs. 10–7 and 10–8 and Color Plates 29–31) has been transformed by the volcanoes, and is the youngest surface we have observed in our solar system. Gravitational forces from the other Galilean satellites

Io

FIGURE 10–6 The volcanoes of Io can be seen erupting on this photograph. The first volcano was discovered by Linda Morabito, an engineer at the Jet Propulsion Laboratory, where the scientists had gathered to receive the images from space. At the time, she was studying images that had been purposely overexposed to bring out the stars so that they could be used for navigation. She saw the large volcanic cloud barely visible on the right limb. The plume extends upward for about 250 km and is scattering sunlight toward us. The bright spot just leftward of Io's terminator is a second volcanic plume projecting above the dark surface into the sunlight.

FIGURE 10–7 A mosaic of Io with a resolution of 8 km, showing volcanic features and no craters. The heart-shaped region at lower center surrounds an erupting volcano named for Pele, the Hawaiian volcano goddess. Its shape changed substantially between this Voyager 1 image and the Voyager 2 images.

FIGURE 10–8 A volcanic caldera, 50 km in diameter, on Io. Dark flows over 100 km long extend from it.

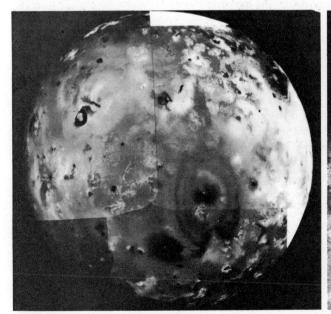

Io

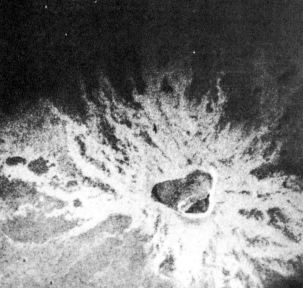

Io

FIGURE 10–9 The surface of Europa, which is very smooth, is covered by this complex array of streaks. Few impact craters are visible.

FIGURE 10–10 This mosaic of Ganymede, Jupiter's largest satellite, shows numerous impact craters, many with bright ray systems. The large crater at upper center is about 150 km across. The mountainous terrain at lower right is a younger region, as is the grooved terrain at bottom center.

Europa

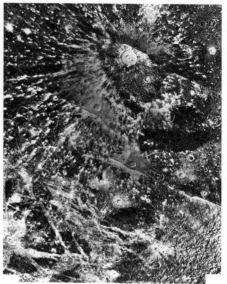

Ganymede

Io's surface, orange in color and covered with strange formations, led Bradford Smith of the University of Arizona, the head of the Voyager imaging team, to remark that "It's better looking than a lot of pizzas I've seen."

Moon	Diameter (km)
Amalthea	270 × 165 × 153
Io	3,632
Europa	3,126
Ganymede	5,276
Callisto	4,820

pull Io slightly inward and outward from its orbit around Jupiter, flexing it. This squeezing and unsqueezing creates heat from friction, which presumably heats the interior and leads to the vulcanism. The surface of Io is covered with sulfur and sulfur compounds, including frozen sulfur dioxide, and the atmosphere is full of sulfur dioxide. It certainly wouldn't be a pleasant place to visit.

Europa (Color Plate 32), the brightest of Jupiter's Galilean satellites, has a very smooth surface and is covered with narrow dark stripes (Fig. 10–9). This suggests that the surface we see is ice. The markings may be fracture systems in the ice, like fractures in the large fields of sea ice near the Earth's north pole. Few craters are visible, suggesting that the ice was soft enough below the crust to close in the craters. Either internal radioactivity or a gravitational heating like that inside Io may have provided the heat to soften the ice. Because Europa possibly has liquid water and extra heating, some scientists consider it a worthy location to check for signs of life.

The largest satellite, Ganymede (Color Plate 33), shows many craters (Fig. 10–10) alongside weird grooved terrain (Fig. 10–11). Ganymede is larger than Mercury but less dense. It contains large amounts of water-ice surrounding a rocky core. But an icy surface is as hard as steel in the cold conditions that far from the Sun, so retains the craters from perhaps 4 billion years ago. The grooved terrain is younger.

Ganymede also shows lateral displacements, where grooves have slid sideways, like those that occur in some places on Earth (for example, the San Andreas fault in California). Ganymede is the only place besides the Earth where such faults have been found. Thus further studies of Ganymede may help our understanding of terrestrial earthquakes.

Callisto (Color Plate 34) has so many craters (Fig. 10–12) that its surface is the oldest of Jupiter's Galilean satellites. Callisto is probably also covered with ice. A huge bull's-eye formation, Valhalla, contains about 10 concentric rings, no doubt resulting from a huge impact. Perhaps ripples spreading from the impact froze into the ice to make Valhalla.

Voyager also observed Amalthea, formerly thought to be Jupiter's innermost satellite. This small chunk of rock is irregular and oblong in shape,

Ganymede

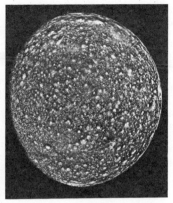

Callisto

FIGURE 10–11 The terrain over large areas of Ganymede is covered with many grooves tens of kilometers wide, a few hundred meters high, and up to thousands of kilometers long. Younger grooves cover older grooves, and offsets sometimes occur along series of grooves.

FIGURE 10–12 Callisto is cratered all over, very uniformly. Several craters have rays or concentric rings. No craters larger than 50 km are visible, from which we deduce that Callisto's crust cannot be very firm. The fact that the limb is so smooth indicates that there is no high relief. Because there are so many craters, the surface must be very old, probably over 4 billion years.

looking like a dark red potato complete with potato "eyes." It is comparable in size to many asteroids, ten times larger than the moons of Mars and ten times smaller than the Galilean satellites. The reddish color may be caused by material splattered from Io. Voyager 2 discovered three previously unknown satellites of Jupiter, making the total of known moons at least 16.

Studies of Jupiter's moons tell us about the formation of the Jupiter system, and help us better understand the early stages of the entire solar system.

10.4 JUPITER'S RING

Though Jupiter wasn't expected to have a ring, Voyager 1 was programmed to look for one just in case; Saturn's ring, of course, was well known, and Uranus' ring had been discovered only a few years earlier. The Voyager 1 photograph indeed showed a wispy ring of material around Jupiter at about 1.8 times Jupiter's radius, inside the orbit of its innermost moon. As a result, Voyager 2 was targeted to take a series of photographs of the ring. From the far side looking back, the ring appeared unexpectedly bright (Fig. 10–13). This brightness probably results from small particles in the ring that scatter the light toward us. Within the main ring, fainter material appears to extend down to Jupiter's surface (Fig. 10–14).

The ring particles may come from Io, or they may come from comet and meteor debris or from material knocked off the innermost moons by meteorites. Whatever their origin, the individual particles probably remain in the ring only temporarily.

In the next chapter, we shall discuss the fundamental ideas about how gravity leads to the formation of planetary rings, in the context of the most famous ring of all: the ring of Saturn.

FIGURE 10–13 Jupiter's ring extends outward from Jupiter's bright limb in this photograph taken by Voyager 2 looking back from the dark side. The lower ring image is cut short by Jupiter's shadow.

FIGURE 10–14 A mosaic of Jupiter's ring from Voyager 2. It is slightly blurred by spacecraft motion, especially in the left-most image.

FIGURE 10–15 The Galileo Probe before launch.

10.5 FUTURE EXPLORATION OF JUPITER

Voyager 1 and Voyager 2 were spectacular successes. The data they provided about the planet Jupiter, its satellites, and its ring, dazzled the eye and revolutionized our understanding.

Jupiter's large mass makes it a handy source of energy to use to send probes to more distant planets. Just as a sling transfers energy to a stone, some of Jupiter's energy can be transferred to a spacecraft through a gravitational interaction.

NASA's Project Galileo will provide one spacecraft to orbit Jupiter and another to drop a probe into Jupiter's clouds. The Orbiter will orbit Jupiter a dozen times in a 20-month period, coming so close to several of Jupiter's moons that pictures will have a resolution 10 to 100 times greater even than those from the Voyagers. The Probe will transmit data for an hour as it falls through the Jovian atmosphere (Fig. 10–15). We expect to lose contact with it after it penetrates the clouds for 130 km or so. The Probe should give us accurate measurements of Jupiter's composition. The comparison to the Sun's composition should help us understand Jupiter's origin. Galileo's launch has been delayed by the space-shuttle tragedy.

The Hubble Space Telescope, in Earth orbit, should be able to take pictures of Jupiter with a resolution approximately equal to that of Voyager. So we should get pictures of Jupiter's clouds every few days for a long time.

KEY WORDS

giant planets, Great Red Spot, radiation belts, Galilean satellites, zones, belts

QUESTIONS

1. Why does Jupiter appear brighter than Mars despite its greater distance from the Earth?

2. Even though Jupiter's atmosphere is very active, the Great Red Spot has persisted for a long time. How is this possible?

3. (a) How did we first know that Jupiter has a magnetic field? (b) What did the recent studies show?

4. It has been said that Jupiter is more like a star than a planet. What facts support this statement? Disagree?

5. What advantages over the 5-m Palomar telescope on Earth did Voyagers 1 and 2 have for making images of Jupiter?

6. What are two types of observations other than photography made from the Voyagers to Jupiter?

7. How does the interior of Jupiter differ from the interior of the Earth?

8. Which moons of Jupiter are icy? Why?

9. Contrast the volcanoes of Io with those of Earth.

10. Compare the surfaces of Callisto, Io, and the Earth's Moon. Explain what this comparison tells us about the ages of features on their surfaces.

11. Using the information given in the text and in Appendices 3 and 4, sketch the Jupiter system, including all the moons and the ring. Mark the groups of moons.

12. Explain how the Hubble Space Telescope might be able to equal Voyager's resolution even though it will be farther from Jupiter.

PROJECTS

1. Using outside reference material, describe in some detail the origins of the names of each of Jupiter's moons in Greek mythology.

2. Over a four-hour interval one night, plot the positions of the Galilean satellites at half-hour intervals.

3. Over a one-week interval, plot the positions of the Galilean satellites from night to night. Using the observations in projects 2 and 3 together, deduce the projection on the sky of the orbits of the individual satellites.

Chapter 11

Saturn and some of its moons, photographed from the Voyager 1 space-craft. Dione is in the foreground, Tethys and Mimas are above it and to Saturn's right, Enceladus and Rhea are to the left, and Titan in its distant orbit is at the top. The symbol for Saturn appears at the top of the facing page.

SATURN: PLANET OF THE BEAUTIFUL RINGS

Saturn is the most beautiful object in our solar system, and possibly even the most beautiful object we can see in the sky (Color Plate 35). The glory of its system of rings makes it stand out even in small telescopes.

Saturn, like Jupiter, Uranus, and Neptune, is a giant planet. Its diameter, without its rings, is 9 times greater than Earth's; its mass is 95 times greater.

The giant planets have low densities. Saturn's is only 0.7 g/cm³, 70 per cent the density of water (Fig. 11–1). The bulk of Saturn is hydrogen molecules and helium. Saturn could have a core of heavy elements, including rocky material, making up 20 per cent of its interior.

The rings extend far out in Saturn's equatorial plane, and are inclined to the planet's orbit. Over a 30-year period, we sometimes see them from above their northern side, sometimes from below their southern side, and at intermediate angles in between. When seen edge on, they are all but invisible (Fig. 11–2).

The rings of Saturn are either material that was torn apart by Saturn's gravity or material that failed to collect into a moon at the time when the planet and its moons were forming. Every massive object has a sphere, called its *Roche limit,* inside of which blobs of matter do not hold together by their mutual gravity. The forces that tend to tear the blobs apart from each other are tidal forces (Section 5.3); they arise because some blobs are closer to the planet than others and are thus subject to higher gravity.

The radius of the Roche limit is usually about 2½ times the radius of the larger body. The Sun also has a Roche limit, but all the planets lie outside it. The natural moons of the various planets lie outside the respective Roche limits. Saturn's rings lie inside Saturn's Roche limit, so it is not surprising that the material in the rings is spread out rather than being collected into a single orbiting satellite.

Saturn has several concentric major rings visible from Earth. The brightest ring is separated from a fainter broad outer ring by an apparent gap called *Cassini's division.* Another ring is inside the brightest ring. We know that the rings are not solid objects, because they move around Saturn at different speeds at different radii.

The rings are thin, for on at least one occasion, stars occulted by the rings could be seen shining dimly through. The rings are relatively much flatter than a phonograph record. Radar waves were first bounced off the rings in 1973. The result of the radar experiments showed that the particles in the rings are probably rough chunks of ice at least a few centimeters and possibly a meter across. Infrared studies show that they are covered with ice.

Like Jupiter, Saturn rotates quickly on its axis, also in about 10 hours. Saturn's delicately colored bands of clouds rotate 10 per cent more slowly at high latitudes than at the poles.

FIGURE 11–1 Since Saturn's density is lower than that of water, it would float, like Ivory Soap, if we could find a big enough bathtub. But it would leave a ring!

Artificial satellites that we send up to orbit around the Earth are constructed of sufficiently rigid materials that they do not break up even though they are within the Earth's Roche limit.

Saturn's moons (Appendix 4) are named after the Titans, in Greek mythology the children and grandchildren of Gaea, the goddess of the Earth, who had been fertilized by a drop of Uranus' blood.

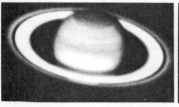

FIGURE 11–2 The rings of Saturn as seen from Earth at various times over several years. They extend in the equatorial plane of Saturn from about 70,000 to over 135,000 km from the planet's center. They are inclined to Saturn's orbit by 27°. (Lowell Observatory photographs)

Saturn gives off radio signals, as does Jupiter, an indication to earthbound astronomers that Saturn also has a magnetic field. And, like Jupiter, Saturn has a source of internal heating.

All the moons known from Earth-based observations, except Titan, range from about 130 km to 1600 km across. Planet-sized Titan, however, is a different kind of body. An atmosphere, including methane, had been detected on Titan by astronomers using terrestrial telescopes. A greenhouse effect is present, making some scientists wonder whether Titan's surface may have been warmed enough for life to have evolved there.

11.1 SATURN FROM SPACE

Voyager 1 reached Saturn in 1980, only a year after Pioneer 11. Voyager 2 arrived in 1981.

11.1a Saturn's Rings

From Pioneer 11, the rings were visible for the first time from a vantage point different from the one we have on Earth. The backlit view obtained by Pioneer 11 (Fig. 11–3) showed that Cassini's division, visible as a dark (and thus apparently empty) band from Earth, appeared bright, and its own

FIGURE 11–3 Though Pioneer 11, which passed Saturn in 1979, gave images far inferior to those from the Voyagers, it gave the first view of Saturn's rings when illuminated by the Sun from the other side. The rings appeared as inverses of the rings as seen from the front. Saturn's moon Rhea is below the planet. Compare the location of light and dark rings with those on the Voyager image opening this chapter. (NASA/Ames; U. of Arizona image processing)

FIGURE 11–4 Saturn and its rings in a mosaic taken by Voyager 1 from a distance of 18 million km, 2 weeks before closest approach. (All Voyager photos courtesy of Jet Propulsion Laboratory/NASA)

dark line of material separates it into inner and outer parts. The brightest ring of Saturn observed from Earth appeared dark on the Pioneer 11 view, presumably because it is too opaque to allow light to pass through it.

The passage of Voyager 1, with its high-resolution cameras, by Saturn was one of the most glorious events of the space program. The structure and beauty of the rings dazzled everyone (Color Plate 36 and Fig. 11–4).

The closer the spacecraft got to the rings, the more rings became apparent. Each of the known rings was actually divided into many thinner rings. By the time Voyager 1 had passed Saturn, we knew of hundreds of rings (Fig. 11–5). Voyager 2 saw still more (Fig. 11–6). Further, a device on Voyager 2 was able to track the change in brightness of a star as it was seen through the rings, and found even finer divisions (Fig. 11–7). The number of these rings (sometimes called "ringlets") is in the hundreds of thousands.

Everyone had expected that collisions between particles in Saturn's rings would make the rings perfectly uniform. But there was a big surprise. As

Studies of the changes in the radio signals from the spacecraft when it went behind the rings showed that the rings are only about 20 m thick, equivalent to the thinness of a phonograph record 30 km across—super long play.

FIGURE 11–5   The complexity of Saturn's rings. About 100 can be seen on this enhanced image, taken by Voyager 1 when it was 8 million km from Saturn. Several narrow rings can even be seen inside the upper right part of the dark Cassini division. A photo of the Earth shows the scale.

A moon discovered by Voyager 1, Saturn's 14th, appears at upper right, just inside the narrow F-ring. It apparently shepherds particles, like a sheep dog, keeping them from falling inward. Another newly discovered moon, just outside the ring, keeps the particles from escaping.

FIGURE 11–6   A close-up from Voyager 2 showing 6,000 km of a major ring, with the narrowest features 15 km wide.

FIGURE 11–7   A device on Voyager 2 followed the changes in brightness of the star delta Scorpii as it passed behind the rings. We see here a computer display of the data from a 50-km-wide section of rings with a resolution of 1 km. Saturn has thousands of such rings.

Figure 11–5

Figure 11–6

Figure 11–7

FIGURE 11–8 Dark radial spokes became visible in the rings as the Voyagers approached. They formed and dispersed within hours. This is a Voyager 2 image, part of a time-lapse series.

FIGURE 11–9 When sunlight scatters forward, the spokes appear bright. This is a Voyager 1 image.

Voyager 1 approached Saturn, we saw that there was changing structure in the rings aligned in the radial direction. The "spokes" look dark from the side illuminated by the Sun (Fig. 11–8), but look bright from behind (Fig. 11–9). This information showed that the particles in the spokes were very small, about 1 micron in size, since only small particles—like terrestrial dust in a sunbeam—reflect light in this way. And it seems that the spoke material is elevated above the plane of Saturn's rings. Since gravity wouldn't cause this, electrostatic forces may be repelling the spoke particles. (You can produce a similar effect by rubbing a rubber comb on cloth; an electrostatic force will then repel a piece of paper. Sometimes, combing your hair makes individual hairs repel each other because an electrostatic force is set up.)

The outer major ring turns out to be kept in place by a tiny satellite orbiting just outside it. And at least some of the rings are kept narrow by "shepherding" satellites (Fig. 11–10) that gravitationally affect the ring material.

Scientists were astonished to find on the Voyager 1 images that the outer ring, the narrow F-ring discovered by Pioneer 11, seems to be made of three braids (Fig. 11–11A). But soon after scientists succeeded in finding explanations for the braided strands, involving the gravity of the pair of newly discovered moons shown in the preceding photograph, Voyager 2 images showed that the rings were no longer intertwined (Fig. 11–11B). No one understands why.

Moon	Diameter (km)
Mimas	392
Enceladus	510
Tethys	1,060
Dione	1,120
Rhea	1,530
Titan	5,150
Hyperion	410 × 220
Iapetus	1,460
Phoebe	220

FIGURE 11–10 The outermost visible ring, the F-ring (which lies outside the A-ring), with its two "shepherding" satellites as seen from Voyager 1. The satellites are about 200 km and 220 km in diameter, respectively.

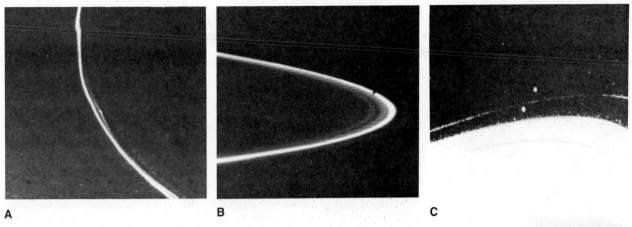

A   B   C

FIGURE 11–11   (A) Two narrow, braided, 10-km-wide rings are visible, as is a broader diffuse component about 35 km wide, on this Voyager 1 image of the F-ring. (B) The Voyager 2 images of the F-ring showed no signs of braiding. (C) Voyager 2 photographed the F-ring as the moons passed each other, but no kinkiness or braiding resulted.

11.1b Saturn Below the Rings

The structure in Saturn's clouds is of much lower contrast than that in Jupiter's clouds. After all, Saturn is colder so it has different chemical reactions. Even so, cloud structure was revealed to the Voyagers at their closest approach. A few circulating ovals similar to Jupiter's Great Red Spot and ovals were detected (Fig. 11–12).

Extremely fast winds, 1800 km/hr—4 times faster than the winds on Jupiter—were measured. On Saturn, the variations in wind speed do not seem to correlate with the positions of belts and zones, unlike the case with Jupiter. Also, as on Jupiter but unlike the case for Earth, the winds seem to be driven by rotating eddies, which in turn get their energy from the planet's interior.

The magnetic field at Saturn's equator is only ⅔ of the field present at the Earth's equator. Remember, though, that Saturn is much larger than the Earth and so its equator is much farther from its center. The total strength of Saturn's magnetic field is 1000 times stronger than Earth's and 20 times weaker than Jupiter's. Saturn has belts of charged particles (Van Allen belts), which are larger than Earth's but smaller than Jupiter's.

11.1c Saturn's Family of Moons

Farther out than the rings, we find the satellites of Saturn (Color Plate 36). Like those of Jupiter, they now have the personalities of independent worlds and are no longer merely dots of light in a telescope.

The largest of Saturn's satellites, Titan, is larger than the planet Mercury and has an atmosphere (Fig. 11–13) with several layers of haze (Fig. 11–14). Studies of how the radio signals faded when Voyager 1 went behind Titan showed that Titan's atmosphere is denser than Earth's. Surface pressure on Titan is 1½ times that on Earth.

Titan's atmosphere is opaque, apparently because of the action of sunlight on chemicals in it, forming a sort of "smog" and giving it its reddish tint. Smog on Earth forms in a similar way. The Voyagers detected nitrogen, which makes up the bulk of Titan's atmosphere. Methane is a minor constituent, perhaps 1 per cent.

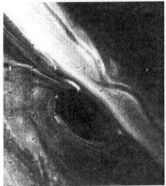

FIGURE 11–12   Time-lapse photos of this large brown spot in Saturn's atmosphere showed circulation in an anti-cyclonic sense, indicating that it is a high-pressure storm. Since such circulation often induces clearing, we may be seeing through an opening in higher clouds to darker underlying clouds. Resolution is 50 km.

Titan

FIGURE 11–13   Titan was disappointingly featureless even to Voyager 1's cameras because of its thick smoggy atmosphere. Its northern polar region was relatively dark.

Titan

FIGURE 11–14 Haze layers can be seen at Titan's limb, with divisions at altitudes of 200, 375, and 500 km.

Mimas

FIGURE 11–15 The impact feature on Mimas, Herschel, is about 130 km in diameter. Its walls are about 5 km high, and the crater may be over 10 km deep. Many of the craters on Mimas are named after characters from the legend of King Arthur, including Lancelot and Gwynevere, Arthur and Merlin.

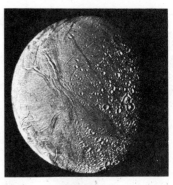

Enceladus

FIGURE 11–16 A computer-enhanced image of Enceladus from Voyager 2. Enceladus resembles Jupiter's Ganymede in spite of being 10 times smaller. Resolution is 2 km on this Voyager 2 mosaic.

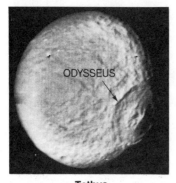

Tethys

FIGURE 11–17 The giant crater on Tethys, visible at lower right, is 500 km across. It is named Odysseus, and other features on Tethys also have names from the *Odyssey*.

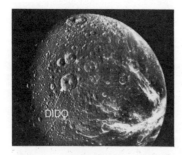

Dione

FIGURE 11–18 Dione, showing impact craters, debris, ridges or valleys, and wispy rays.

The temperature near the surface, deduced from measurements made with Voyager's infrared radiometer, is only about −180°C, somewhat warmed by the greenhouse effect but still extremely cold. This temperature is near that of methane's "triple point," at which it can be in any of the physical states—solid, liquid, or gas. So methane may play the role on Titan that water does on Earth. Parts of Titan may be covered with lakes or oceans of methane mixed with ethane, and other parts may be covered with methane ice or snow.

Some of the organic molecules formed in Titan's atmosphere may rain down on its surface. Thus the surface, hidden from our view, may be covered with an organic crust about a kilometer thick, perhaps partly dissolved in liquid methane. These chemicals are similar to those from which we think life evolved on the primitive Earth. But it is probably too cold on Titan for life to begin.

The surfaces of Saturn's other moons, all icy, are so cold that the ice acts as rigidly as rock and can retain craters. In addition to Titan, four of Saturn's moons are over 1000 km across (Appendix 4). Let us consider the major ones in order from the inside out.

Mimas boasts a huge impact structure, named Herschel, that is 130 km in diameter, nearly ⅓ the diameter of the entire moon (Fig. 11–15). Its central peak is typical of large impact craters on the Earth's Moon and on the terrestrial planets.

Enceladus' surface (Fig. 11–16), has both smooth regions and regions covered with impact craters. The existence of smooth regions suggests that Enceladus' surface was melted comparatively recently and so may be active today. The internal heating may be the result of a gravitational tug of war with Saturn and other moons, as for Jupiter's Io. Some of the patterns in the smooth regions resemble those on Ganymede, which are thought to result from shifting slabs of floating ice.

Tethys also has a large circular feature, called Odysseus, 500 km in diameter—about half Tethys' diameter (Fig. 11–17). Impacts of objects large enough to cause such major features were apparently more frequent than expected. Tethys also has a large valley, Ithaca Chasma, that goes ⅔ of the way around the satellite. The valley could have resulted from the expansion of Tethys as its warm interior froze. Or it could be a fault through the entire object.

Dione, too, shows many impact craters. It also features wisps of material (Fig. 11–18) that may have erupted from inside.

Rhea is heavily cratered on one side; its other side has wispy light markings (Fig. 11–19), like those on Dione. They might be frost from water vapor escaping from cracks in the surface.

Next out is Titan, which we have already discussed.

Hyperion, when photographed by Voyager 2, turned out to have the shape of a disk, like a hamburger.

Strangely, the side of Iapetus (Fig. 11–20) that precedes in its orbit is 5 times darker than the side that trails; it is the color and brightness of coal. The trailing side is the color and brightness of dirty snow. Perhaps the dark material is a hydrocarbon formed by sunlight reacting on some of the methane that wells up, while the bright side is covered by snow and ice. But nobody is sure.

Saturn's outermost satellite, Phoebe, is probably a captured asteroid and will be discussed with asteroids in Section 13.3.

Rhea

FIGURE 11–19 This hemisphere of Rhea shows wispy streaks.

11.2 BEYOND SATURN

Yet another superlative for Voyager 1 was the view it gave us when it looked back on the crescent Saturn (Fig. 11–21). The spacecraft is now travelling up and out of the solar system. Voyager 2's path, on the other hand, passed Uranus (1986) and will pass Neptune (1989).

The Voyagers' trips through the Saturn system exchanged a lot of unknowns for a lot of knowledge plus many specific new mysteries. Monitoring Saturn with the Hubble Space Telescope should give some new information. No spacecraft visits are now planned.

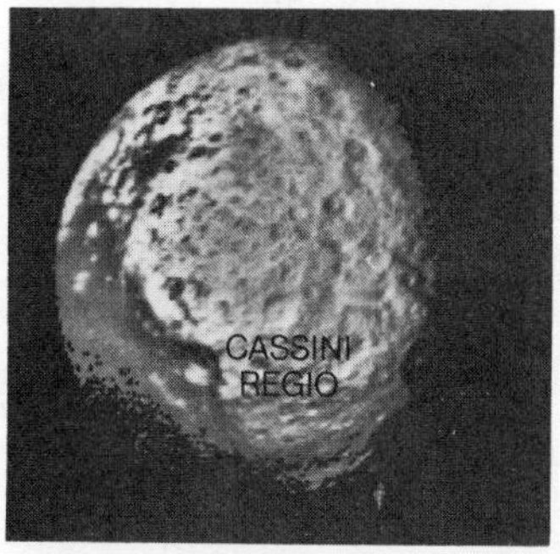

Iapetus

FIGURE 11–20 Iapetus, whose trailing side is 5 times brighter than its leading one. The dark side may have accumulated dust spiralling in toward Saturn or may be covered with hydrocarbons. It could be as black as pitch because it is pitch! A large circular feature is also visible.

FIGURE 11–21 The crescent of Saturn and the planet's rings, taken from 1,500,000 km from the far side of Saturn as Voyager 1 departed. The rings' shadows are visible cutting across the overexposed crescent.

KEY WORDS
Roche limit, Cassini's division

QUESTIONS

1. What are the similarities between Jupiter and Saturn?

2. What is the Roche limit and how does it apply to Saturn's rings?

3. Sketch Saturn's major rings and the orbits of the largest moons, to scale.

4. Describe the major developments in our understanding of Saturn, from ground-based observations to Pioneer 11 to the Voyagers.

5. Why are the moons of the giant planets more appealing for direct exploration by humans than the planets themselves?

6. What holds the material in a narrow ring?

7. Explain how part of the rings can look dark from one side but bright from the other.

8. What did the Voyagers reveal about Cassini's division?

9. What are "spokes" in Saturn's rings and how might they be caused?

10. What have we learned about Titan?

11. Explain how methane on Titan may act similarly to water on Earth.

12. Describe two of Saturn's moons other than Titan.

Chapter 12

A closeup of Miranda, innermost of Uranus's major satellites. This area, only 250-km across, includes a rugged higher-elevation terrain with the 5-km-high cliff of a scarp and a lower, striped terrain. The crater is about 25 km across.

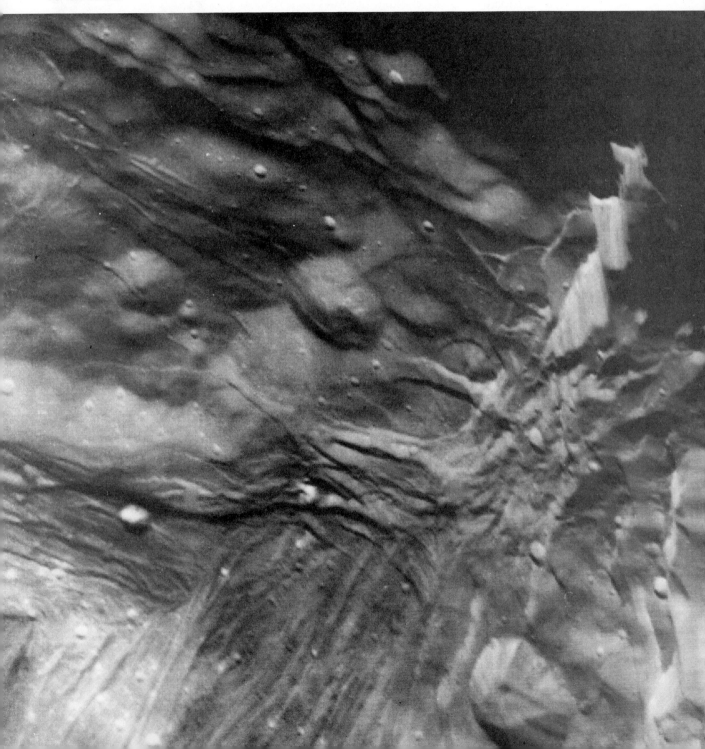

URANUS, NEPTUNE, AND PLUTO: SO NEAR AND YET SO FAR

The two other giant planets beyond Saturn—Uranus (U'ranus) and Neptune—are each about 4 times the diameter of and about 15 times more massive than the Earth. Their albedos are high, which indicates that they are covered with clouds. Like Jupiter and Saturn, Uranus and Neptune don't have solid surfaces, and have atmospheres almost entirely formed of hydrogen and helium. Some of the hydrogen may be in a liquid mantle of water, methane, and ammonia, which surrounds a rocky core, containing mostly silicon and iron. The cores of Uranus and Neptune make up substantial parts of those planets, differing from the relatively more minor cores of Jupiter and Saturn.

The outermost planet, Pluto, is a small, solid body. Its physical features and its orbit are quite different from those of the four giant planets.

In Greek mythology, Uranus was the personification of Heaven and ruler of the world, the son and husband of Gaea, the Earth. Neptune, in Roman mythology, was the god of the sea, and the planet Neptune's trident symbol reflects that origin. Pluto in Greek mythology, was the god of the underworld; some say its symbol also incorporates the initials of Percival Lowell, who sponsored the search that led to its discovery.

12.1 URANUS

Uranus was the first planet to be discovered that had not been known to the ancients. The English astronomer and musician William Herschel reported the discovery in 1781. Actually, Uranus had been plotted as a star on several sky maps during the hundred years prior to Herschel's discovery, but had not been singled out.

Uranus revolves around the Sun in 84 years at an average distance of more than 19 A.U. from the Sun. Uranus appears so tiny that it is not much bigger than the resolution we are allowed by our atmosphere. Uranus is apparently surrounded by thick methane clouds (Fig. 12–1), with a clear atmosphere of molecular hydrogen above them. The trace of methane mixed in with the hydrogen makes Uranus look greenish.

Uranus is so far from the Sun that its outer layers are very cold. Studies of its infrared radiation give a temperature of 58 K. There is no evidence for an internal heat source, unlike the case for Jupiter, Saturn, and Neptune. The recent spacecraft discovery in the ultraviolet of an aurora indicates that Uranus has a magnetic field.

The Voyager 2 spacecraft visited the Uranus system on January 24, 1986. The spacecraft rushed through quickly, since Uranus and its moons are oriented like a bull's-eye in the sky, with their axes of rotation and orbits pointing in the plane of the planets' orbits. The sun is further overhead near one of Uranus's poles than near its equator. It was a surprise to find that both poles, even the one out of sunlight, are about the same temperature. The fact that clouds are drawn out in latitude shows how important rotation is for weather on a planet, more important than whether the sun is overhead. Comparing such a strange weather system with Earth's will help us understand our own weather better.

Scientists on Earth had discovered 9 faint rings around Uranus in 1977. They observed the rings dimming the light from a star (an "occultation") as

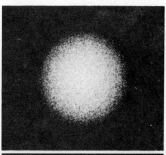

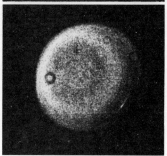

FIGURE 12–1A Uranus, photographed from the ground in blue light, shows no surface markings. **B.** Even from the Voyager 2 spacecraft close to Uranus, few surface features can be seen. The elongated cloud seen in this contrast-enhanced view enabled Uranus's rotation period to be measured. The doughnut is an out-of-focus dust speck.

A

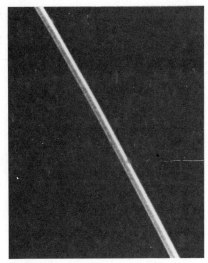

B

FIGURE 12–2 All 9 previously known rings, plus a newly discovered 10th ring (barely visible at left, about midway between the wide epsilon ring and the next ring). This photograph was taken by Voyager 2 in reflected sunlight at a distance of 1.1 million km. (B) The epsilon ring, 100-km wide here, shows a wide bright band, a wide fainter band, and a narrow bright band. The splotches are caused in the imaging process.

the rings moved across the sky in front of the star. The rings have radii 1.7 to 2.1 times the radius of the planet, and are very narrow. This discovery had given the idea of shepherding satellites, since applied to the narrow ringlets subsequently discovered at Saturn. The Voyager 2 flyby provided detailed images of the rings (Fig. 12–2). The backlighted view revealed the presence of dust between the known rings (Fig. 12–3) and showed that smaller dust particles must be swept out of the ring system somehow.

Uranus's five major moons were viewed in detail for the first time by Voyager (Fig. 12–4). They turned out to be very different from each other, with different geologic histories. Voyager also discovered 10 additional moons (Fig. 12–5).

FIGURE 12–3 When the Voyager spacecraft viewed through the rings back toward the sun, the backlighted view revealed new dust lanes between the known rings, which also show in this view. The streaks are stars.

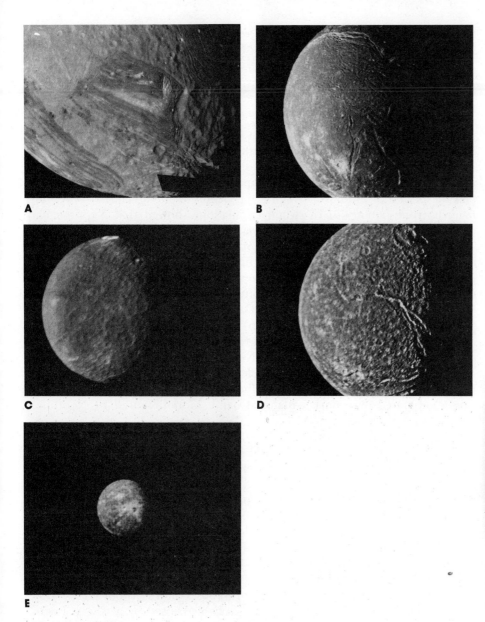

A

B

C

D

E

FIGURE 12–4 (A) Miranda, only 500 km in diameter, shows several different types of geological regions on its surface, in spite of its small size. (B) Ariel, 1,200 km across, is covered with small craters and shows many valleys and scarps. (C) Umbriel, also 1,200 km across, is the darkest of Uranus's moons and shows the least geological activity. The nature of the bright ring is unknown, though it might be a frost deposit around an impact crater. (D) Titania, 1,600 km across, shows so many impact craters that it must have an old surface. The walls of some of the fault valleys may appear bright because of frost deposits. (E) Oberon, also 1,600 km across, was not imaged as well as the other moons. The bright patches may be patterns caused by the formation of an impact crater in an icy surface. (All Voyager images of Uranus courtesy of Jurrie van der Woude, JPL)

FIGURE 12–5 One of the ten moons of Uranus discovered by Voyager 2; it is only 150 km across.

12.2 NEPTUNE

Neptune is even farther from the Sun than Uranus, 30 A.U. compared to about 19 A.U. Neptune takes 165 years to orbit the Sun. Its discovery was a triumph of the modern era of Newtonian astronomy. Mathematicians analyzed the amount that Uranus (then the outermost known planet) deviated from the orbit it would follow if gravity from only the Sun and the other known planets were acting on it. The small deviations could have been caused by gravitational interaction with another planet.

In 1845, soon after he graduated from Cambridge University, John C. Adams (Fig. 12–6) predicted positions for the new planet. But neither the astronomy professor at Cambridge nor the Astronomer Royal made observations to test this prediction. A year later, the French astronomer Urbain Leverrier independently worked out the position of the undetected planet. The French astronomers didn't test his prediction right away either. Leverrier sent his predictions to an acquaintance at Berlin, where a star atlas had

FIGURE 12–6 From Adams's diary, kept while he was in college: "1841, July 3. Formed a design in the beginning of this week, of investigating, as soon as possible after taking my degree, the irregularities in the motion of Uranus which are yet unaccounted for."

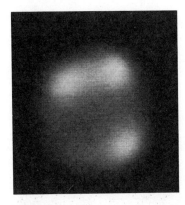

FIGURE 12–7   Galileo's notebook from late December 1612, showing a * marking an apparently fixed star to the side of Jupiter and its moons. The "star," at extreme left, was apparently Neptune. The horizontal line through Jupiter extends 24 Jupiter radii to each side. Jupiter's moons appear as dots on the line.

The episode shows the importance of keeping good lab notebooks, a lesson we should all take to heart.

recently been completed. The Berlin observer, Johann Galle, discovered Neptune within hours by comparing the sky against the new atlas.

Neptune has not yet made a full orbit since it was located in 1846. But it now seems that Galileo observed Neptune in 1613, which more than doubles the period of time over which it has been observed. Galileo's observing records from January 1613 (when calculations showed that Neptune had passed close to Jupiter) show stars that were very close to Jupiter (Fig. 12–7), stars that modern catalogues do not show. Galileo even once noted that one of the "stars" actually seemed to have moved from night to night, as a planet would. The objects that Galileo saw were very close to but not quite exactly where our calculations of Neptune's orbit show that Neptune would have been at that time. Presumably, Galileo saw Neptune, and we can use positions he measured to improve our knowledge of Neptune's orbit.

Neptune, like Uranus, appears greenish in a telescope because of its atmospheric methane. Structure on Neptune has been detected from the ground (Fig. 12–8) on images taken electronically. The images show bright regions in the northern and southern hemisphere that are probably discrete clouds separated by a dark equatorial band. Motion caused by Neptune's rotation can also be seen.

Neptune has two moons (Fig. 12–9). Triton (named after a sea god, son of Poseidon) is large, probably a little larger than our Moon. It is massive enough to have an atmosphere and a melted interior. Triton apparently has a rocky rather than an icy surface, and a thin methane atmosphere; it may have methane frost, too. A lot of nitrogen has been detected, which could make up an ocean there. A second moon, Nereid (the Greek word meaning "sea nymph"), is much smaller, 600 kilometers across. It revolves in a very elliptical orbit with an average radius 15 times greater than Triton's.

Does Neptune have rings, like the other giant planets? There is no obvious reason why it shouldn't. No rings were clearly found at any of several occultations that have been studied. But two observatories reported the same dip at a 1983 occultation; the dip was probably caused by a ring segment. The data correspond to a single ring about 3 of Neptune's radii out from its center. The ring is apparently more similar to Jupiter's insubstantial ring than to the systems of Saturn or Uranus.

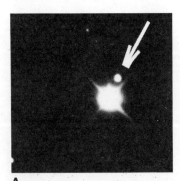

FIGURE 12–8 The use of an electronic solid-state imaging device (a CCD) enabled astronomers to image Neptune in the infrared light that is strongly absorbed by methane gas in Neptune's atmosphere. The bright regions are where non-absorbing clouds of methane ice crystals overlie the methane gas.

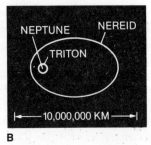

A   **B**

FIGURE 12–9   *(A)* Neptune (showing spikes caused in the telescope) and its nearer satellite, Triton (arrow). *(B)* Nereid's orbit is much larger and more elliptical than that of Triton.

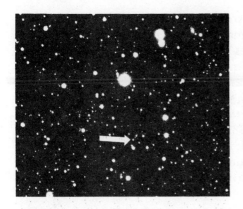 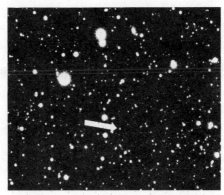

FIGURE 12–10 Small sections of the plates on which Tombaugh discovered Pluto. On February 18, 1930, Tombaugh noticed that one dot among many had moved between January 23, 1930 *(left)*, and January 29, 1930 *(right)*.

Voyager 2 will reach the Neptune system August 24, 1989, and will pass close to Neptune, Triton, and Nereid. If the instruments are still working, these points of light will then be transformed into images of objects easier to comprehend. Scientists can direct the spacecraft as close to Triton as they please, and plan to get images of Triton of higher resolution than of any solar-system object except for the Earth and the Moon.

12.3 PLUTO

Pluto, the outermost known planet, is a deviant. Its orbit is the most out of round and is inclined by the greatest angle with respect to the ecliptic plane, near which the other planets revolve.

Pluto will reach its closest distance from the Sun in 1989. Its orbit is so elliptical that part lies inside the orbit of Neptune. It is now on that part of its orbit, and will remain there until 1999.

The discovery of Pluto was a result of a long search for an additional planet which, together with Neptune, was slightly distorting the orbit of Uranus. Finally, in 1930, Clyde Tombaugh found the dot of light that is Pluto (Fig. 12–10) after a year of diligent study of photographic plates at the Lowell Observatory in Arizona. From its slow motion with respect to the stars from night to night, Pluto was identified as a new planet. Its period of revolution is almost 250 years.

Since Pluto is now within 5 years of its closest approach to the Sun out of its 248-year period, it appears about as bright as it ever does to viewers on Earth. It hasn't been as bright—about magnitude 13.5—for over 200 years. It should be barely visible through even a small telescope under dark-sky conditions.

12.3a Pluto's Mass and Size

Even such basics as the mass and diameter of Pluto are very difficult to determine. It has been hard to deduce the mass of Pluto because to do so requires measuring Pluto's effect on Uranus, a more massive body. (The orbit of Neptune is too poorly known to be of much use.) Moreover, Pluto has made less than one revolution around the Sun since its discovery, thus providing little of its path for detailed study. As recently as 1968, it was concluded that Pluto had 91 per cent the mass of the Earth. It was then realized that our observations were too uncertain to give the mass.

The situation changed drastically in 1978 with the surprise discovery (Fig. 12–11) that Pluto has a satellite. The moon has been named Charon, after the boatman who rowed passengers across the River Styx to Pluto's realm in Greek mythology (and pronounced "Shar′ on," similarly to the name of the discoverer's wife, Charlene). The presence of a satellite allows

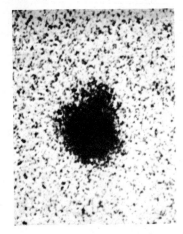

FIGURE 12–11 The discovery image of Charon and Pluto, taken on July 2, 1978. The bump at upper right was interpreted as the satellite. On the basis of photographs like this one, the separation of Charon from Pluto has been estimated. From the separation and the period, a value of 0.2 per cent that of Earth has been derived for Pluto's mass, using Kepler's third law.

Charon may be 1/3 the size of Pluto, and is separated from Pluto by only about 8 Pluto diameters (compared to the 30 Earth diameters that separate the Earth and the Moon). So Pluto/Charon are almost a double-planet system.

The fact that no photographs taken since Charon's discovery show it any more clearly than the picture above illustrates why it had not been previously discovered.

Mutual eclipses of Pluto and Charon began in 1985. For a period of time each year for several years, they occur every 3.2 days. Studying the variations of the total brightness with time is giving us accurate measurements of the sizes and albedos of Pluto and Charon and of the size and orientation of Charon's orbit.

|←1″→|

us to deduce the mass of the planet from Newton's form of Kepler's third law (p. 53). Charon is 5 or 10 per cent of Pluto's mass, and Pluto is only 1/500 the mass of the Earth, ten times less than had been suspected even recently. In this age of space exploration, it is refreshing to see that important discoveries can be made with ground-based telescopes.

Other observations show that Pluto is only about 3,500 km across, about the size of our Moon and much smaller than the Jovian planets.

For limited periods starting in 1985, we have seen Pluto and Charon pass in front of each other as they orbit each other every 6.4 days. When we measure the apparent brightness of Pluto, we are really receiving light from both Pluto and Charon together. Their blocking each other leads to dips in the total brightness we receive from them. Thus we should be able to learn more about their sizes and orbits even though we can't yet see them directly as individual objects (Fig. 12–12). We will have to wait for the Space Telescope to see Pluto and Charon resolved from each other.

12.3b What Is Pluto?

The newer values of Pluto's mass and radius can be used to derive Pluto's density, which turns out to be very low, between 0.5 and 1 times the density of water. Since only ices have such low density, Pluto must be made of frozen materials. Its composition is thus more similar to that of the satellites of the giant planets than to that of Earth or the other inner planets.

Ironically, now that we know Pluto's mass, we calculate that it is far too small to cause the deviations in Uranus' orbit that originally led to Pluto's discovery. Thus the prediscovery prediction was actually wrong. The discovery of Pluto was purely the reward of hard work in conducting a thorough search in a zone of the sky near the ecliptic.

Pluto is not massive enough to retain much of an atmosphere. But a tenuous atmosphere of methane has been detected from Earth in the infrared. The atmosphere could come from methane frost on its surface continually changing to methane gas.

No longer does Pluto, with its moon and its atmosphere, seem so different from the other outer planets. Pluto remains strange in that it is so small next to the giants, and that its orbit is so eccentric and so highly inclined to the ecliptic. The new values of Pluto's mass and density revive the thinking that Pluto may be a former moon of one of the giant planets, probably Neptune, and escaped because of a gravitational encounter with another planet.

If we were standing on Pluto, the Sun would appear over a thousand times fainter than it does to us on Earth. We would need a telescope to see the solar disk, which would be about the same size that Jupiter appears from Earth. No spacecraft to Pluto are planned, but we await observations with the Hubble Space Telescope.

Are there still further planets beyond Pluto? Tombaugh continued his search, and found none. But 1983's Infrared Astronomical Satellite, IRAS, recorded thousands of objects in the infrared. If a 10th planet exists, it would undoubtedly have been recorded, and may one day turn up in the data analysis.

FIGURE 12–12 An image showing Pluto and Charon "resolved" (that is, separated from each other) was obtained for the first time in 1984 by a special observing technique. By taking many short exposures instead of one long one, and suitably analyzing the exposures, the astronomers involved could compensate for the motion of the Earth's atmosphere to get this outstanding image.

QUESTIONS

1. What is strange about the direction of rotation of Uranus?

2. Using Appendix 4, compare the sizes of the moons of Uranus and Neptune with other objects in the solar system.

3. Explain how the occultation of a star can help us discover rings around a planet.

4. Which planets are known to have internal heat sources?

5. What fraction of its orbit has Neptune traversed since it was discovered? Since it was first seen?

6. What fraction of its orbit has Pluto traversed since it was discovered?

7. What evidence suggests that Pluto is not a "normal" planet?

8. In what wavelength range would a cold body like a 10th planet give off the most radiation? How does this affect the chance of discovering such a planet?

9. Summarize the evidence that suggests that Pluto is not a giant planet.

10. Why has the discovery of Charon allowed us to find the mass of Pluto?

Chapter 13

COMETS, METEOROIDS, AND ASTEROIDS: SPACE DEBRIS

Besides the planets and their moons, many other objects are in the family of the Sun. The most spectacular, as seen from Earth, are comets. Bright comets have been noted throughout history, instilling great awe of the heavens.

Asteroids and meteoroids are other residents of our solar system. We shall see how they and the comets are storehouses of information about the solar system's origin.

Comets have long been seen as omens. "When beggars die, there are no comets seen; The heavens themselves blaze forth the death of princes."
Shakespeare, **Julius Caesar**

13.1 COMETS

Every few years, a bright comet fills our sky (Color Plate 13). From a small, bright area called the *head,* a *tail* may extend gracefully over one-sixth (30°) or more of the sky (Fig. 13–1). The tail of a comet is always directed away from the Sun.

Although the tail may give an impression of motion because it extends out only to one side, the comet does not move visibly across the sky as we watch. With binoculars or a telescope, however, an observer can accurately note the position of the head with respect to nearby stars, and detect that the comet is moving at a slightly different rate from the stars as comet and stars rise and set together. Within days or weeks a bright comet will have

The "long hair" that is the tail led to the name **comet,** *which comes from the Greek for "long-haired star,"* **aster kometes.**

FIGURE 13–1 Comet Ikeya-Seki over Los Angeles in 1965, observed from the Mount Wilson Observatory. Photographs like this are taken with ordinary 35-mm cameras; this one was a 32-sec exposure on Tri-X film at f/1.6.

FIGURE 13–2 The head of Halley's Comet in 1910.

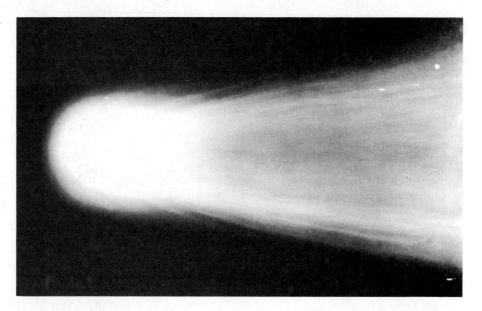

FIGURE 13–2 The head of Halley's Comet in 1910.

FIGURE 13–3 In Comet Mrkos in 1957, the straight *ion tail,* extending toward the top, and the *dust tail,* gently curving toward the right, were clearly distinguished.

become too faint to be seen with the naked eye, although it can be followed for additional weeks with binoculars and then for additional months with telescopes.

Most comets are much fainter than the one we have just described. About a dozen new comets are discovered each year, and most become known only to astronomers. If you should ever discover a comet, and are among the first three to report it to the International Astronomical Union Central Bureau for Astronomical Telegrams, at the Smithsonian Astrophysical Observatory in Cambridge, Massachusetts, it will be named after you.

13.1a The Composition of Comets

At the center of a comet's head is its *nucleus,* which is perhaps a few kilometers across. It is composed of chunks of matter. The most widely accepted theory of the composition of comets, advanced in 1950 by Fred L. Whipple of the Harvard and Smithsonian Observatories, is that the nucleus is like a *dirty snowball.* The nucleus may be made of ices of such molecules as water (H_2O), carbon dioxide (CO_2), ammonia (NH_3), and methane (CH_4), with dust mixed in.

The nucleus itself is so small that we cannot observe it directly from Earth. Radar observations have verified in a few cases that it is a few km across. The rest of the head is the *coma* (pronounced cō′ ma), which may grow to be as large as 100,000 km or so across (Fig. 13–2). The coma shines partly because its gas and dust are reflecting sunlight toward us and partly because gases liberated from the nucleus get enough energy from sunlight to radiate.

The tail can extend 1 A.U. (150,000,000 km), so comets can be the largest objects in the solar system (Color Plates 13 and 17). But the amount of matter in the tail is very small—the tail is a much better vacuum than we can make in laboratories on Earth.

Many comets actually have two tails (Fig. 13–3). The *dust tail* is caused by dust particles released from the ices of the nucleus when they are vaporized. The dust particles are left behind in the comet's orbit, blown slightly away from the Sun by the pressure of sunlight hitting the particles. As a result of the comet's orbital motion, the dust tail usually curves smoothly behind the comet.

The *gas tail* is composed of gas blown out more or less straight behind the comet by the solar wind. As puffs of gas are blown out and as the solar wind varies, the gas tail takes on a structured appearance. Each puff of matter can be seen.

A comet—head and tail together—contains less than a billionth of the mass of the Earth. It has been said that comets are as close as something can come to being nothing.

13.1b The Origin and Evolution of Comets

It is now generally accepted that trillions of incipient comets surround the solar system in a sphere perhaps 50,000 A.U. (almost 1 light year) in radius—the *comet cloud*. The total mass of matter in the cloud may be only 1 to 10 times the mass of the Earth. Occasionally one of the incipient comets leaves the comet cloud, perhaps because gravity of a nearby star has tugged it out of place, and the comet approaches the Sun in a long ellipse. The comet's orbit may be altered if it passes near a Jovian planet. Because the comet cloud is spherical, comets are not limited to the plane of the ecliptic and come in randomly from all angles.

As the comet gets closer to the Sun, the solar radiation begins to vaporize the molecules in the nucleus. The tail forms, and grows longer as more of the nucleus is vaporized. Even though the tail can be millions of km long, it is still so tenuous that only 1/500 of the mass of the nucleus may be lost. Thus a comet may last for many passages around the Sun. But some comets may hit the Sun and be destroyed (Fig. 13–4).

With each reappearance, a comet loses a little mass; eventually, it disappears. We shall see in Section 13.2 that some of the meteoroids are left in its orbit. Some of the asteroids, particularly those that cross the Earth's orbit, may be dead comet nuclei.

Because new comets come from the places in the solar system that are farthest from the Sun and thus coldest, they probably contain matter that is unchanged since the formation of the solar system. So the study of comets is important for understanding the birth of the solar system.

FIGURE 13–4 *(A)* A telescope in orbit that could block out the bright solar disk photographed this comet heading toward the Sun. A white spot the size of the Sun has been added. The comet disappeared behind the disk, which covered a somewhat larger area than the Sun, and did not reemerge. *(B)* Eleven hours after the impact, this cometary material appeared. It may be a splash or it may be part of the comet's tail blown into view. The white spot at upper left is Venus.

FIGURE 13–5 Edmond Halley.

13.1c Halley's Comet

In 1705, the English astronomer Edmond Halley (Fig. 13–5) applied a new method developed by his friend Isaac Newton to determine the orbits of comets from observations of their positions in the sky. He reported that the orbits of the bright comets that had appeared in 1531, 1607, and 1682 were about the same. Because of this, and because the intervals between appearances were approximately equal, Halley suggested that we were observing a single comet orbiting the Sun, and predicted that it would again return in 1758. The reappearance of this bright comet on Christmas night of that year, 16 years after Halley's death, was the proof of Halley's hypothesis (and Newton's method); the comet has since been known as Halley's Comet.

It seems probable that the bright comets reported every 74 to 79 years since 87 B.C. (and possibly even in 240 B.C.) were earlier appearances. The fact that Halley's Comet has been observed at least 13 times endorses the calculations that show that less than 1 per cent of a cometary nucleus's mass is lost at each perihelion passage. Most people only see Halley's Comet once, which makes it a memorable experience.

FIGURE 13–6 Halley's Comet on March 16, 1986, showing structure in the inner 2° of the plasma tail. The structure changes in minutes or hours. (A. Dressler and R. Windhorst, Las Campanas Observatory of the Carnegie Institution of Washington; photographed from Chile)

Halley's Comet went especially close to the Earth during its 1910 return (Fig. 13–6), and the Earth actually passed through its tail. Many people had been frightened that the tail would somehow damage the Earth or its atmosphere, but the tail had no noticeable effect. Even then, most scientists knew that the gas and dust in the tail were too tenuous to harm our environment.

The closest approach of Halley's Comet to the Sun occurred on February 9, 1986. Unfortunately, the Earth was on the opposite side of the Sun at the time, and the comet could not be seen because of the solar glare. A few observations were obtained from the Solar Maximum Mission in Earth orbit, and from the Pioneer Venus Orbiter spacecraft near Venus. When the comet reemerged in late February and in March, spectacular photographs were obtained (Fig. 13–6). The advance knowledge of the appearance of Halley—bright for a comet even if it was not a spectacular object to the eye—allowed detailed observations to be planned for optical, infrared, and radio telescopes. For example, spectroscopy showed many previously undetected ions in the coma and tail.

When Halley's Comet passed through the plane of the Earth's orbit in early March, it was met by an armada of spacecraft. (The United States was conspicuously not represented, because of past cost-cutting.) First, a Soviet pair of spacecraft named Vega arrived as close as 8,000 km from the nucleus. (The "Ve" is from "Venus," where they had visited first, and "Ga" is

Mark Twain was born during the appearance of Halley's Comet in 1835, and often said that he came in with the comet and would go out with it. And so he did, dying during the 1910 return of Halley's Comet.

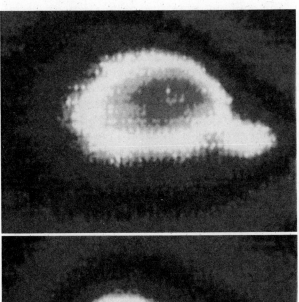

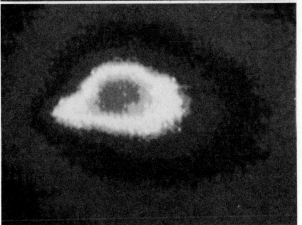

FIGURE 13–7 A & B (A) Two views of Halley's Comet from the Soviet Vega 1, showing a dust jet to the right and then to the left as the spacecraft changed its vantage point. (B) The nucleus and dust jets of Halley's Comet photographed from the European Space Agency's Giotto spacecraft on March 14, 1986. Structure can even been seen on the nucleus's surface. A few seconds after this photograph was taken, the spacecraft was hit with such force by dust that, dramatically, contact with Earth was lost just seconds before its closest approach. (A. courtesy of University of Arizona; B. © 1986 Max-Planck-Institut für Aeronomie, courtesy of Ball Aerospace Corporation)

FIGURE 13–8   Halley's Comet on March 19, 1986, showing the faint, smooth dust tail at top and its complicated plasma tails at bottom. (Donn Reiners; photographed from Hawaii)

from the Soviet word for Halley, there being no "h" in Russian.) The spacecraft studied the dust and gas near the comet, and pinpointed the position of its nucleus (Fig. 13–7A). The position was relayed to the Giotto spacecraft of the European Space Agency. (It was named after the 14th-century Italian artist who included Halley's Comet in a fresco.) Giotto's several instruments studied Halley's gas, dust, and magnetic field from as close as 600 km from the nucleus. The most astounding image was undoubtedly the photograph showing the nucleus itself (Fig. 13–7B). The nucleus turns out to be potato shaped, and about 15 km in its longest dimension, half the size of Manhattan Island. The Vega and Giotto spacecraft confirmed in general the dirty-snowball picture of the nucleus, but the snowball is darker than expected (as black as velvet) and the jets of gas and dust are more localized and stronger than expected. Even at the much greater distance—150,000 km—from the comet, the Japanese spacecraft Suisei that flew by during the same week recorded impacts with dust particles surprisingly having masses as high as a milligram.

From the ground, Halley's Comet was not spectacular. In January, the head and tail were barely visible, even with binoculars. It gave its best show in mid-March (Fig. 13–8). But in March and April, you had to be at a southern latitude to see it well, and most viewing trips were scheduled for early April, when the peak display had been predicted. However, when the comet emerged from the late-March period of bright moon, it had lost much of its tail. It was then visible to both naked eye and to binoculars only as a hazy spot in the sky, with only a trace of tail.

The next appearance of Halley's Comet, in 2061, won't be spectacular again. Not until the one after that, in 2134, will the comet show a long tail. In the meantime, fortunately, we can hope to see many other bright comets—perhaps one every ten years or so—though we may have only a few days notice. Still, the scientific results about Halley's Comet from the ground and from space and of Comet Giacobini-Zinner from a U.S. spacecraft's flyby a few months earlier have given important data to put our understanding of comets on a much firmer basis.

13.2 METEOROIDS

There are many small chunks of matter in interplanetary space, ranging up to tens of meters across. When these chunks are in space, they are called *meteoroids*. When one hits the Earth's atmosphere, friction slows it down and heats it up—usually at a height of about 100 km—until all or most of it is vaporized. Such events result in streaks of light in the sky, which we call *meteors* (popularly known as *shooting stars*). When a fragment of a meteoroid survives its passage through the Earth's atmosphere, the remnant that we find on Earth is called a *meteorite*.

13.2a Types and Sizes of Meteorites

Tiny meteorites less than a millimeter across, *micrometeorites*, are the major cause of erosion on the Moon. Micrometeorites also hit the Earth's upper atmosphere all the time, and remnants can be collected for analysis from balloons or airplanes or from deep-sea sediments. The micrometeorites are thought to be debris from comet tails. They may have been only the size of a grain of sand, and are often sufficiently slowed down that they are not vaporized before they reach the ground.

Space is full of meteoroids of all sizes, with the smallest being most abundant. Most of the small particles, less than 1 mm across, may come from comets. Most of the large particles, more than 1 cm across, may come from collisions of asteroids in the belt around the Sun in which most asteroids are found (Section 13.3).

Most of the meteorites that are found (as opposed to most of those that exist) have a very high iron content—about 90 per cent; the rest is nickel. These *iron meteorites* ("irons," for short) are thus very dense—that is, they weigh quite a lot for their volume.

Most meteorites that hit the Earth are stony in nature, and these *stony meteorites* are often referred to simply as "stones." Because stony meteorites resemble ordinary rocks and disintegrate with weathering, they are not usually discovered unless their fall is observed. That explains why most meteorites discovered at random are irons. But when a fall is observed, most meteorites recovered are stones.

A large terrestrial crater that is obviously meteoritic in origin is the Barringer Meteor Crater in Arizona (Fig. 13–9). It resulted from what was perhaps the most recent large meteorite to hit the Earth, for it was formed only 25,000 years ago.

Recent extrapolation from sky photographs indicates that about 19,000 meteorites hit the Earth each year.

FIGURE 13–9 The Barringer "meteor crater" (actually a meteorite crater) in Arizona. It is 1.2 km in diameter. Dozens of other terrestrial craters are now known, many from aerial or space photographs. The largest may be a depression over 400 km across under the Antarctic ice pack, comparable with lunar craters. Another very large crater, forming Hudson Bay, Canada, is filled with water. Most are either disguised in such ways or have eroded away.

FIGURE 13–10   A bright fireball was observed and photographed on February 5, 1977. Analysis indicated that it probably landed near Innisfree, Alberta, Canada. Scientists flew out to search the snow-covered wheat fields and found this 2-kg meteorite. It is only the third meteorite recovered whose previous orbit around the Sun is known.

FIGURE 13–11   A meteorite landed in 1982 in Wethersfield, Connecticut, coming through the roof and ceiling and bouncing off a bench.

FIGURE 13–12   Scientists picking up meteorites in the Antarctic. The meteorites have probably been long buried in ice, and are made visible when wind erodes the ice. Thousands of meteorites were found in this expedition.

FIGURE 13–13   A meteorite discovered in the Antarctic that seems more like Martian rocks than lunar or terrestrial rocks or other meteorites.

Every few years a meteorite is discovered on Earth immediately after its fall (Fig. 13–10). The chance of a meteorite's landing on someone's house is very small, but it has happened (Fig. 13–11)! Often the positions in the sky of extremely bright meteors are tracked in the hope of finding fresh meteorite falls. The newly discovered meteorites are rushed to laboratories in order to find out how long they have been in space by studying their radioactive elements. Many meteorites have recently been found in the Antarctic, where they have been well-preserved as they accumulated over the years (Fig. 13–12). A few odd Antarctic meteorites seem to have come from the Moon or even from Mars (Fig. 13–13).

The measurements show that the meteoroids were formed up to 4.6 billion years ago, the beginning of the solar system. The abundances of the forms of the elements ("isotopes") in meteorites thus tell us about the solar nebula from which the solar system formed. In fact, up to the time of the Moon landings, meteorites were the only extraterrestrial material we could get our hands on.

13.2c Meteor Showers

Meteors often occur in *showers,* when meteors are seen at a rate far above average. The most widely observed—the Perseids—takes place each summer on about August 12 and the nights on either side of that date. On any clear night a naked-eye observer may see a few *sporadic* meteors an hour, that is, meteors that are not part of a shower. (Just try going out to a field in the country and watching the sky for an hour.) During a shower several meteors may be visible to the naked eye each minute, though this is rare. Meteor showers result from the Earth's passing through the orbits of defunct comets and hitting the meteoroids left behind.

13.3 ASTEROIDS

The nine known planets were not the only bodies to result from the dust cloud that collapsed to form the solar system 4.6 billion years ago. Thousands of *minor planets*, called *asteroids*, also resulted. We detect them by their small motions in the sky relative to the stars (Fig. 13–14).

Most of the asteroids have elliptical orbits between the orbits of Mars and Jupiter, in a zone called the *asteroid belt*. Asteroids are assigned a number in order of discovery and then a name; name and number are often listed together: 1 Ceres, 16 Psyche, and 433 Eros, for example. Asteroids rarely come within a million km of each other, though occasionally collisions do occur, producing meteoroids.

Over 200 known asteroids are larger than 100 km across. The largest asteroids (Fig. 13–15) are the size of some of the moons of the planets. Yet all the asteroids together contain less mass than the Moon. Perhaps 100,000 asteroids could be detected with Earth-based telescopes if we wanted to work on it.

The most accurate way to measure the size of an asteroid is to follow its passage in front of a star. The stars are so far away that the shadow of the

FIGURE 13–14 Asteroids leave a trail on a photographic plate when the telescope is tracking the stars.

Box 13.2 The Extinction of the Dinosaurs	On Earth 65 million years ago, dinosaurs and many other species were extinguished. Evidence has recently been accumulating that these extinctions were sudden and were caused by an Earth-crossing asteroid hitting the Earth. Among the signs is the fact that the rare element iridium is widely distributed around the Earth in a thin layer laid down 65 million years ago, presumably after it was thrown up in the impact. The impact would have raised so much dust into the atmosphere that sunlight could have been shut out for months; plants and animals would not have been able to survive. (We have mentioned toward the end of Section 9.3 the related topic of "nuclear winter.")

Other evidence exists, on the other hand, that the species may have disappeared more gradually, and that an asteroid was therefore not the cause. Research is continuing.

A still newer hypothesis is based on the idea that mass extinctions on Earth took place regularly, with a period of 28 million years. It has been suggested that a faint, undiscovered companion to the Sun comes to the inner part of its orbit with that period, and its gravity then sends a number of comets in toward the Earth where they crash. (Fortunately, the hypothetical star—which has been given the name Nemesis—isn't due back for 15 million years.) Obviously, a lot of verification for both the existence of periodic extinctions and for this theory would have to be found before it can be generally accepted. Some scientists feel that it has already been disproved, while others think it is still reasonable.

FIGURE 13–15   The sizes of the larger asteroids and their relative albedos.

asteroid projected on the Earth in the light of this star is the full size of the asteroid. By timing when the star disappears and reappears, we can measure its diameter along one line across the asteroid. In recent years, groups of amateur astronomers have been making simultaneous observations of occultations from many locations, and so mapping the shape of asteroids.

The Pioneers and Voyagers en route to Jupiter and beyond travelled through the asteroid belt for many months and showed that the amount of dust among the asteroids is not greater than the amount of interplanetary dust in the vicinity of the Earth. So the asteroid belt will not be a hazard for space travel to the outer parts of the solar system.

Saturn's outermost satellite, Phoebe, is probably a captured asteroid rather than a satellite formed in place around Saturn along with the other satellites. So the images from Voyager 2 (Fig. 13–16) reveal an asteroid close up.

Recent studies have led to the conclusion that asteroids are made of different materials from each other, and represent the chemical compositions of different regions of space. The asteroids at the inner edge of the asteroid belt are mostly stony in nature, while the ones at the outer edge are darker (because they contain more carbon). Most of the small asteroids that pass near the Earth belong to the stony group. Three of the largest asteroids belong to the high-carbon group. A third group may be mostly composed of iron and nickel. The differences may be telling us about conditions in the early solar system as it was forming and how the conditions varied with distance from the protosun. Differences in chemical composition also disprove the old theory that the asteroids represent the breakup of a planet that once existed between Mars and Jupiter.

The path of the Galileo spacecraft to Jupiter has been changed to send it near the asteroid 29 Amphitrite in December 1986. The flyby of this large asteroid—200 km in diameter—should give detailed images of its entire surface. The studies of this object, which is located in the asteroid belt, should allow us to verify whether the lines of reasoning we use on ground-based asteroid observations give correct results.

13.3a Special Groups of Asteroids

Aten and *Apollo asteroids* are far from the asteroid belt; their orbits cross that of Earth (Fig. 13–17). The Aten asteroids (Fig. 13–18), further, have orbits whose diameters are currently smaller than the diameter of Earth's orbit. Their orbits probably either did or will also cross the Earth's. In the last decade or so, we have observed a few dozen of all these types of *Earth-crossing asteroids;* there may be 1000 in all. We may be able to send a spacecraft to such a closely approaching asteroid during the next decade.

Earth-crossing asteroids may well be the source of most meteorites, which could be debris of collisions when these asteroids visit the asteroid belt. Eventually, most Earth-crossing asteroids will probably collide with the

FIGURE 13–16   A Voyager 2 view of Saturn's moon Phoebe, taken on September 4, 1981, from a distance of 2.2 million km. Phoebe is 200 km in diameter, twice the size previously believed. Its albedo is only 5 per cent.

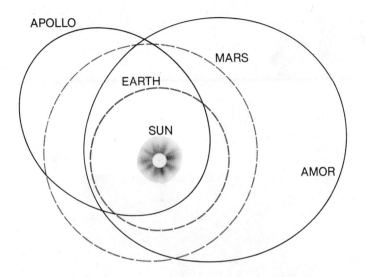

FIGURE 13–17 The orbits of Apollo and Amor, prototypes of classes of asteroids whose orbits approach or cross that of the Earth.

FIGURE 13–18 The streak shows the discovery of the Aten asteroid Ra-Shalom, which is in an Earth-crossing orbit. It was discovered in a search for high-speed objects with the small Schmidt camera at Palomar. Discovered at the time of the Israeli-Egyptian accords, Ra-Shalom was named as a symbol of hope for peace in the Middle East.

Earth. Luckily, we think that only a few dozen of them are greater than 1 km in diameter; statistics show that collisions of this tremendous magnitude should take place every few hundred million years, and could have drastic consequences for life on Earth. Still, this is pretty often on a cosmic scale.

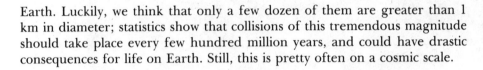

KEY WORDS

comet, head, tail, nucleus, dirty snowball, coma, dust tail, gas tail, comet cloud, meteoroids, meteors, shooting stars, meteorites, micrometeorites, iron meteorites, stony meteorites, showers, sporadic, minor planets, asteroids, asteroid belt, Aten asteroids, Apollo asteroids, Earth-crossing asteroids

QUESTIONS

1. In what part of its orbit does a comet travel head first?
2. Which part of a comet has the most mass?
3. Explain why the comet seen in the photograph opening this chapter showed delicate structure in its tail. Are we observing mainly the dust tail or the gas tail?
4. What is the relation between meteorites and asteroids?

5. Why don't most meteoroids reach the Earth's surface?
6. Why might some meteor showers last only a day while others can last several weeks?
7. Why are meteorites important in our study of the solar system?
8. Compare asteroids with the moons of the planets.

OBSERVING PROJECT

Observe the next meteor shower. No instruments will be necessary; the naked eye is best.

Chapter 14

LIFE IN THE UNIVERSE: HOW CAN WE SEARCH?

We have discussed the nine planets and the dozens of moons in the solar system, and have found most of them to be places that seem hostile to terrestrial life forms. Yet some locations—Mars, with its signs of ancient running water, and perhaps Titan or Europa—allow us to convince ourselves that life may have existed there in the past or might even be present now or develop in the future.

In our first real attempt to search for life on another planet, the Viking landers carried out biological and chemical experiments on Mars. The results showed that there is probably no life on Mars.

Since it seems reasonable that life, as we know it, anywhere in the Universe would be on planetary bodies, let us first discuss the chances of life arising elsewhere in our solar system. Then we will consider the chances that life has arisen in more distant parts of our galaxy or elsewhere in the Universe.

14.1 THE ORIGIN OF LIFE

It would be very helpful if we could state a clear, concise definition of life, but unfortunately that is not possible. Biologists state several criteria that are ordinarily satisfied by life forms—reproduction, for example. Still, there exist forms on the fringes of life—viruses, for example, which need a host organism in order to reproduce—and scientists cannot always agree whether some of these things are "alive" or not.

In science fiction, authors sometimes conceive of beings that show such signs of life as the capability for intelligent thought, even though the being may share few of the other criteria that we ordinarily recognize (Fig. 14–1). But we can make no concrete deductions if we allow such wild possibilities, so scientists prefer to limit the definition of life to forms that are more like "life as we know it."

This rationale implies, for example, that extraterrestrial life is based on complicated chains of molecules that involve carbon atoms. Life on Earth is governed by deoxyribonucleic acid (DNA) and ribonucleic acid (RNA), two carbon-containing molecules that control the mechanisms of heredity. Chemically, carbon is able to form "bonds" with several other atoms simultaneously, which makes these long carbon-bearing chains possible. In fact, we speak of compounds that contain carbon atoms as *organic*.

How hard is it to build up long organic chains? To the surprise of many, an experiment performed in the 1950's showed that making organic molecules was much easier than had been supposed. The experiment showed that if you flash a spark through a glass jar filled with simple molecules like water vapor (H_2O), methane (CH_4), and ammonia (NH_3), organic molecules accumulate in the jar. The simple molecules present at the beginning of the experiment simulate the atmosphere and the sparks simulate the lightning

FIGURE 14–1 "This is the Seven o'Clock News, with Xing Brxt on Jupiter, Klmo Hoxcl on Mars, Ycz Wrabt on Earth . . . ." Drawing by Dana Fradon; © 1980 The New Yorker Magazine, Inc.

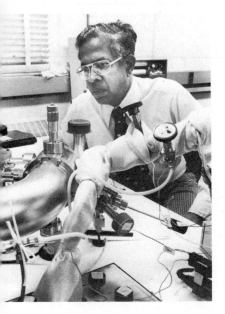

FIGURE 14–2 Cyril Ponnamperuma in his laboratory at the University of Maryland, with some of the equipment he uses to simulate the creation of complicated molecules in gases similar to those in the Earth's early atmosphere.

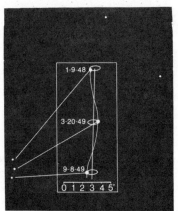

FIGURE 14–3 The image is a composite of three photos of Barnard's star, taken at the Sproul Observatory at intervals of approximately 6 months. The stars at the top have small proper motions and small parallaxes, and appear in the same position on each of the three negatives. Barnard's star, on the other hand, shows its proper motion of over 10 arc sec per year. It also shows an apparent lateral displacement caused by our vantage point, the result of the Earth's orbit around the Sun. The displacement caused by the possible planets does not show at this scale.

that may have existed in the early stages soon after our Earth's formation. The organic molecules that formed were even complex enough to include simple amino acids, the building blocks of life. Modern versions of these experiments have created even more complex organic molecules from a wide variety of simple actions on simple molecules (Fig. 14–2). Scientists have also found extraterrestrial amino acids in two meteorites that had been long frozen in Antarctic ice.

However, mere amino acids or even DNA molecules are not life itself. A jar containing a mixture of all the atoms that are in a human being is not the same as a human being. This is a vital gap in the chain; astronomers certainly are not qualified to say what supplies the "spark" of life.

Still, many astronomers think that since it is not difficult to form complex molecules, life may well have arisen not only on the Earth but also in other locations. Even if life is not found in our solar system, there are so many other stars in space that it would seem that some of them could have planets around them and that life could have arisen there independently of life on Earth.

14.2 OTHER SOLAR SYSTEMS?

We have seen how difficult it was to detect even the outermost planets in our own solar system. At the four-light-year distance of Proxima Centauri, the nearest star to the Sun, any planets around stars would be too faint for us to observe directly. There is hope, though, that the Hubble Space Telescope could see very large ones.

We now have a better chance of detecting such a planet by observing its gravitational effect on the star itself. We study a star's apparent motion across the sky with respect to the other stars and see if the star follows a straight line or appears to wobble. Any wobble would have to result from an object too faint to see that is orbiting the visible star, since the laws of physics hold that objects must travel in straight lines unless forces act on them.

There has long been a report that one of the nearest stars to us, Barnard's star (Fig. 14–3), has such a wobble. But observations from the few observatories that have long time spans of observations conflict, and the matter is now up in the air. The high-precision measurements expected in the next few years from NASA's Hubble Space Telescope and the European Space Agency's Hipparcos should greatly improve the quality of the data.

The indications from the IRAS satellite that solar systems may be forming around stars, as deduced from the excess infrared measured (Fig. 14–4), are not as satisfactory as a direct sighting of a planet or a detection of its gravitational effect. New ground-based observations of a cloud of material orbiting a star are more persuasive (Fig. 14–5). The material present is apparently ices, carbon-rich substances, and silicates. Our Earth formed from such materials, so planets may be forming there. The density of the material indicates, indeed, that planets may be present. Little material is present in the inside portion of the disk; it may have been swept away by the planets. These results surely encourage the point of view that solar systems are common.

Newspaper reports in 1984 that infrared observations detected a planet around each of two nearby stars are controversial, since it is not clear whether the objects are massive planets—50 times the mass of Jupiter—or small stars. After all, the objects' temperature of 1000 K is hot for a planet, but cold for a star. The matter is perhaps one of semantics; the objects may

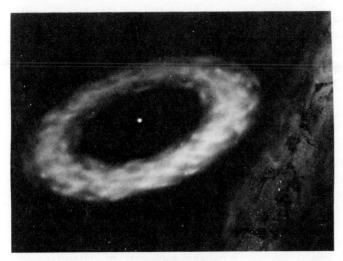

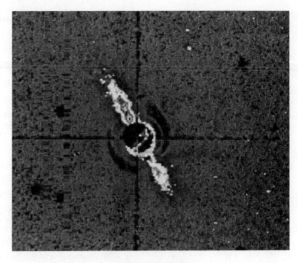

FIGURE 14-4   **FIGURE 14-5**

FIGURE 14-4   An artist's conception of the ring of material around Vega, discovered in the infrared by IRAS. An actual IRAS view of the Milky Way is in the corner of the picture.

FIGURE 14-5   This first picture of what is apparently another solar system was taken with a ground-based telescope. A circular "occulting disk" blocked out the star β (beta) Pictoris; shielding its brightness revealed the material that surrounds it. About 200 times as much mass as the Earth is present in the form of solid material extending 60 billion km from the star. β Pictoris is too far south in the sky to be seen from mid-northern latitudes. The cross of dark lines is from filaments holding up the occulting disk.

be "brown dwarfs," which is just another way of saying objects between planets and stars. In any case, no "solar system"—a system of many planets—has been found in this way.

14.3 THE STATISTICS OF EXTRATERRESTRIAL LIFE

Instead of phrasing one all-or-nothing question about life in the Universe, we can break down the problem into a chain of simpler questions. This procedure was developed by Frank Drake, a Cornell astronomer now at the University of California at Santa Cruz, and extended by, among others, Carl Sagan of Cornell and Joseph Shklovskii, a Soviet astronomer.

14.3a The Probability of Finding Life

First we consider the probability that stars at the centers of solar systems are suitable to allow intelligent life to evolve. For example, the most massive stars evolve relatively quickly (Chapters 19 through 23), and probably stay in the stable state that characterizes the bulk of their lifetimes for too short a time to allow intelligent life to evolve.

Second, we ask what the chances are that a suitable star has planets. Most scientists think that the chances are probably pretty high.

Third, we need planets with suitable conditions for the origin of life. A planet like Jupiter might be ruled out, for example, because it lacks a solid surface and because its surface gravity is high. Also, planets probably must be in orbits in which the temperature does not fluctuate too much.

Fourth, we have to consider the fraction of the suitable planets on which life actually begins. This is the biggest uncertainty, for if this fraction is zero (with the Earth being a unique exception), then we get nowhere with this

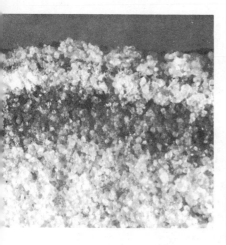

FIGURE 14–6 Lichens growing within a rock in the apparently inhospitable Antarctic. (A lichen is a symbiotic association of fungi and algae.) Within about 1 cm of the rock's top, we find a black zone, mostly fungi, that provides shading, a white lichen layer, and then a green band of algae. If plants could survive here, where temperatures plunge to $-100°C$ in the winter, could they survive on Mars?

FIGURE 14–7 From *The Day the Earth Stood Still.* (© 1951 Twentieth Century–Fox Film Corporation. All rights reserved.) The spaceman *(rear),* a robot named Gort *(front),* and an American friend. The movie described a mission to Earth sent in order to restrain our warlike nature.

entire line of reasoning. Still, the discovery that amino acids can be formed in laboratory simulations of primitive atmospheres, and the discovery of complex molecules in interstellar space, indicate to many astronomers that it is easy to form complicated molecules. And life is found in a wide range of extremes, including under rocks in the Antarctic and near sulfur vents under the ocean (Fig. 14–6).

If we want to have conversations with aliens, we must have a situation where not just life but intelligent life has evolved (Fig. 14–7). We cannot converse with algae or paramecia. Furthermore, the life must have developed a technological civilization capable of interstellar communication. These considerations reduce the probabilities somewhat, but it has still been calculated—with many weakly justified assumptions—that there are likely to be technologically advanced civilizations within a few hundred light years of the Sun.

Now one comes to the important question of the lifetime of the technological civilization itself. We now have the capability of destroying our civilization either dramatically in a flurry of hydrogen bombs or more slowly by, for example, altering our climate or increasing the level of atmospheric pollution. It is a sobering question to ask whether the lifetime of a technological civilization is measured in decades, or whether all the problems that we have—political, environmental, and otherwise—can be overcome, leaving our civilization to last for millions or billions of years. That an Earth-crossing asteroid will eventually impact the Earth with major consequences seems statistically guaranteed on a timescale of a billion years.

We can try to estimate (to guess, really) answers for each of these simpler questions within our chain of reasoning. We can use these answers together to get an answer to the larger question of the probability of extraterrestrial life. Reasonable assumptions lead to the conclusion that there may be millions or billions of planets in this galaxy on which life may have evolved. The nearest one may be within dozens of light years of the Sun. On this basis, many if not most professional astronomers have come to feel that intelligent life probably exists in many places in the Universe. Carl Sagan's latest estimate is that a million stars in the Milky Way Galaxy may be supporting technological civilizations.

A reaction to this view has recently started. Evaluating the above with a more pessimistic set of estimates can lead to the conclusion that we earthlings are alone in our galaxy.

One reason for doubting that the Universe is teeming with life is the fact that extraterrestrials have not established contact with us. Where are they all? To complicate things further, is it necessarily true that if intelligent life evolved, they would choose to explore space or to send out messages?

Philip Morrison has pointed out that we know three facts: (1) no radio signals from afar have been detected; (2) no extraterrestrials are known on Earth; and (3) we are here. He also noted that we now probably have a fourth fact: that there is no life on Mars.

What about self-sustaining colonies voyaging through space for generations? They need not travel close to the speed of light if families are aboard. Even if colonization of space took place at only 1 light year per century, the entire galaxy would still have been colonized in less than 1 per cent of its lifetime, equivalent to 1 week of a human life. The fact that we do not find extraterrestrial life here indicates that our solar system has not been colonized, which in turn indicates that life has not arisen elsewhere. Our descendants may turn out to be the colonizers of the galaxy!

A

B

C

FIGURE 14–8 *(A)* Voyager 2 being prepared for launch. The workers are loading a gold-plated copper record bearing two hours' worth of Earth sounds and a coded form of photographs. The sounds include a car, a steamboat, a train, a rainstorm, a rocket blastoff, a baby crying, animals in the jungle, and greetings in various languages. Musical selections include Bach, Beethoven, rock, jazz, and folk music. *(B)* 116 slides are included on the record. One of them is this slide, which I took when in Australia for an eclipse. It shows Heron Island on the Great Barrier Reef in Australia, in order to illustrate an island, an ocean, waves, beach, and signs of life. *(C)* In the movie *2010*, the Russian spacecraft *Leonov* (left) was stationed a safe distance from the derelict American spacecraft *Discovery* (right), which was tumbling dangerously in a decaying orbit between Jupiter and its moon Io. (From the MGM release *2010*. © 1984 MGM/UA Entertainment Co.)

14.4 INTERSTELLAR COMMUNICATION

What are the chances of our visiting or being visited by representatives of these civilizations? Pioneers 10 and 11 and Voyagers 1 and 2 are even now carrying messages out of the solar system in case an alien interstellar traveller should happen to encounter these spacecraft (Fig. 14–8).

We can hope even now to communicate over interstellar distances by means of radio signals—it takes four years for light or radio waves to reach the nearest star, and it takes even longer for messages to reach other stars. We have known the basic principles of radio for only a hundred years. We are, without thinking about it, sending out signals into space on the normal broadcast channels. A wave bearing the voice of Caruso is expanding into space, and at present is 50 light years from Earth. And once a week a new episode of "Dallas" is carried into the depths of the Universe. Military radars are even stronger.

In 1974 the giant radio telescope at Arecibo, Puerto Rico, was upgraded. At the rededication ceremony, a powerful signal was sent out into space, bearing a message from the people on Earth (Fig. 14–9). It was directed at the globular cluster M13 in the constellation Hercules, on the theory that the presence of 300,000 closely packed stars in that location would increase the chances of our signal being received by a civilization around one of them. But the travel time of the message (at the speed of light) is 24,000 years to M13, so we certainly could not expect to have an answer before twice 24,000, or 48,000 years, have passed. If anybody (or any**thing**) is observing our Sun when the signal arrives, the radio brightness of the Sun will increase by 10 million times for a 3-minute period. A similar signal, if received from a distant star, could be the giveaway that there is intelligent life there.

FIGURE 14–9 *(A)* The message sent to M13, plotted out and with a translation into English added. *(B)* The message was sent as a string of 1679 consecutive characters, in 73 groups of 23 characters each (two prime numbers). There were two kinds of characters, each represented by a frequency; the two kinds of characters are reproduced here as 0's and 1's. The basic binary-system count at upper right is provided with a position-marking square below each number.

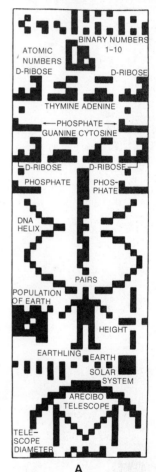
A

B

FIGURE 14–10 Ozma and Dorothy mount the magic stairway. (From *Glinda of Oz* by L. Frank Baum, illustrated by John R. Neill, copyright 1920). Two radio searches for signals from extraterrestrial life were named after Ozma.

How would you go about trying to find out if there was life on a distant planet? If we could observe the presence of abundant oxygen molecules, we might conclude that life was present. The Space Telescope should be able to detect the infrared band of radiation from molecular oxygen on a planet orbiting alpha Centauri.

The most promising way to detect the presence of life at great distances appears to be a search for radio waves. But it would be too overwhelming a task to listen for signals at all frequencies in all directions at all times. One must make some reasonable guesses about how to proceed, choosing frequencies of some cosmic importance. After a very brief search called Ozma (Fig. 14–10) after the queen of the Land of Oz, Ozma II made a search of 600 nearby stars. More recently, over a dozen groups have made radio searches. Nothing has turned up—yet.

Many astronomers feel that carrying out these brief searches was worthwhile, since we had no idea of what we would find. But now many feel that the early promise has not been justified, and the chance that there is extraterrestrial life seems much lower.

More and more astronomers in both the United States and in the Soviet Union are interested in "**c**ommunication with **e**xtra**t**errestrial **i**ntelligence" (CETI), and in the simpler and less expensive "**s**earch for **e**xtra**t**errestrial **i**ntelligence" (SETI).

FIGURE 14–11 C3PO, R2D2, Luke, and Princess Leia gaze outward toward a distant galaxy. © Lucasfilm, Ltd. (LFL) 1980. All rights reserved. From the motion picture *The Empire Strikes Back*, courtesy of Lucasfilm, Ltd.

FIGURE 14–12   A giant spacecraft visiting the closing ceremonies of the 1984 Olympics in Los Angeles. It was actually suspended from a helicopter.

14.5 UFO'S AND THE SCIENTIFIC METHOD

But why, you may ask, if most astronomers accept the probability that life exists elsewhere in the Universe, do they not accept the idea that unidentified flying objects (UFO's) represent visitation from these other civilizations (Fig. 14–11)? The answer to this question leads us not only to explore the nature of UFO's but also to consider the nature of knowledge and truth. The discussion that follows is a personal view, but one that is shared by many scientists.

The most common UFO is a light that appears in the sky that seems unlike anything manmade or any commonly understood natural phenomenon. UFO's may appear as a point or extended, or may seem to vary. But the observations are usually anecdotal, are not controlled as in a scientific experiment, and are not accessible to sophisticated instrumentation.

14.5a UFO's

First of all, most of the sightings of UFO's that are reported can be explained in terms of natural phenomena. Astronomers are experts on strange effects that the Earth's atmosphere can display, and many UFO's can be explained by such effects. Every time Venus shines brightly on the horizon, for example, in my capacity as local astronomer I get a lot of telephone calls from people asking me about the UFO. A planet or star low on the horizon can seem to flash red and green because of atmospheric refraction. Atmospheric effects can affect radar waves as well as visible light.

FIGURE 14–13   The two UFO's are really lights shining through the hole in a phonograph record viewed obliquely. The background was added later.

Sometimes other natural phenomena—flocks of birds, for example—are reported as UFO's. One should not accept explanations that UFO's are flying saucers from other planets before more mundane explanations—including hoaxes, exaggeration, and fraud—are exhausted (Fig. 14–12 and 14–13).

For many of the effects that have been reported, the UFO's would have been defying well-established laws of physics. Where are the sonic booms, for example, from rapidly moving UFO's? Scientists treat challenges to laws of physics very seriously, since our science and technology is based on them.

Most professional astronomers feel that UFO's can be so completely explained by natural phenomena that they are not worthy of more of our time. Furthermore, some individuals may ask why we reject the identification of UFO's with flying saucers, when that explanation is "just as good an explanation as any other." Let us go on to discover what scientists mean by "truth" and how that applies to the above question.

The distance record for round-about signalling was held by a signal from JPL in California that was sent to Voyager 1 near Jupiter, returned to a tracking station in Australia, sent back to JPL in Pasadena, and relayed to a science museum in Hutchinson, Kansas. There it went to a laser that set off a photocell that ignited an explosive charge for the ground-breaking. More recently, a signal was relayed via Pioneer II; the roundtrip took 8 hours.

FIGURE 14–14 When an observation at the 1919 eclipse verified one of Einstein's predictions, his theory was immediately widely accepted. Here we see the equipment used at the eclipse in Brazil in 1919 to test Einstein's theory; it provided some of the key data.

14.5b Of Truth and Theories

At every instant, we can explain what is happening in a variety of ways. When we flip a light switch, for example, we assume that the switch closes an electric circuit in the wall and allows the electricity to flow. But it is certainly possible, although not very likely, that the switch activates a relay that turns on a radio that broadcasts a message to an alien on Mars. The Martian then sends back a telepathic message to the electricity to flow, and the light goes on. The latter explanation sounds so unlikely that we don't seriously consider it. We would even call the former explanation "true," without qualification.

We regard as "true" the simplest explanation that satisfies all the data we have about any given thing. This principle is known as *Occam's Razor;* it is named after a 14th-century British philosopher who originally proposed it. Without this rule, we would always be subject to such complicated doubts that we would accept nothing as known.

Science is based on Occam's Razor, though we don't usually bother to think about it. Sometimes something we call "true" might be more accurately described as a theory. The scientific method is based on hypotheses and theories. A *hypothesis* is an explanation that is advanced to explain certain facts. When it is shown that the hypothesis actually explains most or all of the facts known, then we call it a *theory*. We usually test a theory by seeing whether it can predict things that were not previously observed, and then by trying to confirm whether the predictions are valid.

An example of a theory is the Newtonian theory of gravitation, which for many years explained almost all the planetary motions. Only a small discrepancy in the orbit of Mercury, as we describe in Section 18.6, remained unexplained. In 1916, Einstein presented a general theory of relativity as a better explanation of gravitation. His theory made certain predictions about the positions of stars during total solar eclipses. When his predictions were verified (Fig. 14–14), widespread acceptance of his theory came immediately.

Is Newton's theory "true"? Yes, in most regions of space. Is Einstein's theory "true"? We say so, although we may also think that one day a new theory will come along that is more general than Einstein's in the same way that Einstein's is more general than Newton's.

How does this view of truth tie in with the previous discussion of UFO's? Scientists have assessed the probability of UFO's being flying saucers from other worlds, and most have decided that the probability is so low that the possibility is not even worth considering. We have better things to do with our time and with our national resources. We have so many other, simpler explanations of the phenomena that are reported as UFO's that when we apply Occam's Razor, we call the identification of UFO's with extraterrestrial visitation "false." UFO's may be unidentified, but they are probably not flying, nor for the most part are they objects.

Occam's Razor, sometimes called the Principle of Simplicity, is a razor in the sense that it is a cutting edge that allows a distinction to be made among theories.

KEY WORDS

organic, Occam's Razor, theory, hypothesis

QUESTIONS

1. Why is carbon vital for life?

2. Describe two ways that the Hubble Space Telescope will aid in the search for planets around distant stars.

3. What infrared measurements have proved relevant to the search for life?

4. Choose numbers for the probabilities of each of the paragraphs of Section 14.3a, and calculate the number of intelligent civilizations that may exist among our gal-axy's trillion stars by multiplying these probabilities.

5. Why can't we hope to carry on a conversation at a normal rate with extraterrestrials on a distant star?

6. How have we already "given away" our existence to distant civilizations?

7. Describe how Einstein's general theory of relativity serves as an example of the scientific method.

PART III

Studying the Stars

Many prominences appear around the sun, and a giant eruptive prominence ejects matter westward (toward the left). This hydrogen-light photograph, taken from a high mountain in Hawaii, looks past the dark "occulting disk" that hides the sun's surface.

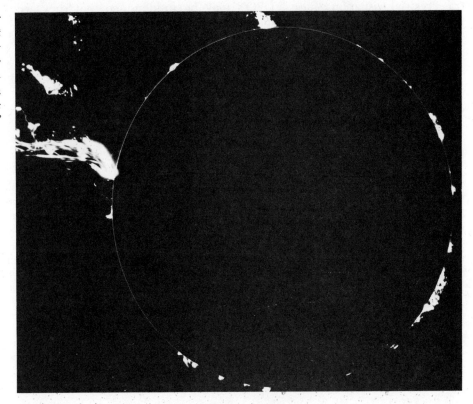

When we look up at the sky at night, most of the objects we see are stars. In the daytime, we see the sun, which is itself a star. The moon, the planets, even a comet may give a beautiful show, but, however spectacular, they are only minor actors on the stage of observational astronomy.

From the center of a city, we may not be able to see very many stars, because city light scattered by the earth's atmosphere makes the light level of the sky brighter than most stars. All together, about 6000 stars are bright enough to be seen with the unaided eye under good observing conditions.

When we look at the constellations, we are observing only the directions of the stars and not whether the stars are physically very close together. Nor do the constellations tell us anything about the nature of the stars. Astronomers tend not to be interested in the constellations because the constellations do not give useful information about how the universe works. After all, the constellations are optical-illusions—they merely tell the directions in which stars lie. Most astronomers want to know why and how. Why is there a star? How does it shine? Such studies of the workings of the universe are called

144

astrophysics. Almost all modern-day astronomers are astrophysicists as well, for they not only make observations but also think about their meaning.

Some properties of the stars that we can see with the naked eye tell us about the natures of the stars themselves. For example, some stars seem blue-white in the sky, while others appear slightly reddish. From information of this nature, astronomers are able to determine the temperatures of the stars. We can sort stars into categories and tell how far they are away and how they move in space.

Most stars occur in pairs or in larger groups, and we shall discuss types of groupings and what we learn from their study in Chapter 17. Binary stars are important, for example, in finding the masses of stars. Some stars vary in brightness, a property that astronomers use to tell us the scale of the universe itself. The study of star clusters leads to our understanding of the ages of stars and how they evolve.

In Chapter 18 we study an average star in detail. It is the only star we can see close up—the sun. The phenomena we observe on the sun take place on other stars as well, as has recently been verified by telescopes in space.

In recent years, the heavens have been studied in other ways besides observing the ordinary (visible) light that is given off by many astronomical objects. Other forms of radiation—radio waves, x-rays, gamma rays, ultraviolet, and infrared—and interstellar particles called cosmic rays are increasingly studied from the earth's surface or from space. Many a contemporary astronomer—even many who consider themselves observers rather than theoreticians—has never looked through an optical telescope. Still, astronomy began with optical studies, and our story begins there. This part of the book will tell us how we study stars, and what they teach us. Part IV will take up the life stories of stars.

Chapter 15

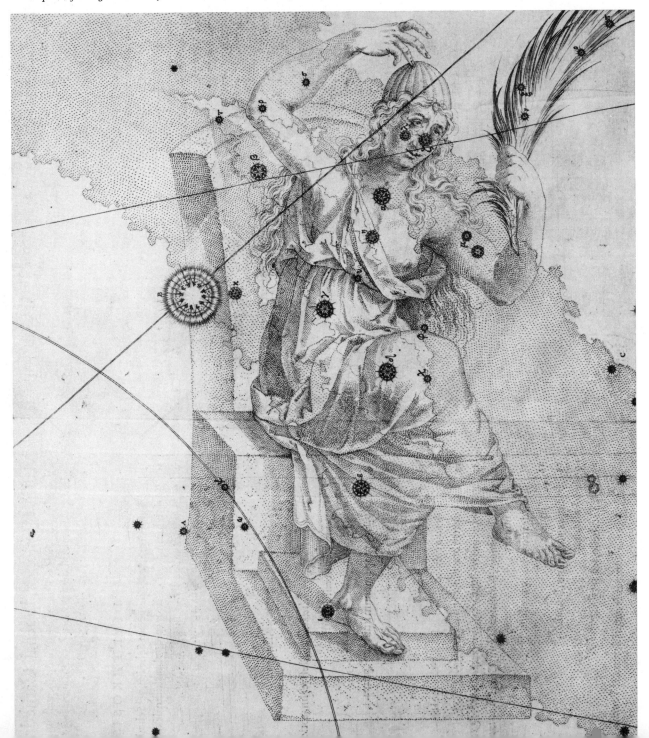

Cassiopeia, from Johann Bayer's Uranometria, *first published in 1603.*

OBSERVING STARS: WHAT WE SEE

The thousands of stars in the sky that we see with our eyes, and the millions more that telescopes reveal, are glowing balls of gas. Their bright surfaces send us the light that we see. Though we learn a lot about a star from studying its surface, we can never see through to a star's interior where the important action goes on.

For the next few chapters, we will discuss the surfaces of stars and what they tell us. Only when we finish this useful study will we go on, in Part IV, to discuss the stellar interiors.

15.1 COLORS AND TEMPERATURES

When you heat an iron poker in a fire, it begins to glow and then becomes red hot. If we could make it hotter still, it would become white hot. To understand the temperature of the poker or of a hot gas, we break its light down into its component colors. A graph of color versus the energy at each color is called a *spectrum,* as is the actual display of colors spread out (Color Plate 5). A dense gas or a solid gives off a *continuous spectrum,* that is, light changing smoothly in intensity from one color to the next.

We have seen that, technically, each color corresponds to light of a specific *wavelength.* This "wave theory" of light is not the only way we can consider light, but it does lead to very useful and straightforward explanations. Light that is about 4500 angstroms (where an angstrom [Å] is one ten-billionth of a meter) is blue light. Light that is about 6000 Å is yellow light. Light that is about 6500 Å is red light. The entire visible region of the spectrum is in this range of wavelength. We can remember the colors we perceive from the name of the friendly fellow ROY G. BIV: Red Orange Yellow Green Blue Indigo Violet, going from longer to shorter wavelengths.

The outer layers of stars come close to emitting radiation as *black bodies,* that is, as ideal emitters and absorbers of radiation. We can approximate the visible radiation from the outer layer of a star from a "black-body curve," a graph of the amount of energy emitted at each wavelength. A different black-body curve corresponds to each temperature (Fig. 15–1). The curves shown are for the same volume of gas at different temperatures. Note that as the temperature increases, the gas gives off more energy at every wavelength. Note also that the wavelength at which most energy is given off (where the curve "peaks") is farther and farther toward the blue as the temperature increases.

Astronomers rely heavily on the quantitative expression of these two radiation laws to measure the temperatures of stars. By simply measuring the brightness of a star at two known wavelengths, and comparing the relative brightness to the black-body curves for different temperatures, astronomers can take the temperature of the star.

The actual amount of radiation a star gives off depends both on how hot its surface is and how large its surface is. But the shape of the black-body curve doesn't change with a star's size.

Astronomers often measure the wavelength of light in angstroms (abbreviation Å, after A. J. Ångstrom, a Swedish astronomer of the last century). Violet light is about 4000 Å and red light is about 6500 Å.

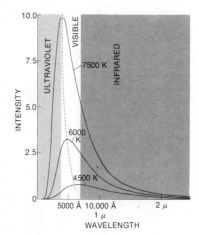

FIGURE 15–1 The intensity of radiation for different stellar temperatures. The dotted line shows how the peak of the curve shifts to shorter wavelengths as temperature increases; this is known as Wien's displacement law.

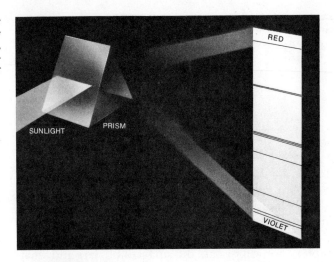

15.2 SPECTRAL LINES

When Joseph Fraunhofer looked in detail at the spectrum of the sun in the early 1800's, he noticed that the continuous range of colors in the sun's light was crossed by dark lines (Fig. 15–2 and Color Plate 5). He saw a dozen or so of these "Fraunhofer lines"; we have since mapped millions.

The dark Fraunhofer lines turn out to be from relatively cool gas absorbing radiation from behind it. We thus say that they are *absorption lines*. (Note that "absorption" is spelled with a "p," not with a "b.") Atoms of each of the chemical elements in the gas absorb light at a certain set of wavelengths. By seeing what wavelengths are absorbed, we can tell what elements are in the gas and the proportion of them present.

How does the absorption take place? To understand it, we have to study processes inside the atoms themselves. Atoms are the smallest particles of a given chemical element. For example, all hydrogen atoms are alike, all iron atoms are alike, and all uranium atoms are alike. As was discovered in 1911, atoms contain relatively light particles, which we call *electrons,* orbiting relatively massive central objects, which we call *nuclei.* The nucleus has a positive electric charge and electrons have negative electric charges. The formation of spectral lines depends chiefly on the electrons. We discuss the nuclei and further atomic properties in Section 19.3.

Why aren't the electrons pulled into the nucleus? In 1913, Niels Bohr (Fig. 15–3) made a suggestion that some arbitrary rule kept the electrons in their orbits. His suggestion was the beginning of the *quantum theory.* The theory incorporates the idea that light consists of individual packets—*quanta* of energy—which had been worked out a decade earlier. We can think of the quanta of energy as particles of light, which are called *photons.* For some purposes, it is best to consider light as waves, while for others it is best to consider light as particles (that is, as photons). Quantum theory was elaborated into a detailed structure called *quantum mechanics* in 1926 and the years following. Though we will not discuss quantum mechanics here, we should mention that it is one of the major intellectual advances of the 20th century. Regretfully, in order to get on with our discussion of stars, we will leave quantum mechanics for you to study elsewhere.

Let us return to absorption lines. We mentioned that they are formed as light passes through a gas (Fig. 15–4). The atoms in the gas take up some

FIGURE 15–3 Niels Bohr and his wife, Margrethe.

SOURCE   SPECTRUM

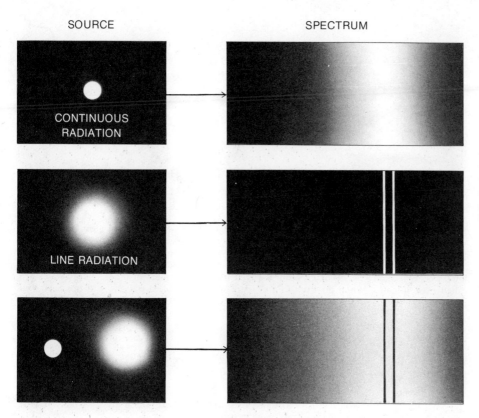

CONTINUOUS
RADIATION

LINE RADIATION

FIGURE 15–4 The left column shows sources of continuous or spectral-line radiation. We look at them from the right (that is, looking through the spectral-line source toward the source of continuous radiation in the bottom case). The wide right column shows the spectra we see.

The top row shows a source that emits continuous radiation. The middle row shows a gas that by itself has an emission line spectrum. When we look through cooler gas at a hotter source, we get an absorption line (bottom row).

of the light at specific wavelengths. The energy of this light goes into giving the electrons in those atoms more energy. But energy can't pile up in the atoms. If we look at the atoms from the side, so that they are no longer seen in silhouette against a background source of light, we would see the atoms giving off just as much energy as they take up (Fig. 15–5). Essentially, they

FIGURE 15–5 When we view a source that emits a continuum through a vapor that emits emission lines, we may see absorption lines at the wavelengths that correspond to the emission lines. Each of the inset boxes shows a graph of intensity (in the vertical direction) against wavelength (in the horizontal direction). Each graph of the spectrum shown is the one you would measure if you were at the tip of the arrow looking back along the arrow. A peak in the spectrum is an emission line, and a dip is an absorption line. Note that the view from the right shows an absorption line, while from any other angle the vapor in the center appears to be giving off an emission line. Only when you look through one source silhouetted against a hotter source do you see any absorption lines.

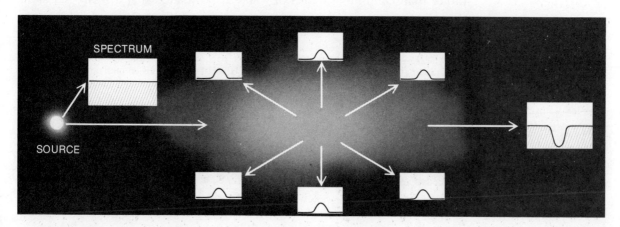

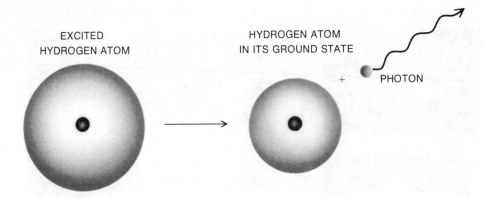

EXCITED
HYDROGEN ATOM

HYDROGEN ATOM
IN ITS GROUND STATE

+   PHOTON

FIGURE 15–6 For many purposes, it is sufficient to think of a hydrogen atom as though it was made of billiard-ball objects. Actually, we can only know that the electron has a certain probability of being in any given place, so current calculations deal with an "electron cloud." Since the hydrogen atom has only a single electron, it is a particularly simple case to study. The lowest possible energy state of an atom is called its ground state. All other energy states are called excited states. When an atom in an excited state gives off a photon, it drops back to a lower energy state, perhaps even to the ground state. We see the photons as an emission line.

give off this energy at the same set of wavelengths. These wavelengths are the *emission lines*, wavelengths at which there is an abrupt jump in the intensity of light.

All stars have absorption lines. They are formed, basically, as light from layers just inside the star's surface passes through atoms right at the surface (though detailed models are much more complicated). Emission lines occur only in special cases for stars. We see absorption lines on the sun's surface, but we can see emission lines by looking just outside the sun's edge, where only dark sky is the background. Extended regions of gas called "nebulae" give off emission lines, as we will explain later, because they are not silhouetted against background sources of light.

Bohr, in 1913, presented a model of the simplest atom—hydrogen—and showed how emission and absorption lines occur in it. Stars often show the spectrum of hydrogen. Laboratories on earth also produce the spectrum of hydrogen; the same set of spectral lines shows in both emission and in absorption. Hydrogen consists of a single central particle in its nucleus with a single electron moving around it (Fig. 15–6). Bohr's model for the hydrogen atom explained why hydrogen has only a few spectral lines (Fig. 15–7), rather than continuous bands of color. It postulated that a hydrogen atom

FIGURE 15–7 The hydrogen lines in the visible part of the spectrum, with their wavelengths shown in angstroms. Hα (H-alpha) is in the red, Hβ (H-beta), Hγ (H-gamma), and Hδ (H-delta) are in the blue, and the other lines are in the violet. You can see that the lines form a series that converges to the left—to the shorter wavelengths.

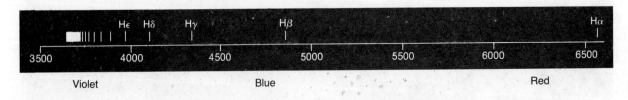

Hε Hδ   Hγ   Hβ   Hα

3500   4000   4500   5000   5500   6000   6500

Violet   Blue   Red

Color Plate 37 (top): The solar corona at the 1980 total eclipse, photographed by the author's expedition to India. Note the many streamers typical of solar maximum.

Color Plate 38: (top) The flash spectrum of the solar corona at the 1977 total solar eclipse. The crescents are chromospheric spectral lines. The horizontal streak is an overexposed spectrum of a Baily's bead. (bottom) The spectrum during totality shows the green and red emission lines from the corona and the emission lines of prominences dotted around the solar limb, and the reflection of clouds from the grating making the spectrum. (Dennis di Cicco photos; Williams College expedition)

Color Plate 39 (top): An erupting prominence on the sun, photographed on August 9, 1973, in the ultraviolet radiation of ionized helium at 304 Å with the Naval Research Laboratory's slitless spectroheliograph aboard Skylab. Because of the technique used, radiation at nearby wavelengths appears as a background. A black dot visible on the solar surface slightly to the left of the base of the prominence may be the site of the flare that caused the eruption. Supergranulation is visible on the solar disk in the light of He II, and a macrospicule shows on the extreme right. (NRL/NASA photo)

Color Plate 40 (bottom): Contours of equal intensity computed for another eruptive prominence observed in the radiation of ionized helium with the NRL instrument aboard Skylab. Each color represents a different strength of emission. This view shows the prominence 90 minutes after its eruption, when the gas had moved 500,000 km from the solar surface. (NRL/NASA photo)

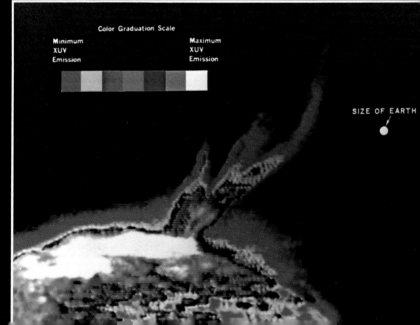

Color Graduation Scale

Minimum
XUV
Emission

Maximum
XUV
Emission

SIZE OF EARTH

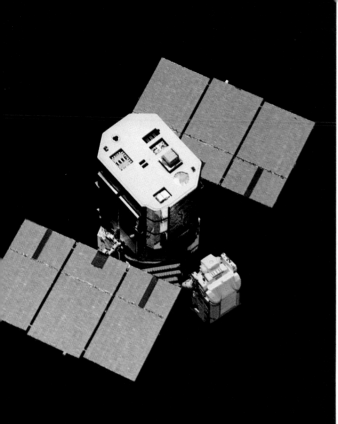

Color Plate 41: A Space Shuttle astronaut made an excursion to the damaged Solar Maximum Mission, but could not attach himself to it (left). Later, the spacecraft was retrieved by the remote arm and attached to the Space Shuttle, where it was repaired (right).

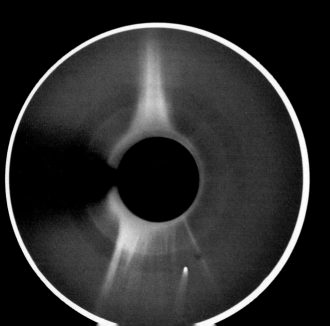

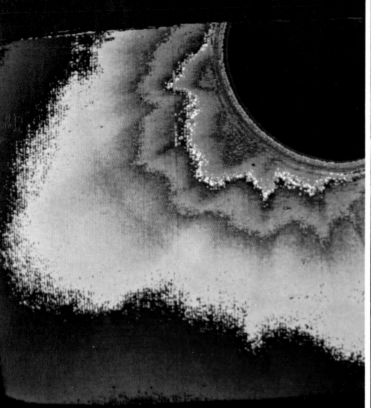

HAO SMM CORONAGRAPH/POLARIMETER
DOY 103   UT= 1837

HAO SMM CORONAGRAPH/POLARIMETER
DOY 103   UT= 1416 POL=0

Color Plate 44: Two false-color views of the corona in visible light with the coronagraph aboard Solar Maximum Mission. The different colors show the intensities, which correspond roughly to the density of the gas present. Blue is the densest and yellow is the least dense. The occulting disk, which is the quarter-circle in the corner of each image, is 1.75 times the diameter of the sun. It is immediately surrounded by a series of colored rings that are an optical effect in the system. The image on the right shows a sharp-edged narrow feature known as a coronal spike. (Courtesy of Lewis House, Ernest Hildner, William Wagner, and Constance Sawyer — HAO/NCAR/NSF and NASA)

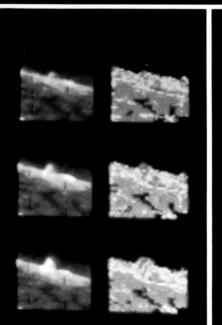
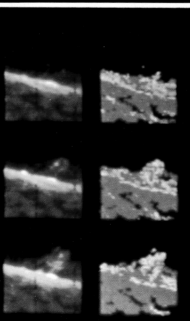

Color Plate 45 (left): A solar flare observed on April 30, 1980, from the Solar Maximum Mission. The sequence on the left, at 2.5-minute intervals, shows the radiation of three-times ionized carbon at 1548 Å. The sequence on the right shows line-of-sight velocities found from the Doppler shift. Red represents material moving away and blue represents approaching material.

We see a jet of hot gas arising from the base of a magnetic loop and filling the top of the loop. Some of the gas may even go over the top of the loop. The velocity picture supports this last idea since it shows material coming toward us at the top of the loop. A bright flare was detected in x-rays 3 minutes after the last frame shown.

Color Plate 46 (right): A similar sequence of events observed from the Solar Maximum Mission on its next orbit 90 minutes later. Each picture here is separated by 7.5 minutes. Although at first glance the material appears to be rising, detailed measurement gives a different picture. Actually features are not moving but new ones could give off or take up energy only in one of a fixed set of amounts, just as you can climb up stairs from step to step but cannot float in between. The position of the electron relative to the nucleus determines the amount of energy in a hydrogen atom. So the electron can only jump from *energy level* to energy level, and not hover in between values of the fixed set of energies allowable. When the electron jumps from a higher energy level to a lower one, a photon is given off. Many photons together make an emission line. When light hits an electron and makes it jump to a higher energy level, photons are absorbed and we see an absorption line. Each change in energy corresponds to a fixed wavelength; the higher the energy, the shorter the wavelength. Photons of blue light have higher energy than photons of red light.

15.3 THE SPECTRAL TYPES OF STARS

Hydrogen's spectrum is particularly simple. Atoms with more electrons have more possible energy levels so more jumps between energy levels are possible. They thus have more complicated sets of spectral lines.

When we look at a variety of stars, we see many different sets of spectral lines, usually from many elements mixed in together in the star's outer layers. Hydrogen is often prominent, though often iron or calcium lines are also present.

At the turn of·the century, Annie Cannon at the Harvard Observatory classified hundreds of thousands of stars by their spectra (Fig. 15–8). She first classified them by the strength of their hydrogen lines, defining stars with the strongest lines as "spectral type A," stars of slightly weaker hydrogen lines as "spectral type B," and so on. It was soon realized that hydrogen lines were strongest at some particular temperature, and were weaker at hotter temperatures (because the electrons escaped completely from the atom) or at cooler temperatures (because the electrons did not so often reach the energy levels other than their lowest possible). Rearranging the spectral types in order of temperature—from hottest to coolest—gave: O B A F G K M. Generations of astronomy students have memorized the order using the mnemonic "Oh, be a fine girl, kiss me."

FIGURE 15–8   Part of the computing staff of the Harvard College Observatory in about 1917. Annie Jump Cannon, fifth from the right, classified over 500,000 spectra in the decades following 1896. Her catalogue, called the Henry Draper catalogue after the benefactor who made the investigation possible, is still in use today. Many stars are still known by their HD (Henry Draper catalogue) numbers, such as HD 176387. Also in this photograph is Henrietta S. Leavitt, fifth from the left, whose work on Cepheid variable stars we shall discuss later.

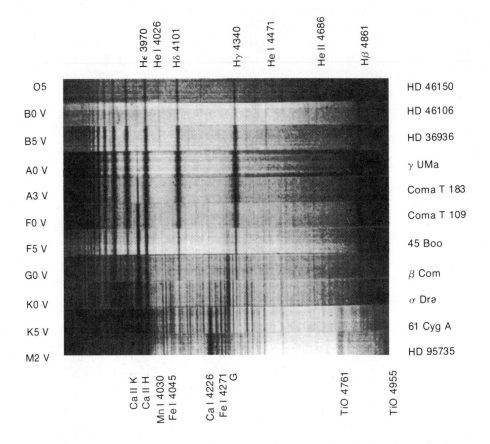

FIGURE 15–9 Spectral types. Note that all the stellar spectra shown have absorption lines. The hydrogen lines shown are Hβ (H-beta), Hγ (H-gamma), Hδ (H-delta), and Hε (H-epsilon). They form an easy-to-distinguish series of lines from right to left in spectral type A0; most of the lines farther left are other members of the series of hydrogen's spectral lines.

The Roman numeral "V" indicates that the stars are normal ("dwarf") stars. The numbers following the letters in the left-hand column represent a subdivision of the spectral types into tenths; the step from B8 to B9 is equal to the adjacent step from B9 to A0. The abbreviations of the elements forming the lines and the individual wavelengths (in angstroms) of the most prominent lines appear at the top and the bottom of the graph. The names of the stars whose spectra are shown appear at the right.

Looking at a set of stellar spectra in order (Fig. 15–9) shows how the hydrogen lines are strongest at spectral type A, which corresponds roughly to 10,000°K. A pair of spectral lines from calcium becomes strong in spectral type G, like the sun, at about 6,000°K. In the coolest stars, those of spectral type M, the temperatures are so low (only about 2000°K) that molecules can survive, and we see complicated sets of spectral lines from them. On the other extreme, the hottest stars, of spectral type O, reach 50,000°K. Astronomers subdivide the spectral types into tenths; thus, we have A7, A8, A9, F0, F1, F2, etc. The first thing an astronomer does when studying a star is to determine its spectral type and thus its temperature.

We have known that stars are mainly made of hydrogen and helium only since the 1920's. When Cecilia Payne (later known as Cecilia Payne-Gaposchkin) at Harvard first suggested the idea during that decade, based on her analyses of stellar spectra, it seemed impossible. It took years before astronomers accepted it.

KEY WORDS

spectrum, continuous spectrum, wavelength, black bodies, absorption lines, electrons, nuclei, quantum theory, quanta, photons, quantum mechanics, emission lines, energy level

QUESTIONS

1. What are two differences between a star and a planet?

2. The sun's spectrum reaches its maximum intensity at a wavelength of 5600 Å. Would the spectrum of a star whose temperature is twice that of the sun peak at a longer or a shorter wavelength? In what part of the spectrum might that be?

3. A law of radiation (Wien's law) states that when gas is heated by any factor (that is, becomes a number of times hotter), the wavelength at which its spectrum peaks becomes shorter by the same factor. The sun's temperature is about 5800 K and its spectrum peaks at 5600 Å. An O star's temperature may be 40,000 K. At what wavelength does its spectrum peak?

4. For the O star of Question 3, in what part of the spectrum does its spectrum peak? Can the peak be observed with the Palomar telescope? Explain.

5. One black body peaks at 2000 Å. Another, of the same size, peaks at 10,000 Å. Which gives out more radiation at 2000 Å? Which gives out more radiation at 10,000 Å?

6. Discuss the relation of energy levels and spectral lines. Explain why absorption and emission lines of the same element have the same wavelength.

7. Star A appears to have the same brightness through a red and a blue filter. Star B appears brighter in the red than in the blue. Star C appears brighter in the blue than in the red. Rank these stars in order of increasing temperature.

8. What is the difference between continuous radiation and an absorption line? Continuous radiation and an emission line? Draw a spectrum that shows both continuous radiation and absorption lines. Can you draw absorption lines without continuous radiation? Can you draw emission lines without continuous radiation? Explain.

9. Does the spectrum of the solar surface show emission or absorption lines?

10. Make up your own mnemonic for the spectral types.

Chapter 16

The Big Dipper, seen here through an aurora borealis, is part of the constellation Ursa Major, the Big Bear. The middle star in the handle of the dipper (the tail of the bear) can be seen as double.

HOW FAR ARE
THE STARS?

To judge how far away something is, you might use your binocular vision. Your brain interprets the slightly different images from your two eyes to give a nearby object some three-dimensionality and to assess its distance. Also, we often unconsciously judge how far away an object is by assessing its size compared with the sizes of other objects. Psychologists have made oddly shaped rooms in which the eye and brain are fooled.

But the stars are so far away that they are all points, so we cannot judge their size. And our eyes are much too close together to give us binocular vision. Only for the nearest stars can we reliably measure their distance fairly directly. In this chapter, we will see how this is done. We will also discuss how this ability helps us categorize stars.

16.1 TRIANGULATING TO THE NEARBY STARS

Though our eyes aren't far enough apart to give binocular vision, we can get a sort of binocular vision by taking advantage of our location on a moving platform: the earth. At six month-intervals, we move entirely across the earth's orbit. Since the average size of the earth's orbit—150 million kilometers (93 million miles)—is called 1 Astronomical Unit (A.U.), we move by 2 A.U.

This distance is enough to give us a slightly different perspective on the nearest few thousand stars. These stars appear to shift very slightly against the background of more distant stars (Fig. 16–1). The nearest star—known as Proxima Centauri—appears to shift by only the diameter of a dime at a distance of 2 km! It turns out to be about 4.2 light years away. (It is in the southern constellation of Centaurus, and is not visible from the U.S.) By calculating the length of the long side of a giant triangle—by "triangulating"—we can measure distances in this way out to a few hundred light years. But our galaxy is perhaps 100,000 light years across, so this is less than 1 per cent of the diameter of the galaxy. And for stars a few hundred light years away, our values are very uncertain.

The situation will improve somewhat in a few years with the launch by the European Space Agency of a satellite named Hipparcos. Hipparcos is designed to measure distances accurately by the above method. It will survey hundreds of thousands of stars all over the sky. Also, the Hubble Space Telescope will be able to improve the triangulation to a very small number of stars, but it will do these to a greater degree than Hipparcos.

16.2 ABSOLUTE MAGNITUDES

Automobile headlights appear faint when they are far away but can almost blind us when they are up close. Similarly, stars that are inherently the same

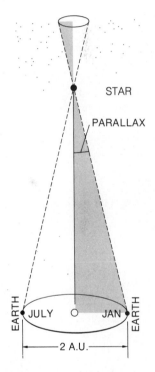

FIGURE 16–1 The nearer stars seem to be slightly displaced with respect to the farther stars when viewed from different locations in the earth's orbit. We measure the "parallax angle"—half the angle the star appears to move against the background of more distant stars when viewed from opposite sides of the earth's orbit. We know that the short side of the lower shaded triangle is the earth's distance from the sun. From the angle and the short side, we can calculate the length of the long side of the triangle, which is the star's distance.

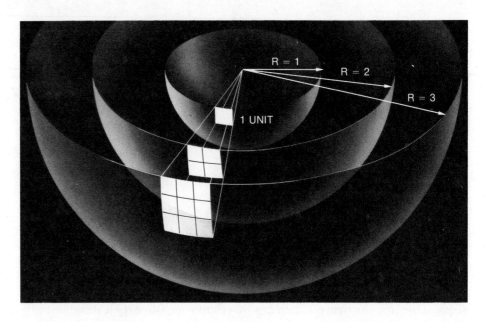

FIGURE 16–2   The inverse-square law. Radiation passing through a sphere twice as far away as another sphere has spread out so that it covers $2^2 = 4$ times the area; n times farther away it covers n^2 times the area.

brightness **appear** to be of different brightnesses to us, depending on their distances. We have previously described (in Section 3.3) how we give their brightnesses in "apparent magnitude."

To tell how inherently bright stars are, astronomers have set a specific distance at which to compare stars. The distance, which is a round number in the special units astronomers use for triangulation, comes out to be about 32.6 light years. We calculate how bright a star would appear on the magnitude scale if it were 32.6 light years away from us. We call this value its *absolute magnitude.*

For a star that is actually at the standard distance, its absolute magnitude and apparent magnitude are the same. For a star that is farther from us than that standard distance, we would have to move it closer to us to get it to that standard distance. This would make its magnitude brighter (a lower value numerically), so its absolute magnitude is brighter than its apparent magnitude.

The method is particularly valuable, since astronomers have alternative ways of finding a star's absolute magnitude. Then by comparing its absolute magnitude and its apparent magnitude (which can easily be measured at a telescope), they calculate how far away the star must be.

The method works because we understand how a star's energy spreads out with distance (Fig. 16–2). The energy from a star changes with the square of the distance; the energy decreases as the distance increases. Since one value goes up as the other goes down, it is an inverse relationship. We call it the *inverse-square law.*

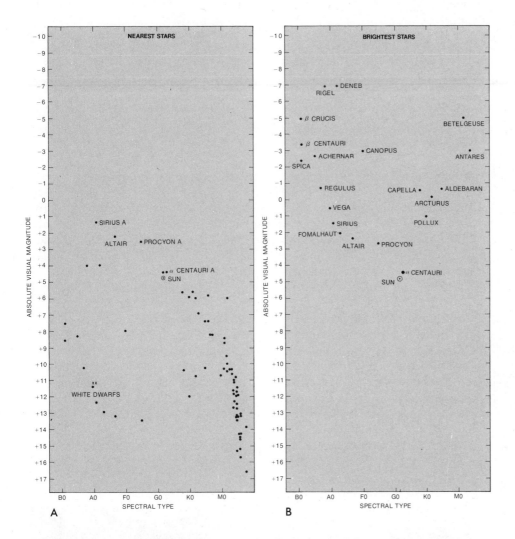

FIGURE 16–3   Color-magnitude diagrams *(A)* for the nearest stars in the sky and *(B)* for the brightest stars in the sky. The brightness scale is given in absolute magnitude—the apparent brightness a star would have at a standard distance. Because the effect of distance has been removed, the intrinsic properties of the stars can be compared directly on such a diagram.

Note that none of the nearest stars is intrinsically very luminous. Also, the brightest stars in the sky are, for the most part, intrinsically very luminous, even though they are not usually the very closest to us.

16.3 COLOR-MAGNITUDE DIAGRAMS

What if we plot a graph that has the temperature of stars on the horizontal axis (x-axis) and the brightness of the stars on the vertical axis (y-axis)? Since the temperature can be determined by measuring the color, and the brightness can be expressed in magnitudes, we then have a *color-magnitude diagram*.

If we plot such a diagram for the nearest stars (Fig. 16–3A), we find that most of them fall on a narrow band that extends downward (fainter) from the sun. If we plot such a diagram for the brightest stars we see in the sky, we see that the stars are more scattered, but that all are brighter than the sun (Fig. 16–3B).

FIGURE 16-4 Henry Norris Russell and his family circa 1917.

FIGURE 16-5 Ejnar Hertzsprung in the 1930's.

The idea of such plots came to two astronomers at about the same time in the early years of this century. Henry Norris Russell (Fig. 16–4), at Princeton in the United States, plotted the absolute magnitudes. Ejnar Hertzsprung (Fig. 16–5), in Denmark, plotted apparent magnitudes but did them for a group of stars that were all of the same distance. He could do this by considering a cluster of stars (Fig. 16–6). The two methods came to the same thing. Color-magnitude diagrams are often known as Hertzsprung-Russell diagrams or as H-R diagrams.

When we graph quite a lot of stars, or put together both nearby and farther stars (Fig. 16–7), we can see clearly that most stars fall in a narrow band across the color-magnitude diagram. This band is called the *main sequence*. Normal stars in the longest-lasting phase of their lifetimes are on the main sequence. The position of a star on the graph does not change much during this time; the sun stays at more-or-less the same position on the main sequence for about 10 billion years. Stars on the main sequence are called *dwarfs*. The sun is a dwarf.

A few stars are brighter than other stars of their same color (shown perhaps as spectral type or as temperature). Since the same amount of gas at the same temperature gives off the same amount of energy, these stars must be bigger than the main-sequence stars. They are thus called *giants* or even *supergiants*. The reddish star Betelgeuse in the shoulder of the constellation Orion is a supergiant.

A few stars are fainter than dwarfs of the same color. These fainter stars are called *white dwarfs*. Do not confuse white dwarfs with normal dwarfs. White dwarfs are smaller than dwarfs. The sun is 1.4 million kilometers across, while the white dwarf Sirius B (a companion of the bright star Sirius) is only about 10 thousand kilometers across, roughly the size of the earth.

FIGURE 16-6 A physical grouping of stars in space: a "globular cluster" of perhaps 1,000,000 stars (Section 17.3). We see M15 in the constellation Pegasus.

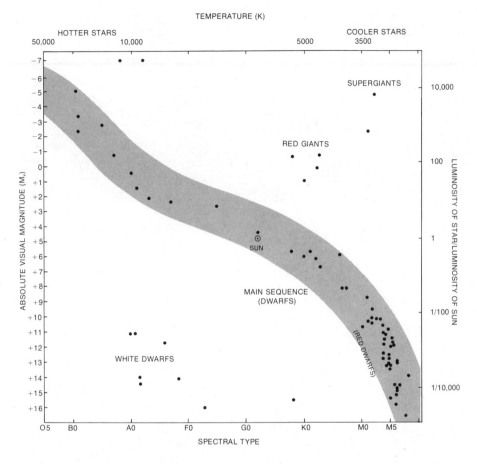

FIGURE 16–7 The color-magnitude diagram, with the stars of both parts of Figure 16–3 included. The spectral-type axis *(bottom)* is equivalent to the temperature axis *(top)*. The absolute-magnitude axis *(left)* is equivalent to the luminosity axis *(right)*.

We shall see later that giants, supergiants, and white dwarfs are later stages of life for stars.

16.4 THE MOTIONS OF STARS

The stars are so far away that they hardly move across the sky. Only for the nearest stars can we detect any such motions, which are called *proper motions*. Only for the most precise work do astronomers have to take the effect of the proper motions over decades into account. The Hipparcos satellite will improve our measurements of many proper motions.

Astronomers can actually measure motions toward and away from us much better than they can measure motions from side to side. A motion toward or away from us on a line joining us and a star (a radius) is called a *radial velocity*. A radial velocity shows up as a *Doppler shift*, a change of the spectrum in wavelength.

Doppler shifts in sound are more familiar to us. You can easily hear the pitch of a car's engine drop as the car passes us. We are hearing the wavelength of its sound waves increase as the object passes us and begins to re-

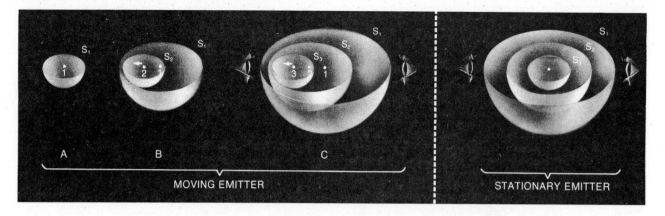

FIGURE 16–8   The Doppler shift explained. An object emits waves of radiation that can be represented by spheres representing the peaks of the wave, each centered on the object and expanding. In the left side of the drawing, the emitter is moving in the direction of the arrow. In part *A,* we see that the peak emitted when the emitter was at point 1 becomes a sphere (labelled S) around point 1, though the emitter has since moved toward the left. In part *B,* some time later, sphere S has continued to grow around point 1 (even though the emitter is no longer there), and we also see sphere S_2, which shows the position of the peaks emitted when the emitter had moved to point 2. Sphere S_2 is thus centered on point 2, even though the emitter has continued to move on. In *C,* still later, yet a third peak of the wave has been emitted, S_3, this time centered on point 3, while spheres S_1 and S_2 have continued to expand.

For the case of the moving emitter, observers who are being approached by the emitting source (those on the left side of the emitter, as shown on the left side of part *C*) see the three peaks drawn coming past them relatively bunched together (that is, at shorter intervals of time, which is the same as saying at a higher frequency), as though the wavelength were shorter, since we measure the wavelength from one peak to the next. This corresponds to a wavelength farther to the blue than the original, a *blueshift.* Observers from whom the emitter is receding (those on the right side of the emitter, as shown on the right side of part *C*) see the three peaks coming past them with decreased frequency (at increased intervals of time), as though the wavelength were longer. This corresponds to a color farther to the red, a *redshift.*

Contrast the case at the extreme right, in which the emitter does not move and so all the peaks are centered around the same point. No redshifts or blueshifts arise.

Note that once the light wave is emitted, it travels at a constant speed ("the speed of light," 3×10^{10} cm/s, equivalent to seven times around the earth in a second), so that a shorter (lower) wavelength corresponds to a higher frequency, and a longer (higher) wavelength corresponds to a lower frequency.

cede (Fig. 16–8). A similar effect takes place with light when we observe light that was emitted by an object that is moving toward or away from us. The effect, though, is not obvious to the eye; sensitive devices are necessary to detect the Doppler shift in light.

When an object is moving away from us, its spectrum is shifted slightly to longer wavelengths. Since red is at the long-wavelength end of the visible, we say that the object's light is *redshifted.* When an object is moving toward us, its spectrum is *blueshifted,* that is, shifted toward shorter wavelengths. Even when an object's radiation is beyond the red, we still say it is redshifted whenever the object is receding.

The fraction of its wavelength by which light is redshifted or blueshifted is the same as the fraction of the speed of light that the object is moving. (That is, the wavelength shift is proportional to the speed of the object.) So astronomers can measure an object's radial velocity by measuring the wavelength of a spectral line and comparing the wavelength to a similar spectral line measured from a stationary source on earth (Fig. 16–9). Stars in our galaxy have only small Doppler shifts, which shows that they are travelling

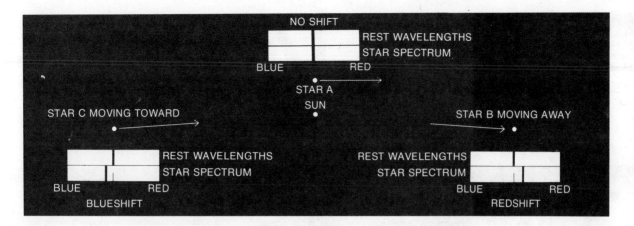

FIGURE 16–9 The Doppler effect. In each pair of spectra, the position of the spectral line in the laboratory is shown on top and the position observed in the spectrum of the star is shown below it. Lines from approaching stars appear blueshifted, lines from receding stars appear redshifted, and lines from stars that are moving transverse to us are not shifted because the star has no velocity toward or away from us. A short vertical line marks the unshifted position on the spectra showing shifts.

less than 1 per cent of the speed of light. We shall see, when we discuss galaxies, that the situation is very different for the universe as a whole. Doppler shifts turn out to be the key to our understanding of the past and future of our universe.

KEY WORDS

absolute magnitude, inverse-square law, color-magnitude diagram, main sequence, dwarfs, giants, supergiants, white dwarfs, proper motions, radial velocity, Doppler shift, redshifted, blueshifted

QUESTIONS

1. If a star that is 100 light years from us appears to be 10th magnitude, would its absolute magnitude be a larger or a smaller number?

2. If the sun were 10 times farther from us than it is, how many times less light would we get from it?

3. Sketch a color-magnitude diagram, and distinguish between dwarfs and white dwarfs.

4. Compare the temperature of white dwarfs with dwarfs, and explain their relative brightness.

5. Compare the temperature of red giants with red dwarfs, and explain their relative brightness.

6. Does measuring Doppler shifts depend on how far away an object is?

7. Two stars have the same velocity in space in the same direction, but star A is 100 light years from the earth and star B is 300 light years from the earth. (a) Which has the larger velocity measured by a Doppler shift? (b) Which has the larger proper motion?

8. How might Space Telescope help in determining proper motions?

Chapter 17

M13, a globular cluster in the constellation Hercules.

STARS TOGETHER: BINARIES, VARIABLES, AND CLUSTERS

Though the sun probably exists as an isolated star in space, most stars are part of pairs or groups. If our planet were part of a double-star system, we might see two or more suns rising. And if our sun were part of a cluster of stars, the nighttime sky would be ablaze with bright points of light.

Astronomers take advantage of stellar pairs and groups to find out how much mass stars contain and how old some stars are. They take advantage of other stars that vary in brightness to find out how far away those stars are. In this chapter, we will see how.

17.1 BINARY STARS

If you look up at the handle of the Big Dipper, you might be able to see, even with your naked eye, that the middle star (Mizar) has a fainter companion (Alcor). Native Americans called these stars a horse and rider. But Alcor and Mizar are not noticeably revolving around each other as a result of their mutual gravity. They may have some link, though. In a few cases, *optical doubles*—chance apparent associations—occur.

When you look at Mizar through a telescope, you can see that it is split in two. These two stars **are** revolving around each other, and are thus known as a *visual binary*. We see visual binaries best when the stars are relatively far away from each other (Fig. 17–1).

The two components of Mizar are known as Mizar A and Mizar B. If we look at the spectrum of either one (Fig. 17–2), we can see that over a period of days, the spectral lines seem to split and come together again. The spectrum shows that two stars are actually present in Mizar A, and another two present in Mizar B. They are *spectroscopic binaries*. Even when only the

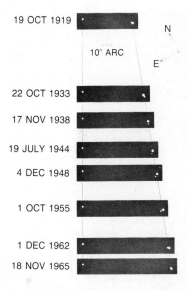

19 OCT 1919

N

10" ARC

E

22 OCT 1933

17 NOV 1938

19 JULY 1944

4 DEC 1948

1 OCT 1955

1 DEC 1962

18 NOV 1965

FIGURE 17–1 The two components of the visual binary Krüger 60 are seen to orbit each other with a period of about 44 years in this series of photographs taken at the Leander McCormick Observatory in Virginia and at the Sproul Observatory in Pennsylvania. We also see the proper motion of the Krüger 60 system, as it moves farther away from the single star at the left.

Albireo (β [beta] Cygni) contains a B star and a K star, which make a particularly beautiful pair because of their different colors.

FIGURE 17–2 Two spectra of Mizar (ζ (zeta) Ursae Majoris) taken 2 days apart show that it is a spectroscopic binary. The lines of both stars are superimposed in the upper stellar absorption spectrum (top arrow), but are separated in the lower spectrum (bottom arrow) by 2 Å, which corresponds to a relative velocity of 140 km/sec. Emission lines from a laboratory source are shown at the extreme top and extreme bottom (vertical white lines) to provide a comparison with a source at rest that has lines at known wavelengths.

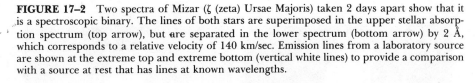

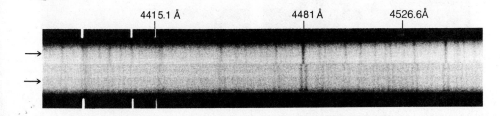

4415.1 Å   4481 Å   4526.6Å

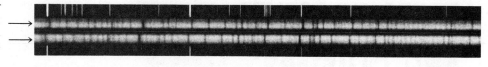

FIGURE 17–3 Two spectra of Castor B (α (alpha) Geminorum B) taken at different times show a Doppler shift. Thus the star is a spectroscopic binary, even though lines from only one of the components can be seen. The comparison spectrum of a laboratory source appears at the top and bottom. Note that emission lines from this laboratory source are in the same horizontal positions at extreme top and bottom, while the absorption lines of the stellar spectra (marked with arrows) are shifted laterally (that is, in wavelength) with respect to each other.

spectrum of one star is detectable (Fig. 17–3), we can still tell that it is part of a spectroscopic binary if the spectral lines shift back and forth in wavelength over time.

Yet another type of binary star is detectable when the light from a star periodically dims because one of the components passes in front of (eclipses) the other (Fig. 17–4). In *eclipsing binaries* like this one, we can often determine enough about the stars' orbits around each other that we can calculate how much mass the stars must have to stay in those orbits. We can measure a star's mass directly mainly for eclipsing binaries, so we know only a few dozen stellar masses. It turns out that, for main-sequence stars, the brighter stars have more mass. The most massive stars contain about 60 times as much mass as the sun. The least massive stars contain about 1/15th the mass of the sun.

Eclipsing binaries are detected by us only because they happen to be aligned so that one star passes between the earth and the other star. Thus all binaries would be eclipsing if we could see them at the proper angle. The same binary system might be seen as different types depending on the angle at which we happen to view the system (Fig. 17–5). When we discuss the evolution of stars, we will see that matter often flows from one member of a binary system to another. This interchange of matter can change a star's evolution very drastically. And the flowing matter can heat up by friction to the very high temperatures at which x-rays are emitted. In recent years, our satellite observatories above the earth's atmosphere have detected many such sources of celestial x-rays.

One more type of double star can be detected when we observe a wobble in the path of a star as its proper motion takes it slowly across the sky (Fig. 17–6). The laws of physics hold that a mass must move in a straight line unless affected by a force. The wobble off a straight line tells us that an invisible object must be pulling the visible star off to the side. Measurement of the positions and motions of stars is known as astrometry, so these stars are called *astrometric binaries*. A similar method is being used to try to detect planets around distant stars. Results are inconclusive so far, though perhaps Hipparcos and the Hubble Space Telescope will find planets in this way.

FIGURE 17–4 The shape of the light curve of an eclipsing binary depends on the sizes of the components and the angle from which we view them. At lower left, we see the light curve that would result for star B orbiting star A, as pictured at upper left. When star B is in the positions shown with subscripts at top, the regions of the light curve marked with the same subscripts result. The eclipse at B_4 is total. At right, we see the appearance of the orbit and the light curve for star D orbiting star C, with the orbit inclined at a greater angle than at left. The eclipse at D_4 is partial. From earth, we observe only the light curves, and use them to determine what the binary system is really like, including the inclination of the orbit and the sizes of the objects.

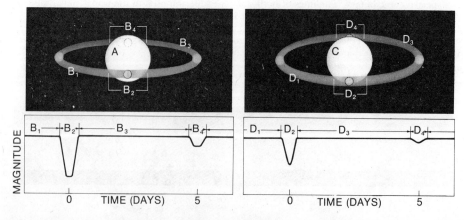

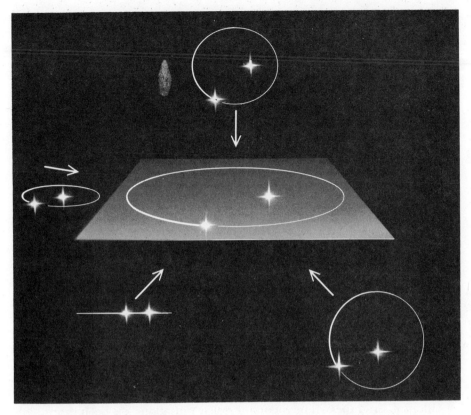

FIGURE 17–5 The appearance and the Doppler shift of the spectrum of a binary star depend on the angle from which we view the binary. From far above or below the plane of the orbit, we might see a visual binary, as shown on the top and at the lower right of the diagram. From close to but not exactly in the plane of the orbit, we might see only a spectroscopic binary, as shown at left. (The stars appear closer together, so might not be visible as a visual binary.) From exactly in the plane of the orbit, we would see an eclipsing binary, as shown at lower left.

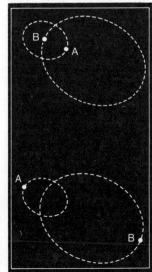

FIGURE 17–7 The light curve for Mira, the first variable star to be discovered; similar long-period variables are known as Mira variables. Dates are given at top in Julian days (J. D.), elapsed time since January 1, 4713 B.C.

17.2 VARIABLE STARS

Many stars vary in brightness over hours, days, or months. Eclipsing binaries are one example of such *variable stars,* but many other types of stars are actually individually changing in brightness. Many professional and amateur astronomers follow the *light curves* of such stars, that is, graphs of how their brightness changes over time.

A very common type of variable star changes slowly in brightness with a period of months or up to a couple of years. The first of these *long-period variables* to be discovered was the star Mira in the constellation Cetus, the Whale. At its maximum brightness, it is 3rd magnitude, quite noticeable to the naked eye. At its minimum brightness, it is far below naked-eye visibility (Fig. 17–7). Such stars are supergiants whose outer layers actually change in temperature.

FIGURE 17–6 Sirius A and B, an astrometric binary. From studying the motion of Sirius A (often called, simply, Sirius) astronomers deduced the presence of Sirius B before it was seen directly. Sirius B's orbit is larger because Sirius A is a more massive star.

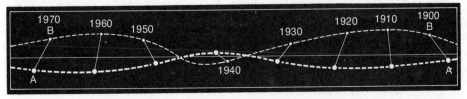

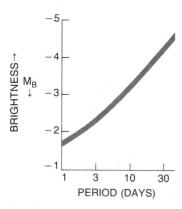

A particularly important type of variable star for astronomers are the *Cepheid variables,* of which the star delta Cephei is a prime example (Fig. 17–8). A strict relation exists linking the period over which the star varies with the star's absolute magnitude (which measures its intrinsic brightness). This relation (Fig. 17–9) is the key to the importance of the stars. After all, we can observe a star's light curve easily, with just a telescope. From comparing the period of a newly observed Cepheid with the known relation, we can tell the absolute magnitude of the star. From comparing the absolute magnitude with the star's measured apparent magnitude, we can tell how far away the star is. Cepheids are supergiant stars, bright enough that we can detect them in some of the nearby galaxies. Finding the distance to these Cepheids gives us the distances to the galaxies they are in. This is the most accurate method we have of finding the distances to galaxies, and so is at the basis of most of our measurements of the size of the universe. The relation was discovered in the early years of this century by Henrietta Leavitt (Fig. 17–10) in the course of her study of stars in the Magellanic Clouds (Fig. 17–11). She knew only that all the stars in each of the Magellanic Clouds were about the same distance from us, so that she could compare their apparent magnitudes without worrying about distance effects. The Magellanic Clouds turn out to be the nearest galaxies to our own.

FIGURE 17–9   The period-luminosity relation for Cepheid variables.

Cepheid variables are actually changing in size, which leads to their variations in brightness. The theory of stellar pulsations has been thoroughly worked out. Stars of a related type, *RR Lyrae variables,* have shorter periods than regular Cepheids. RR Lyrae variables are mostly found in clusters of stars (see the next section). Since they all have the same average absolute magnitude, observing an RR Lyrae star and comparing its apparent magnitude with its absolute magnitude enables us to tell quickly how far away the cluster is.

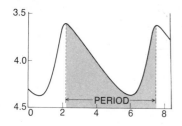

FIGURE 17–10   Henrietta S. Leavitt, who worked out the period-luminosity relation, at her desk at the Harvard College Observatory in 1916.

FIGURE 17–11   From the southern hemisphere, the Magellanic Clouds are high in the sky. In this view from Perth, Australia, the Large Magellanic Cloud is about 20° above the horizon and the Small Magellanic Cloud is twice as high. They are not quite this obvious to the naked eye. Canopus, the second brightest star in the sky, appears at the lower left.

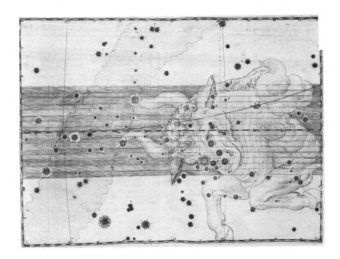

FIGURE 17–12 Taurus, the Bull, from the star atlas of Bayer (1603).

17.3 STAR CLUSTERS

The face of Taurus, the Bull (Fig. 17–12) is outlined by a V of stars, which are close together out in space. On Taurus's back rides another group of stars (Fig. 17–13 and Color Plate 48), the Pleiades (Plee'-a-deez). Both are examples of *open clusters*, groupings of dozens or a few hundred stars. We can often see 6 of the Pleiades with the naked eye, but binoculars reveal dozens more and telescopes reveal the rest. Open clusters are irregular in shape. They are found near the Milky Way, which means that they are in the plane of our galaxy.

In the constellation Hercules, as shown in the picture opening this chapter, we see a cluster of stars with spherical symmetry. This *globular cluster* (see also Color Plate 47) looks like a fuzzy mothball when seen through a small telescope. Larger telescopes reveal many individual stars. Globular clusters (Fig. 17–14) contain tens of thousands or hundreds of thousands of stars, many more than open clusters. Globular clusters appear far above and below the Milky Way; they form a spherical *halo* around the center of our galaxy. We find their distances from earth by studying RR Lyrae stars in them.

When we study the spectra of stars in clusters to find out what elements are in them, we find out that all the stars are about 90 per cent hydrogen with almost all the rest helium (that is, 90 per cent of the number of atoms are hydrogen). Under 1 per cent of the atoms are made of elements heavier than helium, even in the open clusters. But the abundances of these "heavy elements" in the stars in globular clusters are ten times lower. We now think that the globular clusters formed about 10–18 billion years ago, early in the life of the galaxy, from the original gas in the galaxy. This gas had a very low abundance of elements heavier than hydrogen and helium. As time

The Pleiades are often known as the Seven Sisters, after the seven daughters of Atlas who were pursued by Orion and who were given refuge in the sky. That one is missing—the Lost Pleiad—has long been noticed. Of course, a seventh star is present (and hundreds of others as well), although too faint to be plainly seen with the naked eye. The Pleiades seem to be riding on the back of Taurus.

FIGURE 17–13 The Pleiades, a galactic cluster. The six brightest stars are visible to the naked eye.

 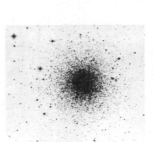 

FIGURE 17–14 Three views of the globular cluster NGC 288. From left to right, we see a plate taken with a blue-sensitive emulsion, a plate taken with a red-sensitive emulsion, and a plate taken with an infrared-sensitive emulsion.

FIGURE 17–15 *(A)* A color-magnitude diagram showing two open clusters superimposed. We see that the fainter stars of galactic clusters are on the main sequence, while the brighter stars are above and to the right of the main sequence. Almost all stars in the younger clusters, like h and χ (chi) Persei, follow the main sequence, while more members of older clusters, like M67, have had time to evolve toward the red-giant region. The lower parts of both h and χ Persei and M67 are on the main sequence. By observing the point where the cluster turns off the main sequence, we deduce the length of time since its stars were formed, i.e., the age of the cluster.

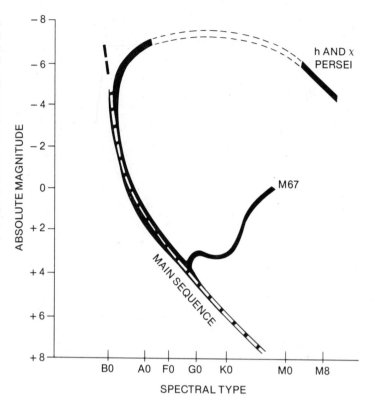

FIGURE 17–16 An x-ray burster before, during, and after its 50-second burst. Imaged with the Einstein observatory.

passed, stars in our galaxy "cooked" lighter elements into heavy elements through fusion in their interiors, died, and spewed off these heavy elements into space. The stars in open clusters formed later on from enriched gas. Thus the open clusters have higher abundances of these heavier elements.

We can tell the ages of clusters by looking at their color-magnitude diagrams. For the youngest open clusters, almost all the stars are on the main sequence. But the stars at the upper-left part of the main sequence—which are relatively hot, bright, and massive—live on the main sequence for much shorter times than the cooler stars, which are fainter and less massive. The points representing their colors and magnitudes thus begin to lie off the main sequence (Fig. 17–15). Since we can calculate the main-sequence lifetimes of stars of different colors, we can tell the age of an open cluster by which stars do not appear on the main sequence. The ages of open clusters range from "only" a few million years up to billions of years.

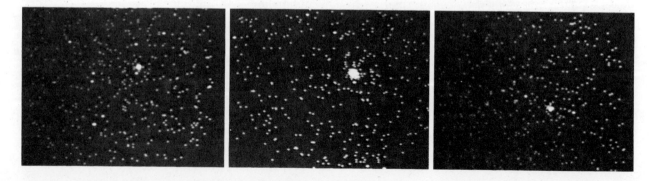

BOX 17.1 Star Clusters	*Open Clusters*	*Globular Clusters*
	No regular shape	Shaped like a ball, stars more closely packed toward center
	Many young stars	All old stars
	C-m diagrams have long main sequences	C-m diagrams have short main sequences
	Where stars leave the main sequence tells the cluster's age	All clusters have the same H-R diagram and thus the same age
	Stars have similar composition to sun	Stars have lower abundances of heavy elements than sun
	Hundreds of stars per cluster	10,000–1,000,000 stars per cluster
	Found in galactic plane	Found in galactic halo

Though different open clusters have different-looking color-magnitude diagrams because of this effect, all globular clusters have essentially the same color-magnitude diagram. Thus all the globular clusters are of essentially the same age. Few of their stars are on the main sequence, indicating that all globular clusters are very old—10 billion to 18 billion years old. Their ages set a lower limit for the age of the universe.

Observations from the Einstein x-ray observatory have detected bursts of x-rays coming from a few regions of sky, including a half-dozen globular clusters (Fig. 17–16). We think that among the many stars in those clusters, some have lived their lives, died, and are attracting material from nearby companions to flow over to them. We are seeing x-rays from this material as it "drips" onto the dead star's surface and heats up. Clusters thus turn out to have many surprises.

KEY WORDS

optical doubles, visual binary, spectroscopic binary, eclipsing binary, astrometric binaries, variable stars, light curves, long-period variables, Cepheid variables, RR Lyrae variables, open clusters, globular clusters, halo

QUESTIONS

1. Sketch the orbit of a double star that is simultaneously a visual, an eclipsing, and a spectroscopic binary.

2. (a) Assume that an eclipsing binary contains two identical stars. Sketch the intensity of light received as a function of time. (b) Sketch to the same scale another light curve to show the result if both stars were much larger while the orbit stayed the same.

3. Explain briefly how observations of a Cepheid variable in a distant galaxy can be used to find the distance to the galaxy. Why can't we use triangulation instead?

4. Why is it more useful to study the color-magnitude diagram of a cluster of stars instead of one for stars in a field chosen at random?

5. Relate the masses of main-sequence stars to their positions on the color-magnitude diagram.

6. Sketch the path in the sky of the visible (with a solid curve) and invisible (with a dotted curve) components of an astrometric binary.

7. What are the Magellanic Clouds? How did we find out the distances to them?

8. What are two distinctions between open and globular clusters?

Chapter 18

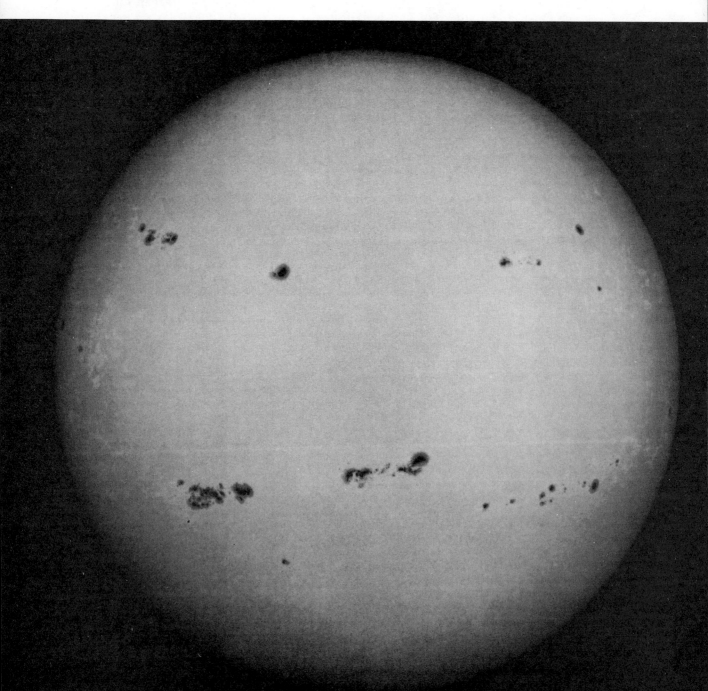

The sun, observed in white light in November 9, 1979, near the most recent solar maximum.

OUR STAR: THE SUN

Not all stars are far away; one is close at hand. By studying the sun, we not only learn about the properties of a particular star but also can study processes that undoubtedly take place in more distant stars as well. We will first discuss the *quiet sun*, the solar phenomena that appear every day. Afterwards, we will discuss the *active sun*, solar phenomena that appear non-uniformly on the sun and vary over time.

18.1 BASIC STRUCTURE OF THE SUN

We think of the sun as the bright ball of gas that appears to travel across our sky every day. We are seeing only one layer of the sun, part of its atmosphere. The layer that we see is called the *photosphere,* which simply means the sphere from which the light comes (from the Greek "photos," light). As is typical of many stars, about 94 per cent of the atoms and nuclei in the outer parts are hydrogen, about 5.9 per cent are helium, and a mixture of all the other elements makes up the remaining one-tenth of one per cent. The overall composition of the interior is not very different.

The sun is a typical star, in the sense that stars much hotter and much cooler, and stars intrinsically much brighter and much fainter, exist. Radiation from the photosphere peaks (is strongest) in the middle of the visible spectrum; after all, our eyes evolved over time to be sensitive to that region of the spectrum because the greatest amount of the solar radiation is emitted there. If we lived on a planet orbiting an object that emitted mostly x-rays, we, like Superman, might have x-ray vision.

Beneath the photosphere is the solar *interior*. All the solar energy is generated there at the solar *core*, which is about 10 per cent of the solar diameter at this stage of the sun's life. The temperature there is about 15,000,000 K.

The photosphere is the lowest level of the *solar atmosphere* (Fig. 18–1). Though the sun is gaseous through and through, with no solid parts, we still use the term atmosphere for the upper part of the solar material because it is relatively transparent.

Just above the photosphere is a jagged, spiky layer about 10,000 km thick, only about 1.5 per cent of the solar radius. This layer glows colorfully pinkish when seen at an eclipse, and is thus called the *chromosphere* (from the Greek *chromos,* color). Above the chromosphere, a ghostly-white halo called the *corona* (from the Latin, crown) extends tens of millions of kilometers into space. The corona is continually expanding into interplanetary space and in this form is called the *solar wind.* The earth is bathed in the solar wind.

FIGURE 18–1 The parts of the solar atmosphere and interior. The solar surface is depicted as it appears through light equally filtered across the spectrum (called *white light*), and through filters that pass only light of certain elements in certain temperature stages. These specific wavelengths, counterclockwise from the top, are the Hα line of hydrogen (which appears in the red), the K line of ionized calcium (which appears in the part of the ultraviolet that passes through the earth's atmosphere and can be seen with the naked eye), and the line of ionized helium in the extreme ultraviolet (which can be observed only from rockets and satellites; it is less than 1/10 the shortest wavelength that passes through the atmosphere).

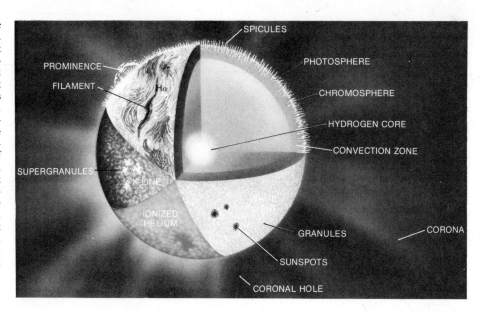

18.2 THE PHOTOSPHERE

The sun is a normal star of spectral type G2, which means that its surface temperature is about 5800 K. But the sun is the only star close enough to allow us to study its surface in detail. We can resolve parts of the surface about 700 km across, about the distance from Boston to Washington, D.C.

Sometimes we observe the sun in *white light*—all the visible radiation taken together. When we study the solar surface in white light with 1 arc second resolution, we see a salt-and-pepper texture called *granulation* (Fig. 18–2). The effect is similar to that seen in boiling liquids on earth. But the granules are about the same size as the limit of our resolution, so are difficult to study. NASA plans to launch SOT—the Solar Optical Telescope—about 1990; it will allow astronomers to resolve features 10 times smaller than we can now.

The spectrum of the solar photosphere, like that of other G stars, is a continuous spectrum crossed by absorption lines (Fig. 18–3). Hundreds of

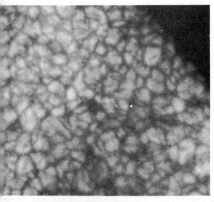

FIGURE 18–2 The small cells separated by dark boundaries are the granules; they are the limit of detail of what we can see from the ground. When Solar Optical Telescope is launched, perhaps in the early 1990's, we hope for an increase in resolution by a factor of 10.

TABLE 18–1   The Most Common Elements on the Sun

	SYMBOL	ATOMIC NUMBER
For each 1,000,000 atoms of hydrogen, there are	H	1
63,000 atoms of helium	He	2
690 atoms of oxygen	O	8
420 atoms of carbon	C	6
87 atoms of nitrogen	N	7
45 atoms of silicon	Si	14
40 atoms of magnesium	Mg	12
37 atoms of neon	Ne	10
32 atoms of iron	Fe	26
16 atoms of sulfur	S	16
3 atoms of aluminum	Al	13
2 atoms of calcium	Ca	20
2 atoms of sodium	Na	11
2 atoms of nickel	Ni	28
1 atom of argon	Ar	18

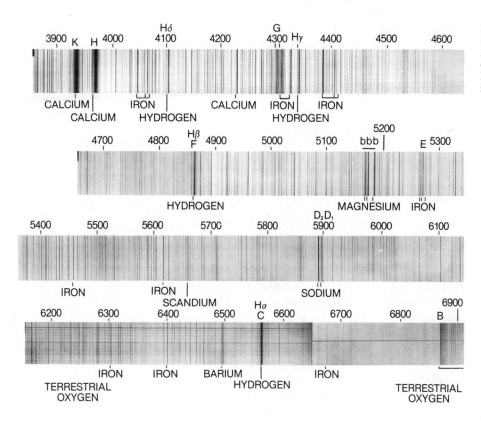

FIGURE 18–3 Part of the solar spectrum in the visible from 3900 Å to 6900 Å. It is continuous radiation crossed by dark Fraunhofer lines. The colors are shown in Color Plate 5.

thousands of these absorption lines, which are also called Fraunhofer lines, have been photographed and catalogued. They come from most of the chemical elements. Iron has many lines in the spectrum. The hydrogen lines are strong but few in number.

From the spectral lines, we can figure out the percentages of the elements. None of the elements other than hydrogen and helium makes up as much as one-tenth of one per cent of the number of hydrogen atoms (Table 18–1).

18.3 THE CHROMOSPHERE

Under high resolution, we see that the chromosphere is not a spherical shell around the sun but rather is composed of small "spicules" (Fig. 18–4). These jets of gas rise and fall, and have been compared in appearance to blades of grass or burning prairies.

Spicules are more-or-less cylinders of about 700 km across and 7000 km tall. They seem to have lifetimes of about 5 to 15 minutes, and there may be approximately half a million of them on the surface of the sun at any given moment.

FIGURE 18–4 Spicules at the solar limb. They are only about 700 km in diameter, about the same size as the granules.

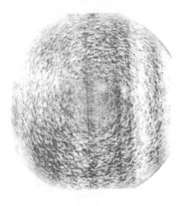

FIGURE 18–5 Supergranulation is best visible in an image like this one, in which the velocities on the sun are shown. Dark areas are receding and bright areas are approaching us. The supergranules are each less than 1 mm across in this reproduction. Because matter flows from the center to the edge of each supergranule, the Doppler shift makes one side appear dark and the other bright on this velocity image.

Studies of velocities on the sun showed the existence of large organized cells of matter on the surface of the sun called *supergranulation* (Fig. 18–5). Supergranulation cells look somewhat like polygons of approximately 30,000 km diameter. Each supergranulation cell may contain hundreds of individual granules. Matter seems to well up in the middle of a supergranule and then slowly move horizontally across the solar surface to the supergranule boundaries. The matter then sinks back down at the boundaries.

Chromospheric matter appears to be at a temperature of 10,000 to 15,000 K, somewhat higher than the temperature of the photosphere. New ultraviolet spectra of distant stars recorded by the International Ultraviolet Explorer have shown unmistakable signs of chromospheres in stars of spectral types like the sun. Thus by studying the solar chromosphere we are also learning what the chromospheres of other stars are like.

18.4 THE CORONA

During total solar eclipses, when first the photosphere and then the chromosphere are completely hidden from view, a faint white halo around the sun becomes visible. This corona (Fig. 18–6) is the outermost part of the solar atmosphere, and extends throughout the solar system. Close to the solar limb, the corona's temperature is about 2,000,000 K.

Even though the temperature of the corona is so high, the actual amount of energy in the solar corona is not large. The temperature quoted is actually a measure of how fast individual particles (electrons, in particular) are moving. There aren't very many coronal particles, even though each particle has a high velocity. The corona has less than one-billionth the density of the earth's atmosphere, and would be considered to be a very good vacuum in a laboratory on earth. For this reason, the corona serves as a unique and valuable celestial laboratory in which we may study gaseous "plasmas" in a near vacuum. Plasmas are gases separated into positively and negatively charged particles and can be shaped by magnetic fields. We are

FIGURE 18–6 The total solar eclipse of June 11, 1983, photographed from East Java, Indonesia. The photograph was taken through a filter that is denser at its center than at its edges so as to reduce the bright inner corona. This allows us to see the much fainter streamers of the outer corona in the same photograph, a factor of 10,000 in brightness. Bright prominences are visible on the limb.

FIGURE 18–7 From a few mountain sites, the innermost corona can be photographed without need for an eclipse. The coronagraph at the Haleakala Observatory of the University of Hawaii on Maui took these observations on March 8, 1970, the day after a total eclipse, with their coronagraph. Five exposures were superimposed to improve the image.

trying to learn how to use magnetic fields on earth to control plasmas in order to provide energy through nuclear fusion (Section 19.4).

Photographs of the corona show that it is very irregular in form. Beautiful long *streamers* extend away from the sun in the equatorial regions. The shape of the corona varies continuously and is thus different at each successive eclipse. The structure of the corona is maintained by the magnetic field of the sun.

The corona is normally too faint to be seen except at an eclipse of the sun because it is fainter than the everyday blue sky. But at certain locations on mountain peaks on the surface of the earth, the sky is especially clear and dust-free, and the innermost part of the corona can be seen (Fig. 18–7).

Several manned and unmanned spacecraft have used devices that made a sort of inferior artificial eclipse to photograph the corona hour by hour in visible light (Color Plates 41 to 44). These satellites studied the corona to much greater distances from the solar surface than can be studied with coronagraphs on earth. (They cannot see the inner part, which we can still study best at eclipses.) Among the major conclusions of the research is that the corona is much more dynamic than we had thought. For example, many blobs of matter were seen to be ejected from the corona into interplanetary space (Fig. 18–8), one per day or so.

The visible region of the coronal spectrum, when observed at eclipses, shows continuous radiation, absorption lines, and emission lines. The emission lines do not correspond to any normal spectral lines known in laboratories on earth or on other stars, and for many years their identification was one of the major problems in solar astronomy. The lines were even given the name of an element: coronium. (After all, the element helium was first discovered on the sun.) In the late 1930's, it was discovered that the emission lines arose in atoms that had lost over a dozen electrons each. This was the major indication that the corona was very hot (and that coronium doesn't exist). The corona must be very hot indeed, millions of degrees, to have enough energy to strip that many electrons off atoms.

The gas in the corona is so hot that it emits mainly x-rays, photons of high energy. The photosphere, on the other hand, is too cool to emit x-rays. As a result, when photographs of the sun are taken in the x-ray region of the spectrum (from satellites), they show the corona and its structure (Fig. 18–9).

Detailed examination of the x-ray images shows that most, if not all, the radiation appears in the form of loops of gas joining separate points on the solar surface. We must understand the physics of coronal loops in order to understand how the corona is heated. It is not sufficient to think in terms of a uniform corona, since the corona is obviously not so uniform.

The x-ray image also shows very dark areas at the sun's north pole and extending downward across the center of the solar disk. These dark locations are *coronal holes*, regions of the corona that are particularly cool and quiet. The density of gas in those areas is lower than the density in adjacent areas.

FIGURE 18–8 This example of solar ejection was photographed from the manned Skylab spacecraft. Here we see two pictures superimposed. The eruption at the extreme right, photographed with the coronagraph, is the response of the corona to a lower-level eruption, photographed in the ultraviolet and superimposed on the central dark region occulted by a disk in the middle of the coronagraph.

FIGURE 18–9 An x-ray photograph of the sun taken from Skylab on June 1, 1973. The dark region across the center is a coronal hole. It is joined with a coronal hole at the north pole.

FIGURE 18–10 A sunspot, showing the dark *umbra* surrounded by
the lighter, mottled *penumbra.*
Granulation is visible in the surrounding photosphere.

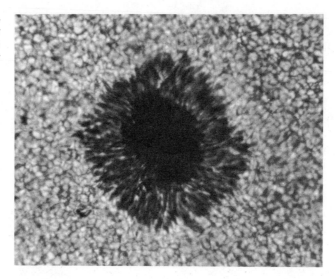

There is usually a coronal hole at one or both of the solar poles. Less
often, we find additional coronal holes at lower solar latitudes. The regions
of the coronal holes seem very different from other parts of the sun. The
solar wind flows to earth mainly out of the coronal holes, so it is important
to understand the coronal holes to understand our environment in space.

18.5 SUNSPOTS AND OTHER SOLAR ACTIVITY

Many solar phenomena vary with an 11-year cycle, which is called the *solar
activity cycle.* The most obvious are *sunspots* (Fig. 18–10), which appear relatively dark when seen in white light. Sunspots appear dark because they are
giving off less radiation than the photosphere that surrounds them. Thus
they are relatively cool, since cooler gas radiates less than hotter gas. Actually, if we could somehow remove a sunspot from the solar surface and
put it off in space, it would appear bright against the dark sky; a large one
would give off as much light as the full moon seen from earth.

A sunspot includes a very dark central region, called the *umbra* from the
Latin for "shadow" (plural: *umbrae*). The umbra is surrounded by a *penumbra*
(plural: *penumbrae*), which is not as dark.

To understand sunspots we must understand magnetic fields. When
iron filings are put near a simple bar magnet on earth, the filings show a

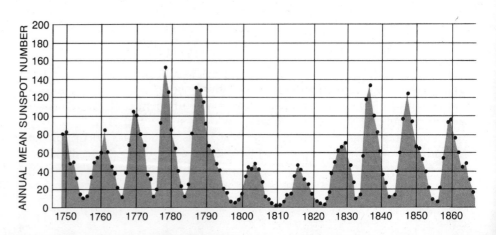

FIGURE 18–11 *(A)* Lines of force from a bar magnet are outlined by iron filings sprinkled on a paper covering the magnet. One end of the magnet is called a north pole and the other is called a south pole. Similar poles ("like poles")—a pair of norths or a pair of souths—repel each other and unlike poles (1 north and 1 south) attract each other. Lines of force go between opposite poles. *(B)* The magnetic field on November 9, 1979, the day that the white-light photograph that opens this chapter was taken. One polarity appears as relativley dark and the opposite polarity as relatively light. Note how the regions of strong fields correspond to the positions of sunspots. Notice also how the two members of a pair of sunspots have polarity opposite to each other, and how the polarity that comes first as the sun rotates is different above and below the equator.

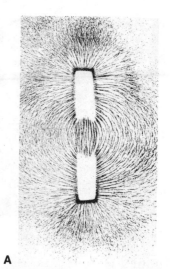

A

pattern (Fig. 18–11). The magnet is said to have a north pole and a south pole, and the magnetic field linking them is characterized by what we call *magnetic-field lines* (after all, the iron filings are spread out in what look like lines). The earth (as well as some other planets) has a magnetic field that has many characteristics in common with that of a bar magnet. The structure seen in the solar corona, including the streamers, shows matter being held by the sun's magnetic field.

The strength of the solar magnetic field shows up in spectra. George Ellery Hale showed, in 1908, that the sunspots are regions of very high magnetic-field strength on the sun, thousands of times more powerful than the earth's magnetic field. Sunspots usually occur in pairs, and often these pairs are part of larger groups. In each pair, one sunspot will be typical of a north magnetic pole and the other will be typical of a south magnetic pole.

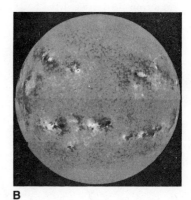

B

Magnetic fields are able to restrain matter—this is the property we are trying to exploit on earth to contain superheated matter sufficiently long to allow nuclear fusion to take place for energy production. The strongest magnetic fields in the sun occur in sunspots. The magnetic fields in sunspots keep energy from being carried upward to the surface. As a result, sunspots are cool and dark. The parts of the corona above active regions are hotter and denser than the normal corona. Presumably the energy is guided upward by magnetic fields. These locations are prominent in radio or x-ray maps of the sun.

Sunspots were discovered in 1610, independently by Galileo and by others. In about 1850, it was realized that the number of sunspots varies with an 11-year cycle (Fig. 18–12), the *sunspot cycle*. Every 11-year cycle, the north magnetic pole and south magnetic pole on the sun reverse; what had been a north magnetic pole is then a south magnetic pole and vice versa. So it is 22 years before the sun returns to its original state, making the real period of the solar-activity cycle 22 years.

The last maximum of the sunspot cycle—the time when there is the greatest number of sunspots—took place in late 1979. Shortly thereafter,

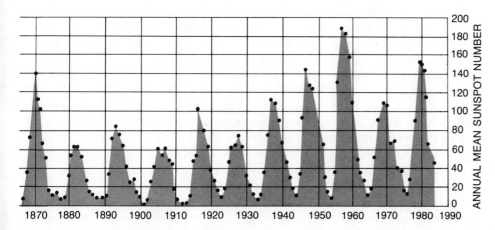

FIGURE 18–12 The 11-year sunspot cycle is but one manifestation of the solar activity cycle.

A

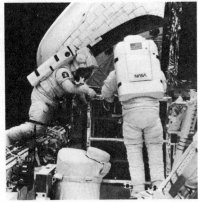

B

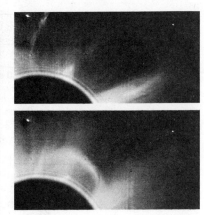

C

FIGURE 18–13 *(A)* Space-shuttle astronauts retrieved the Solar Maximum Mission spacecraft with the remote arm and attached it to the cargo bay. *(B)* The astronauts were able to repair or replace almost all the defective systems. *(C)* The eruption of a large prominence led to this transient event in the corona, observed from the coronagraph aboard Solar Maximum Mission. 56 minutes elapsed between the two frames. (Courtesy of Lewis House, Ernest Hilder, William Wagner, and Constance Sawyer—HAO/NCAR/NSF and NASA)

NASA launched a satellite called the Solar Maximum Mission to study solar activity at that time (Fig. 18–13 and Color Plates 41 and 43 through 46).

Careful studies of the solar activity cycle are now increasing our understanding of how the sun affects the earth. Although for many years scientists were skeptical of the idea that solar activity could have a direct effect on the earth's weather, scientists presently seem to be accepting more and more the possibility of such a relationship.

An extreme test of the interaction may be provided by the interesting probability that there were no sunspots at all on the sun from 1645 to 1715! This period, called the *Maunder minimum*, was largely forgotten until recently. An important conclusion is that the solar activity cycle may be much less regular than we had thought.

Much of the evidence for the Maunder minimum is indirect, and has been challenged, as has the specific link of the Maunder minimum with colder climate during that period. It would be reasonable for several mechanisms to affect the earth's climate on this time scale, rather than only one.

Precise measurements made from the Solar Maximum Mission have shown that the total amount of energy flowing out of the sun varies slightly, by up to 0.002 (0.2 per cent). The dips in energy seem to correspond to the existence of large sunspots. Astronomers are now trying to figure out what happens to the blocked energy.

18.5a Flares

Violent activity sometimes occurs in the regions around sunspots. Tremendous eruptions called *solar flares* (Fig. 18–14) can eject particles and emit radiation from all parts of the spectrum into space. These solar storms begin in a few seconds and can last up to four hours. Temperatures in the flare can reach 5 million kelvins, even hotter than the quiet corona. Flare particles that are ejected reach the earth in a few hours or days and can cause disruptions in radio transmission, cause the aurorae (Color Plate 26) and even cause surges on power lines. Because of these solar-terrestrial relationships, high priority is placed on understanding solar activity and being able to predict it. Observing flares in the ultraviolet and x-ray region was one of the Solar Maximum Mission's major goals. The radio emission of the sun also increases at the time of a solar flare.

No specific theory for explaining the eruption of solar flares is generally accepted. But it is agreed that a tremendous amount of energy is stored in the solar magnetic fields in sunspot regions. Something unknown triggers the release of the energy.

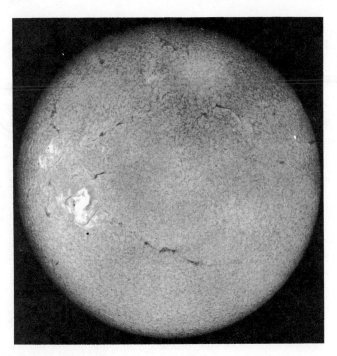

FIGURE 18-14 One of the largest solar flares in decades occurred on August 7, 1972, and led to power blackouts, short-wave radio blackouts, and aurorae. The whole sun is shown at the peak of the flare, which is the bright region to the left of center. The picture was taken through a filter passing only the light of hydrogen; flares are usually not visible in white light.

18.5b Filaments and Prominences

Studies of the solar atmosphere through filters that pass only hydrogen radiation also reveal other types of solar activity. Dark *filaments* are seen threading their way for up to 100,000 km across the sun in the vicinity of sunspots. They mark the locations of zero magnetic field that separate regions of positive and negative magnetic field.

When filaments happen to be on the sun's edge, they project into space, often in beautiful shapes. Then they are called *prominences* (Fig. 18–15). Prominences can be seen with the eye at solar eclipses, and glow pinkish at that time because of their emission in hydrogen and a few other spectral lines. They can be observed from the ground even without an eclipse, if a filter that passes only light emitted by hydrogen gas is used. Prominences appear to be composed of matter in a condition of temperature and density similar to matter in the quiet chromosphere, somewhat hotter than the photosphere. Sometimes prominences can hover above the sun, supported by magnetic fields, for weeks or months. Other prominences change rapidly (Color Plates 39 and 40).

FIGURE 18–15 A solar prominence. Gas can remain in this shape, supported by the sun's magnetic field, for weeks or months.

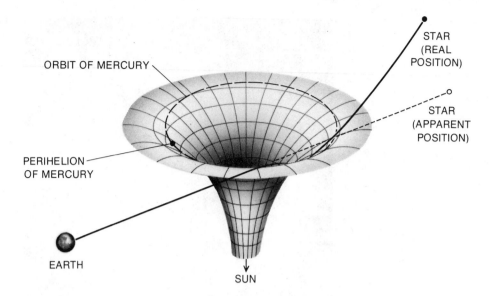

FIGURE 18–17   Under the general theory of relativity, the presence of a massive body essentially warps the space nearby. This can account for both the bending of light near the sun and the advance of the perihelion point of Mercury by 43 arc sec per century (the diameter of the moon in 4,200 years) more than would otherwise be expected. The diagram shows how a two-dimensional surface warped into three dimensions can change the direction of a "straight" line that is constrained to its surface; the warping of space is analogous, although with a greater number of dimensions to consider. The effect is similar to a golfer putting on a warped green. Though the ball is hit in a straight line, we see it appear to curve.

18.6 THE SUN AND THE THEORY OF RELATIVITY

The intuitive notion we have of gravity corresponds to the theory of gravity advanced by Isaac Newton in 1687. We now know, however, that Newton's theory and our intuitive ideas are not sufficient to explain the universe in detail. Theories advanced by Albert Einstein in the first decades of this century now provide us with a more accurate understanding.

The sun, as the nearest star to the earth, has been very important for testing some of the predictions of Albert Einstein's theory of gravitation, which is known as the general theory of relativity. The theory, which Einstein advanced in final form in 1916, could be checked by three observational tests that depended on the presence of a large mass like the sun for experimental verification.

First, Einstein's theory showed that the closest point to the sun of Mercury's elliptical orbit would move slightly around the sky over centuries. Such a movement had already been noted. Calculations with Einstein's theory accounted precisely for the amount of the movement. It was a plus for Einstein to have explained it, but the test of a scientific theory is really whether it predicts new things rather than whether it explains old ones.

The second test arose from a major new prediction of Einstein's theory: light from a star would act as though it were bent toward the sun by a very small amount (Fig. 18–16). We on earth, looking back past the sun, would see the star from which the light was emitted as though it were shifted slightly away from the sun (Fig. 18–17). Only a star whose radiation grazed the edge of the sun would seem to undergo the full deflection; the effect diminishes as one considers stars farther away from the solar limb. To see the effect, one has to look near the sun at a time when the stars are visible,

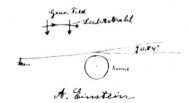

FIGURE 18–16 The prediction in Einstein's own handwriting of the deflection of starlight by the sun, taken from a letter from Einstein to the astronomer George Ellery Hale at Mt. Wilson Observatory. "Lichtstrahl" is "light ray." Einstein asked if the effect could be measured without an eclipse, to which Hale replied negatively. The drawing came from a date when Einstein had developed only an early version of the theory, which gave a predicted value half of his later predicted value (0.84 seconds of arc, where one second of arc is 1/3600 degree).

FIGURE 18-18 The first American report of the events that made Einstein world famous. From *The New York Times* of November 9, 1919.

and this could be done only at a total solar eclipse.

The effect was verified at the total solar eclipse of 1919. Scientists hailed this confirmation of Einstein's theory, and from the moment of its official announcement, Einstein was recognized by scientists and the general public alike as the world's greatest scientist (Fig. 18-18).

Similar observations have been made at subsequent eclipses, though they are very difficult to make. Fortunately, the effect is constant through the spectrum and the test can now be performed more accurately by observing how the sun bends radiation from radio sources, especially quasars. The results agree with Einstein's theory to within 1 per cent, enough to make the competing theories very unlikely.

The third test was to verify the prediction of Einstein's theory that strong gravity would cause the spectrum to be redshifted. This effect is very slight for the sun, but has been barely detected. It has best been verified for extremely dense stars in which mass is very tightly packed together (a type of star described in Section 20.3).

As a general rule, scientists try to find theories that not only explain the data that are at hand, but also make predictions that can be tested. This is an important part of the *scientific method*. Because the bending of radiation was a prediction of the general theory of relativity that had not been anticipated, its verification was a more convincing proof of the theory's validity than the theory's ability to explain the known shift in Mercury's orbit.

ECLIPSE SHOWED GRAVITY VARIATION

Diversion of Light Rays Accepted as Affecting Newton's Principles.

HAILED AS EPOCHMAKING

British Scientist Calls the Discovery One of the Greatest of Human Achievements.

Copyright, 1919, by The New York Times Company.
Special Cable to THE NEW YORK TIMES.
LONDON, Nov. 8.—What Sir Joseph Thomson, President of the Royal Society, declared was " one of the greatest—perhaps the greatest—of achievements in the history of human thought " was discussed at a joint meeting of the Royal Society and the Royal Astronomical Society in London yesterday, when the results of the British observations of the total solar eclipse of May 29 were made known.

There was a large attendance of astronomers and physicists, and it was generally accepted that the observations were decisive in verifying the prediction of Dr. Einstein, Professor of Physics in the University of Prague, that rays of light from stars, passing close to the sun on their way to the earth, would suffer twice the deflection for which the principles enunciated by Sir Isaac Newton accounted. But there was a difference of opinion as to whether science had to face merely a new and unexplained fact or to reckon with a theory that would completely revolutionize the accepted

KEY WORDS

quiet sun, active sun, photosphere, interior, core, solar atmosphere, chromosphere, corona, solar wind, white light, granulation, supergranulation, streamers, coronal holes, solar activity cycle, sunspots, umbra (umbrae), penumbra (penumbrae), magnetic-field lines, sunspot cycle, Maunder minimum, solar flares, filaments, prominences, scientific method

QUESTIONS

1. Sketch the sun, labelling the interior, the photosphere, the chromosphere, the corona, sunspots, and prominences.

2. Draw a graph showing the sun's temperatures starting with the core and going upward through the corona.

3. Define and contrast a prominence and a filament.

4. Why are we on earth particularly interested in coronal holes?

5. List three phenomena that vary with the solar activity cycle.

6. Why can't we observe the corona every day from earth's surface?

7. How do we know that the corona is hot?

8. Describe relative advantages of ground-based eclipse studies and of satellite studies of the corona.

9. Describe the sunspot cycle.

10. In what tests of general relativity does the sun play an important role? Describe the current status of these investigations.

TOPICS FOR DISCUSSION

Discuss the relative importance of solar observations (a) from the ground, (b) at eclipses, (c) from satellites.

PART IV
The Life Cycle of Stars

The Cygnus loop, the remnant of a supernova. The remainder of the star that explodes as a supernova contracts to become either a neutron star or a black hole, depending on how much mass is left.

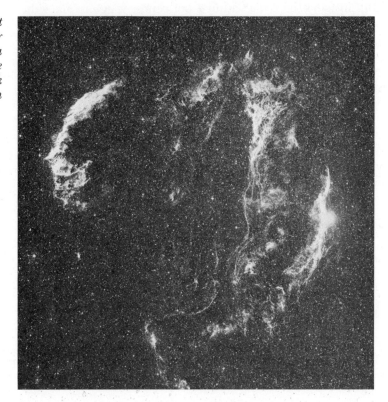

We have seen how the color-magnitude diagram for a cluster of stars allows us to deduce how old the cluster is. Our human lifetimes are very short compared to the billions of years that a typical star takes to form, live its life, and die. Thus our hope of understanding the life history of an individual star depends on studying large numbers of stars, for presumably we will see them at different stages of their lives.

Though we can't follow an individual star from cradle to grave, studying many different stars shows us enough stages of development to allow us to write out a stellar biography. Chapters 19 through 23 are devoted to such life stories. Similarly, we could study the stages of human life not by watching some individual's aging, but rather by studying people of all ages who are present in a city on a given day.

In Chapter 19, we shall study the birth of stars, and observe some places

Color Plate 47 (top): M15, a globular cluster in the constellation Pegasus, is visible to the naked eye as a hazy patch. It contains about 100,000 stars. (Canada-France-Hawaii Telescope)

Color Plate 48 (bottom): The Pleiades, M45, is an open cluster in the constellation Taurus, the Bull. Reflection nebulae are visible around many of the brightest stars. The spikes are artifacts caused in the telescope. (© 1985 Royal Observatory, Edinburgh)

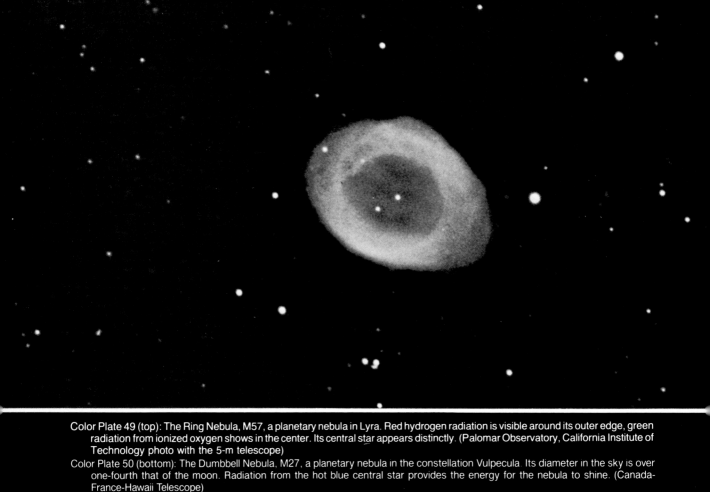

Color Plate 49 (top): The Ring Nebula, M57, a planetary nebula in Lyra. Red hydrogen radiation is visible around its outer edge, green radiation from ionized oxygen shows in the center. Its central star appears distinctly. (Palomar Observatory, California Institute of Technology photo with the 5-m telescope)

Color Plate 50 (bottom): The Dumbbell Nebula, M27, a planetary nebula in the constellation Vulpecula. Its diameter in the sky is over one-fourth that of the moon. Radiation from the hot blue central star provides the energy for the nebula to shine. (Canada-France-Hawaii Telescope)

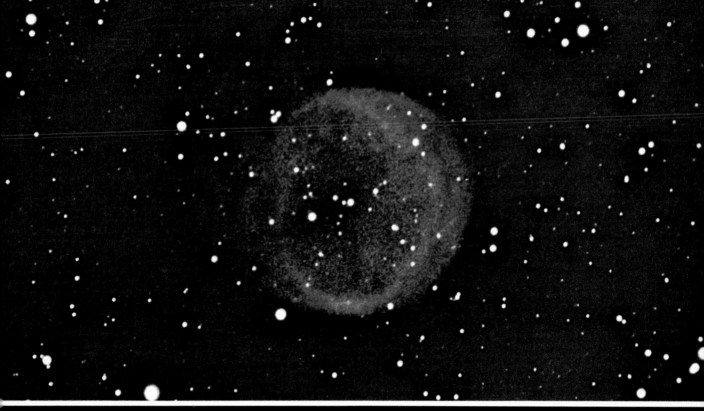

Color Plate 51 (top) A planetary nebula, NGC 6781, in the constellation Aquila (Palomar Observatory, California Institute of Technology photo with the 1 2-m Schmidt camera)

Color Plate 52 (bottom, left) A radio image of the Cassiopeia A supernova remnant, observed at a wavelength of 20 cm with the VLA (Courtesy of Philip E Angerhofer, Richard A Perley, Bruce Balick, and Douglas Milne with the VLA of NRAO)

Color Plate 53 (bottom, right) An x-ray image of the Cassiopeia A supernova remnant, observed with the Einstein Observatory The bright ring is thought to be the region associated with the expanding shock front No pulsar has been detected (Courtesy of S. S Murray and colleagues at the Harvard-Smithsonian Center for Astrophysics)

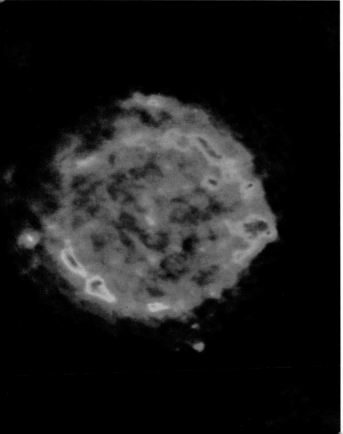

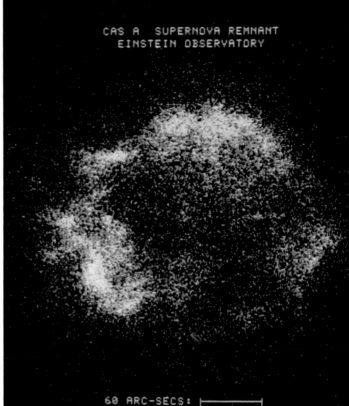

CAS A SUPERNOVA REMNANT
EINSTEIN OBSERVATORY

60 ARC-SECS:

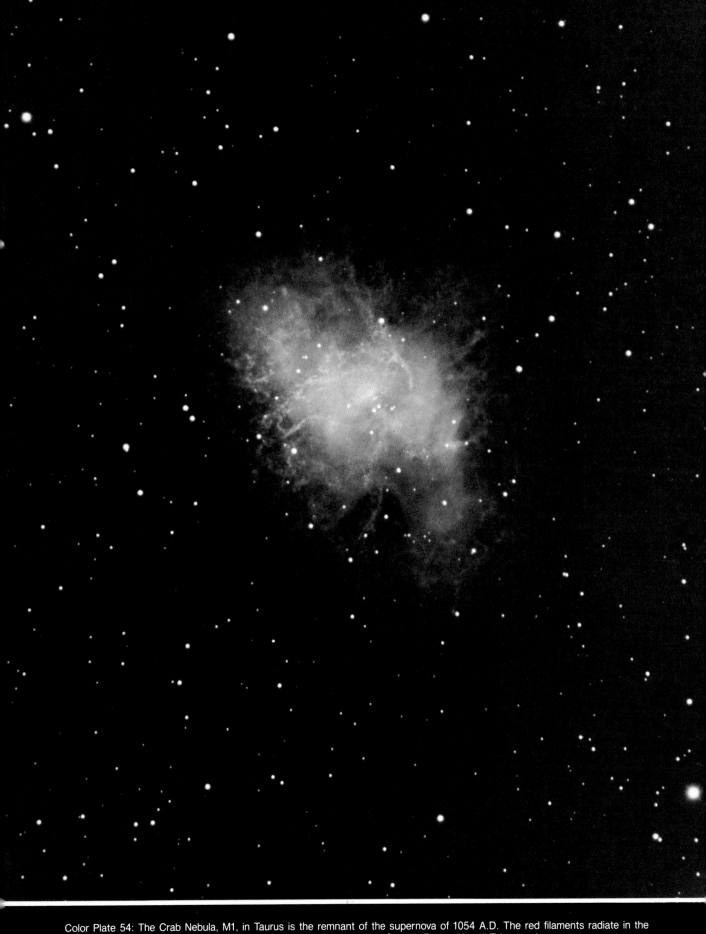

Color Plate 54: The Crab Nebula, M1, in Taurus is the remnant of the supernova of 1054 A.D. The red filaments radiate in the hydrogen lines; the white continuum is from synchrotron radiation. (Canada-France-Hawaii Telescope)

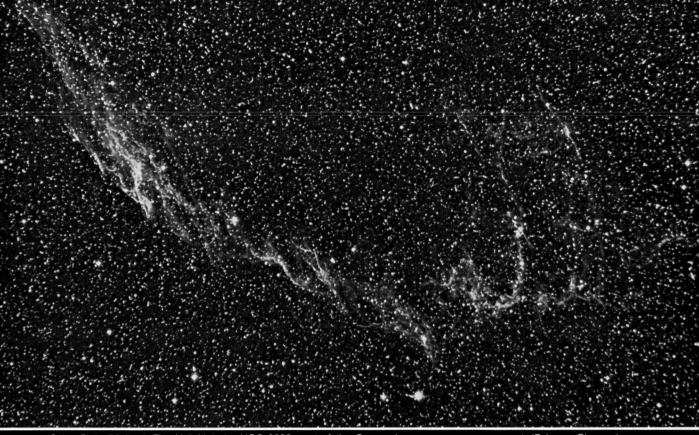

Color Plate 55 (top): The Veil Nebula, NGC 6992, part of the Cygnus loop, a supernova remnant. (Palomar Observatory, California Institute of Technology photo with the 1.2-m Schmidt camera)

Color Plate 56 (bottom): The Orion Nebula, M42, is glowing gas excited by the Trapezium, four hot stars. The nebula contains stars in formation; it is 25 ly across and 400 parsecs away. (Palomar Observatory, California Institute of Technology photo with the 5-m telescope)

in the sky where we expect stars to be born very soon, possibly even in our own lifetimes. We shall then consider the properties of stars during the long, stable phase in which they spend most of their histories.

In the following chapters, we go on to consider the ways in which stars end their lives. Stars like the sun sometimes eject shells of gas that glow beautifully; the part of such a star that is left behind then contracts until it is as small as the earth (see Chapter 20). Sometimes when a star is newly visible in a location where no star was previously known to exist—a "nova"—we are seeing the interaction of a dead solar-type star with a companion.

Sometimes newly visible stars rival whole galaxies in brightness. Then we may be seeing the spectacular death throes of massive stars or the end of the second life of a low-mass star (see Chapter 21). The study of what happens to stars after that explosive event is currently a topic of tremendous interest to many astronomers. Some of the stars become pulsars (Chapter 22). Other stars, we think, may even wind up as black holes, objects that are invisible and therefore difficult—though not impossible—to detect (see Chapter 23).

When we write a stellar biography from the clues that we get from observation, we are acting like detectives in a novel. As new methods of observation become available, we are able to make better deductions, so the extension of our senses throughout the spectrum has led directly to a better understanding of stellar evolution. Our work in the x-ray part of the spectrum, for example, has told us more about tremendously hot gases like those that result from an exploding star. X-ray astronomy is also intimately connected with our current exciting search for a black hole.

We can set up divisions, using terminology from the sport of boxing, depending on the mass of stars. We use the name featherweight star for collapsing gas that is less than about 7 per cent as massive as the sun; it never reaches stardom, and becomes a brown dwarf. Stars more massive than that but containing less than about four solar masses eventually lose some of their mass and become white dwarfs; we call these lightweight stars. Heavyweight stars, which contain greater than eight solar masses, explode as one of the types of supernovae. Some wind up as neutron stars, which we may detect as pulsars or in x-ray binaries. Other heavyweight stars, after they become supernovae, wind up withdrawing from the universe in the form of black holes. The fate of middleweight stars, between 4 and 8 solar masses, is less clear.

Let us start in the next chapter with the formation and the main lifetime of stars. The following three chapters will then discuss the death of stars.

Chapter 19

The dark region is a globule that is so opaque that it blocks our view of the stars behind. Crossing the left (north) boundary are knots of gas ejected from a young star; they glow as they interact with gas flowing out of the star. The star that ejected them, visible only in the infrared, is near HH46. It has apparently ejected HH46 and HH47A&B in a line slanting toward us. We are seeing a "bipolar ejection" from a young star of the type known as T Tauri. Part of another globule is at far left.

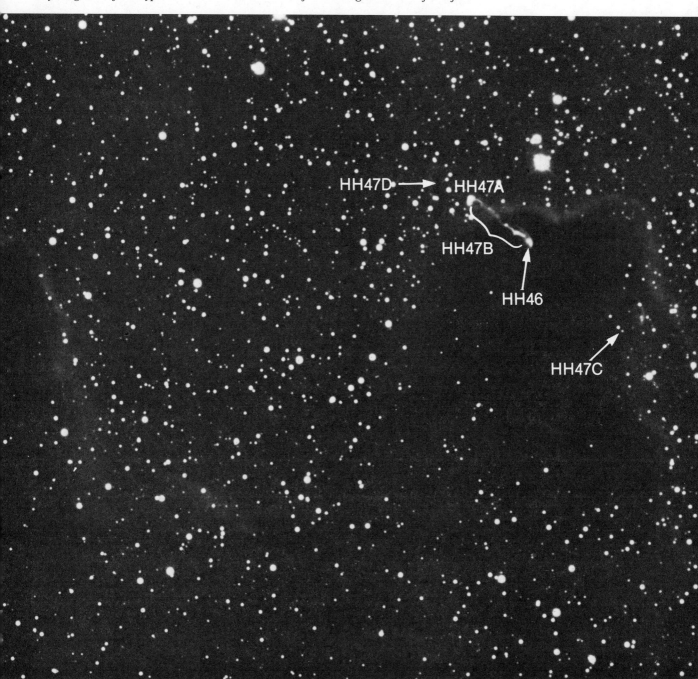

HOW STARS SHINE

Even though individual stars shine for a relatively long time, they are not eternal. Stars are born out of gas and dust that may exist within a galaxy; they then begin to shine brightly on their own. Though we can observe only the outer layers of stars, we can deduce that the temperatures at their centers must be millions of kelvins. We can even deduce what it is deep down inside that makes the stars shine.

We begin this chapter by discussing the birth of stars. We then consider the processes that go on inside a star during its life on the main sequence. Finally, we begin the story of the evolution of stars when they finish this stage of their lives. The following four chapters will continue the story of what is called *stellar evolution*.

19.1 STARS IN FORMATION

The birth of a star begins with a region of gas and dust. The dust (tiny solid particles) may have escaped from the outer atmospheres of giant stars. The gas and dust from which stars are forming is best observed in the infrared and radio regions of the spectrum (Chapter 25).

Consider a region that reaches a higher density than its surroundings, perhaps from a random fluctuation in density or—in a leading theory of why galaxies have spiral arms—because a wave of compression passes by. Or a star may explode nearby (a "supernova") sending out a shock wave that compresses gas and dust. In any case, once a region gains higher than average density, gravity keeps the gas and dust contracting. As they contract, energy is released, and some of that energy heats the matter. We have already seen that such gravitational energy was released in the early solar system.

Such a not-quite-yet-formed star is called a "protostar" (from the prefix of Greek origin meaning "primitive"). At first, the protostar brightens, since its outer layers are heating up. While this is occurring, the central part of the protostar continues to contract and its temperature also rises. The higher temperature results in a higher pressure, which pushes outward more and more strongly. By this time, the dust has vaporized and the gas has become opaque, so that energy emitted from the central region does not escape directly. Eventually the outward force resulting from the pressure balances the inward force of gravity for the central region. This balance is the key to understanding stable stars.

Theoretical analysis shows that the dust surrounding the stellar embryo we call a protostar should absorb much of the radiation that the protostar emits. The radiation from the protostars should heat the dust to temperatures that cause it to radiate primarily in the infrared. Infrared astronomers

FIGURE 19–1 T Tauri itself is embedded in a Herbig-Haro object. The H-H object has changed in brightness considerably in the last century, first fading from view and then brightening. This probably results from changes in the angle at which T Tauri illuminated the cloud. The H-H object has now been constant in brightness for decades. The image of a second H-H object is merged with the image of T Tauri on this plate.

H¹ NUCLEUS
PROTON

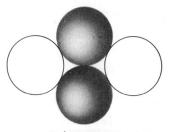

He⁴ NUCLEUS
ALPHA PARTICLE

FIGURE 19–2 The nucleus of hydrogen's most common form is a single proton while the nucleus of helium's most common form consists of two protons and two neutrons.

have found many objects that are especially bright in the infrared but that have no known optical counterparts. The IRAS satellite, which orbited in 1983, has discovered so many of these that we now think that about one star forms each year in our Milky Way Galaxy. These objects seem to be located in regions where the presence of a lot of dust and gas and other young stars indicates that star formation might be going on.

In the visible part of the spectrum, several classes of stars that vary erratically in brightness are found. One of these classes, called *T Tauri stars*, contains stars as massive as or less massive than the sun. The star that typifies the class—T Tauri itself (Fig. 19–1)—is named with a capital Latin letter, as are the first variable stars to be named in a constellation. The visible radiation of T Tauri stars can vary by as much as several magnitudes; presumably, these stars have not quite settled down to a steady and reliable existence.

It came as a surprise in 1982 when T Tauri turned out to have a companion observable only in the infrared. The companion's temperature is only 800 K (less than half the temperature of a star). The companion may be embedded inside and taking up matter from a disk of gas around T Tauri. The companion might even be too small to be a star, which would shine on its own by the fusion process we discuss in the next section. If so, it would be more like a giant planet in formation. It is, in any case, 30 times farther from T Tauri than Jupiter is from our sun. The companion of T Tauri may be just as good a "brown dwarf" as the objects described in Section 14.2.

19.2 ENERGY SOURCES IN STARS

If stars got all their energy from contracting, they would not shine for very long on an astronomical time scale—only about 30 million years. Yet we know that rocks 4 billion years old have been found on earth, so the sun has been around at least that long. Some other source of energy must hold the stars up against their own gravitational pull.

Actually, a protostar will heat up until its central portions become hot enough for *nuclear fusion* to take place, at which time it reaches the main sequence of the color-magnitude diagram. Using this process, which we will soon discuss in detail, the star can generate enough energy inside itself to support it during its entire lifetime on the main sequence. The energy makes the particles in the star move around rapidly. For short, we say that the particles have a "high temperature." The particles exert a *thermal pressure* pushing outward, providing a force that balances gravity's inward pull.

The basic fusion processes in most stars fuse four hydrogen nuclei into one helium nucleus. (Similarly, energy is released when forms of hydrogen undergo fusion in a hydrogen bomb here on earth.) In the process, tremendous amounts of energy are released.

A hydrogen nucleus is but a single proton. A helium nucleus is more complex. It consists of two protons and two neutrons (Fig. 19–2). The mass of the helium nucleus that is the final product of the fusion process is slightly less than the sum of the masses of the four hydrogen nuclei that went into it. A small amount of the mass "disappears" in the process: 0.007 (0.7 per cent) of the mass of the four hydrogen nuclei.

The mass does not really simply disappear, but rather is converted into energy according to Albert Einstein's famous formula $E = mc^2$. Note that c, the speed of light, is a large number (3×10^{10}cm/sec), and c^2 is even larger.

Thus even though m is only a small fraction of the original mass, the amount of energy released is prodigious. The loss of only 0.007 of the central part of the sun, for example, is enough to allow the sun to radiate as much as it does at its present rate for a period of at least ten billion (10^{10}) years. This fact, not realized until 1920 and worked out in more detail in the 1930's, solved the longstanding problem of where the sun and the other stars get their energy.

All the main-sequence stars are approximately 90 per cent hydrogen (that is, 90 per cent of the atoms are hydrogen), so there is lots of raw material to stoke the nuclear "fires." We speak colloquially of "nuclear burning," although, of course, the processes are quite different from the chemical processes that are involved in the "burning" of logs or of autumn leaves. In order to be able to discuss these processes, we must first discuss the general structure of nuclei and atoms.

19.3 ATOMS

As we mentioned in Section 15.2, an atom consists of a small *nucleus* surrounded by *electrons* (Fig. 19–3). Most of the mass of the atom is in the nucleus, which takes up a very small volume in the center of the atom. The effective size of the atom, the chemical interactions of atoms to form molecules, and the nature of spectra are all determined by the electrons.

FIGURE 19–3 An artist's conception of an atom; the nucleus is much tinier than shown with respect to the region that contains the electrons. Most atoms are much more complex than hydrogen or helium. In neutral atoms, the number of electrons is equal to the number of protons.

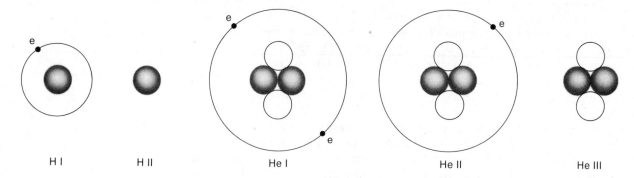

H I   H II   He I   He II   He III

FIGURE 19–4 Hydrogen and helium ions. The sizes of the nuclei are greatly exaggerated with respect to the sizes of the orbits of the electrons.

The nuclear particles with which we need be most familiar are the *proton* and *neutron*. Both these particles have nearly the same mass, 1836 times greater than the mass of an electron, though still tiny (Appendix 2). The neutron has no electric charge and the proton has one unit of positive electric charge. The electrons, which surround the nucleus, have one unit each of negative electric charge. When an atom loses an electron, it has a net positive charge of 1 unit for each electron lost. The atom is now a form of *ion* (Fig. 19–4).

Since the number of protons in the nucleus determines the charge of the nucleus, it also determines the quota of electrons that the neutral state of the atom must have. To be neutral, after all, there must be equal numbers of positive and negative charges. Each *element* (sometimes called "chemical element") is defined by the specific number of protons in its nucleus. The element with one proton is hydrogen, that with two protons is helium, that with three protons is lithium, and so on.

Though a given element always has the same number of protons in a nucleus, it can have several different numbers of neutrons. (The number of neutrons is usually somewhere between 1 and 2 times the number of protons. The lightest elements—hydrogen, which need have no neutrons, helium, and beryllium—are the only exceptions to this rule.) The possible forms of the same element having different numbers of neutrons are called *isotopes*.

For example, the nucleus of ordinary hydrogen contains one proton and no neutrons. An isotope of hydrogen (Fig. 19–5) called deuterium (and sometimes "heavy hydrogen") has one proton and one neutron. Another isotope of hydrogen called tritium has one proton and two neutrons.

Most isotopes do not have specific names, and we keep track of the numbers of protons and neutrons with a system of superscripts and subscripts. The subscript **before** the symbol denoting the element is the number of protons (called the *atomic number*), and a superscript **after** the symbol is the total number of protons and neutrons together (called the *mass number*). For example, $_1H^2$ is deuterium, since deuterium has one proton, which gives the subscript, and an atomic mass of 2, which gives the superscript. Deute-

FIGURE 19–5 Isotopes of hydrogen and helium. $_1H^2$ (deuterium) and $_1H^3$ (tritium) are much rarer than the normal isotope, $_1H^1$. $_2He^3$ is much rarer than $_2He^4$.

$_1H^1$

$_1H^2 = D$
= DEUTERIUM

$_1H^3 = T$
= TRITIUM

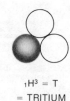

$_2He^3$

$_2He^4$

rium has atomic number equal 1 and mass number equal 2. Similarly, $_{92}U^{238}$ is an isotope of uranium with 92 protons (atomic number = 92) and mass number of 238, which is divided into 92 protons and $238 - 92 = 146$ neutrons.

Each element has only certain isotopes. For example, most naturally occurring helium is in the form $_2He^4$, with a lesser amount as $_2He^3$. Sometimes an isotope is not stable, in that after a time it will spontaneously change into another isotope or element; we say that such an isotope is *radioactive.*

Many other nuclear particles have been detected; some only show up for tiny fractions of a second under extreme conditions in giant particle accelerators ("atom smashers"). The nuclear particles are apparently made up of even more fundamental particles called *quarks.* Quarks often exist in threes. Protons and neutrons as well as many other subatomic particles are made of 3 quarks together (Fig. 19–6). Six kinds of quarks (called "flavors") have been discovered. Each exists in three versions (called "colors"). The flavors are somewhat whimsically known as up, down, strange, charmed, truth, and beauty. The colors are known as red, green, and blue. The study of quarks and subatomic particles is a major activity of contemporary physics.

During certain types of radioactive decay, a particle called a *neutrino* is given off (Fig. 19–7). A neutrino is a neutral particle (its name comes from the Italian for "little neutral one"). Neutrinos have a very useful property for the purpose of astronomy: they do not interact very much with matter. Thus when a neutrino is formed deep inside a star, it can usually escape to the outside without interacting with any of the matter in the star. A photon of radiation, on the other hand, can travel only about 1 cm in a stellar interior before it is absorbed, and it is millions of years before a photon zigs and zags its way to the surface.

The elusiveness of the neutrino not only makes it a valuable messenger—indeed, the only possible direct messenger—carrying news of the conditions inside the sun at the present time, but also makes it very difficult for us to detect on earth. A careful experiment carried out over many years has found only 1/3 the expected number of neutrinos. A new generation of this experiment is now being planned to help us understand this potentially important discrepancy.

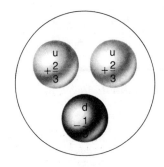

PROTON   +1 CHARGE UNIT

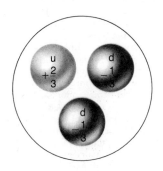

NEUTRON   0 CHARGE UNIT

FIGURE 19–6 The most common quarks are *up (u)* and *down (d);* ordinary matter in our world (including protons and neutrons) is made of them. The *up* quark has an electrical charge of +2/3, and the *down* quark has a charge of −1/3. This fractional charge is one of the unusual things about quarks; prior to their invention (discovery?), it had been thought that all electric charges came in whole numbers. Note how the charge of the proton and of the neutron is the sum of the charges of their respective quarks.

FIGURE 19–7 Neutrinos pass through the earth so easily, that this 1-km-long mound of earth at the Fermi National Accelerator Laboratory near Chicago serves as a filter that lets only neutrinos and no other particles through from a beam the atom smasher there produces.

FIGURE 19–8 The proton-proton chain; e⁺ stands for a positron, ν (nu) is a neutrino, and γ (gamma) is radiation at a very short wavelength.

In the first stage, two nuclei of ordinary hydrogen fuse to become a deuterium (heavy hydrogen) nucleus, a positron (the equivalent of an electron, but with a positive charge), and a neutrino. The neutrino immediately escapes from the star, but the positron soon collides with an electron. They annihilate each other, forming gamma rays. (A positron is an anti-electron, an example of anti-matter; whenever a particle and its antiparticle meet, they annihilate each other.)

Next, the deuterium nucleus fuses with yet another nucleus of ordinary hydrogen to become an isotope of helium with two protons and one neutron. More gamma rays are released.

Finally, two of these helium isotopes fuse to make one nucleus of ordinary helium plus two nuclei of ordinary hydrogen. The protons are numbered to help you keep track of them.

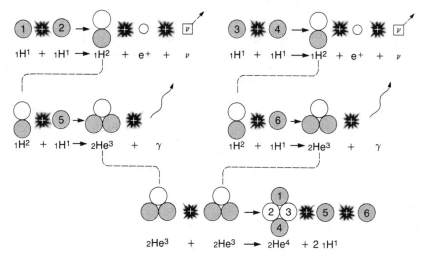

$$_1H^1 + {}_1H^1 \longrightarrow {}_1H^2 + e^+ + \nu \qquad {}_1H^1 + {}_1H^1 \longrightarrow {}_1H^2 + e^+ + \nu$$

$$_1H^2 + {}_1H^1 \longrightarrow {}_2He^3 + \gamma \qquad {}_1H^2 + {}_1H^1 \longrightarrow {}_2He^3 + \gamma$$

$$_2He^3 + {}_2He^3 \longrightarrow {}_2He^4 + 2 \, _1H^1$$

The Greek letters derive from a former confusion of radiation and particles. We know now that α *(alpha) particles, formerly called* α *rays, are helium nuclei;* β *(beta) particles, formerly called* β *rays, are electrons, and* γ *(gamma) rays, as we have seen, are radiation of short wavelength.*

19.4 STELLAR ENERGY CYCLES

Several chains of reactions have been proposed to account for the fusion of four hydrogen atoms into a single helium atom. Hans Bethe, now at Cornell University, suggested some of these procedures during the 1930's. The different chain reactions prevail at different temperatures, so chains that are dominant in very hot stars may be different from the ones in cooler stars.

When the temperature of the center of a star is less than 15 million K, the *proton-proton chain* (Fig. 19–8) dominates. In it, we put in six hydrogens, and wind up with one helium plus two hydrogens. The net transformation is four hydrogens into one helium. But the six protons contained more mass than do the final single helium plus two protons. The small fraction of mass that disappears is converted into an amount of energy that we can calculate with the formula $E = mc^2$. By Einstein's special theory of relativity, mass and energy are equivalent and interchangeable, linked by this equation.

For stellar interiors hotter than that of the sun, the *carbon-nitrogen cycle* (Fig. 19–9) dominates. This cycle begins with the fusion of a hydrogen nucleus with a carbon nucleus. After many steps, and the insertion of four hydrogen nuclei, we are left with one helium nucleus plus a carbon nucleus. Thus as much carbon remains at the end as there was at the beginning, and the carbon can start the cycle again. Again, four hydrogens have been converted into one helium, 0.007 of the mass has been transformed, and an equivalent amount of energy has been released according to $E = mc^2$.

Stars with even higher interior temperatures, above 10^8 K, fuse helium nuclei to make carbon nuclei. The nucleus of a helium atom is called an "alpha particle" for historical reasons. Since three helium nuclei ($_2He^4$) go into making a single carbon nucleus ($_6C^{12}$), the procedure is known as the *triple-alpha process* (Fig. 19–10). A series of other processes can build still heavier elements inside stars. These processes are called *nucleosynthesis*.

The theory of nucleosynthesis can account for the abundances we observe of the elements heavier than helium. Currently, we think that the synthesis of isotopes of hydrogen and helium took place in the first few minutes after the origin of the universe (Chapter 28), and that the heavier elements were formed, along with additional helium, in stars or in supernovae. William A. Fowler of Caltech shared the 1983 Nobel Prize in Physics for his work on nucleosynthesis, including the measurements of the rates of the nuclear reactions that make the stars shine.

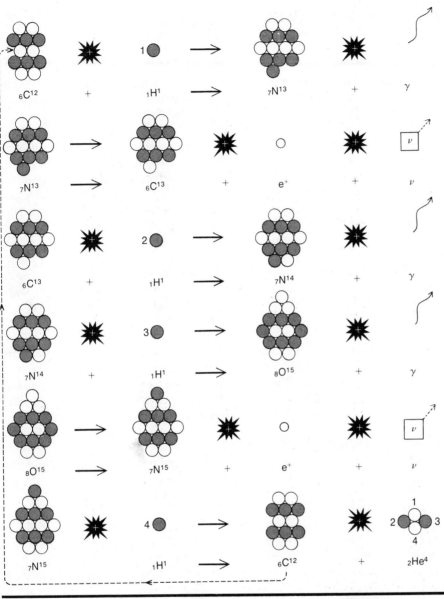

FIGURE 19–9 The carbon-nitrogen cycle. The 4 hydrogen atoms are numbered. Note that the carbon is left over at the end, ready to enter into another cycle.

$_6C^{12}$ + $_1H^1$ ⟶ $_7N^{13}$ + γ

$_7N^{13}$ ⟶ $_6C^{13}$ + e^+ + ν

$_6C^{13}$ + $_1H^1$ ⟶ $_7N^{14}$ + γ

$_7N^{14}$ + $_1H^1$ ⟶ $_8O^{15}$ + γ

$_8O^{15}$ ⟶ $_7N^{15}$ + e^+ + ν

$_7N^{15}$ + $_1H^1$ ⟶ $_6C^{12}$ + $_2He^4$

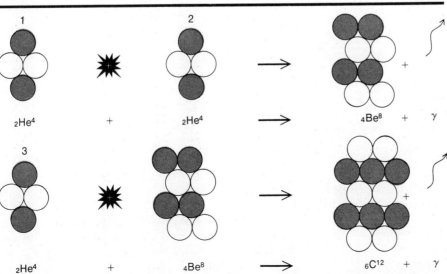

FIGURE 19–10 The triple-alpha process, which takes place only at temperatures above about 10^8 K. Beryllium, $_4Be^8$, is but an intermediate step. The 3 helium nuclei are numbered.

$_2He^4$ + $_2He^4$ ⟶ $_4Be^8$ + γ

$_2He^4$ + $_4Be^8$ ⟶ $_6C^{12}$ + γ

19.5 THE STELLAR PRIME OF LIFE

Let us now fit our knowledge of nuclear processes into astronomy. We last discussed a protostar in a collapsing phase, with its internal temperature rapidly rising.

One of the most common definitions of temperature describes the temperature as a measure of the velocities of individual atoms or other particles. A higher temperature corresponds to higher particle velocities.

For a collapsing protostar, the energy from the gravitational collapse goes into giving the individual particles greater velocities; that is, the temperature rises. For nuclear fusion to begin, atomic nuclei must get close enough to each other so that the force that holds nuclei together, the *strong nuclear force*, can play its part. But all nuclei have positive charges, because they are composed of protons (which bear positive charges) and neutrons (which are neutral). The positive charges on any two nuclei cause an electrical repulsion between them, which tends to prevent fusion from taking place.

However, at the high temperatures typical of a stellar interior, some nuclei have enough energy to overcome this electrical repulsion. They come sufficiently close to each other for the strong nuclear force to take over.

Once nuclear fusion begins, enough energy is generated to raise the pressure greatly. For main-sequence stars, the energy comes mainly from the proton-proton chain for cooler stars and the carbon-nitrogen cycle for hotter ones. The pressure provides a force that pushes outward strongly enough to balance gravity's inward pull. In the center of a star, the fusion process is self-regulating. The star finds a balance between thermal pressure pushing out and gravity pushing in. When we learn how to control fusion in power-generating stations on earth, which currently seems decades off, our energy crisis will be over.

The greater a star's mass, the hotter its core becomes before it generates enough pressure to counteract gravity. The hotter core gives off more energy, so the star becomes brighter, explaining why main-sequence stars of high luminosity have large masses. Thus more massive stars use their nuclear fuel at a much higher rate than less massive stars. Even though the more massive stars have more fuel to burn, they go through it relatively quickly. The next three chapters continue the story of stellar evolution by discussing the fate of stars when they have used up the hydrogen in their cores.

KEY WORDS

stellar evolution, T Tauri stars, nuclear fusion, thermal pressure, $E = mc^2$, nucleus, electrons, proton, neutron, ion, element, isotopes, atomic number, mass number, radioactive, quarks, neutrino, proton-proton chain, carbon-nitrogen cycle, triple-alpha process, nucleosynthesis, strong nuclear force

QUESTIONS

1. Since individual stars can live for billions of years, how can observations taken at the current time tell us about stellar evolution?

2. What is the source of energy in a protostar? At what point does a protostar become a star?

3. Arrange the following in order of development: T Tauri stars; sun; pulsars.

4. Give the number of protons, the number of neutrons, and the number of electrons in: ordinary hydrogen ($_1H^1$), lithium ($_3Li^6$), iron ($_{26}Fe^{56}$).

5. (a) If you remove one neutron from helium, the remainder is what element? (b) Now remove one proton. What is left?

6. What is the major fusion process that takes place in the sun?

7. What forces are in balance for a star to be on the main sequence?

8. What does it mean for the temperature of a gas to be higher?

9. In what form is energy carried away in the proton-proton chain?

10. Why do neutrinos give us different information about the stars than does light?

Chapter 20

The Helix Nebula, the nearest planetary nebula to us, only 400 light years away. Long exposures show that its diameter in the sky is about the same as the moon's, though it is too faint to see with the naked eye. (Photo with the Anglo-Australian 4-m telescope)

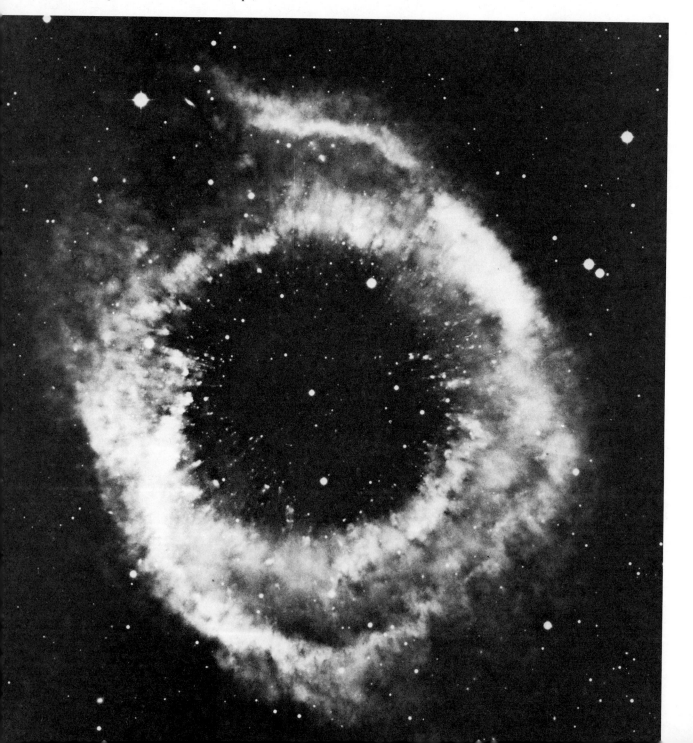

THE DEATH OF THE SUN

The length of time a star stays on the main sequence depends on its mass. The more massive a star is, the shorter its stay on the main sequence. The most massive stars may be there for only a few million years. A star like the sun, on the other hand, is not especially massive and will live on the main sequence for about 10 billion years. Since it has taken over 4 billion years for humans to evolve, it is a good thing that some stars can be this stable for this long.

In this chapter, we will discuss what will happen when the sun dies. In the following chapters, we will discuss the death of more massive stars.

20.1 RED GIANTS

Any star containing about as much mass as the sun or less will have the same fate. As fusion exhausts the hydrogen from its center, its internal pressure will diminish. Gravity will pull the outer layers in, and the core will heat up again. Hydrogen will begin "burning" in a shell around the core. (The process is nuclear fusion, not burning as on earth.) The new energy will cause the outer layers of the star to swell by a factor of 10 or more. It will become very large, so large that when the sun reaches this stage, its diameter will be 10 per cent the size of the Earth's orbit. Its surface will then be relatively cool for a star, only about 2000 K, so it will appear reddish. Such a star is called a *red giant*. Red giants appear at the upper right of color-magnitude diagrams (Fig. 20–1). The sun will be in this stage for about 10 million years, only 0.1 per cent of its lifetime on the main sequence.

Red giants are so bright that we can see them at quite a distance, and a few are therefore among the brightest stars in the sky. Arcturus in Boötes and Aldebaran in Taurus are both red giants.

The core becomes so hot that helium will start fusing into heavier elements (the triple-alpha process), but this stage will last only a brief time. Subsequently, the star becomes smaller and fainter. We wind up with a star whose core is carbon, which is surrounded by shells of helium and hydrogen that are undergoing fusion.

20.2 PLANETARY NEBULAE

As the carbon core contracts and heats up, it generates more energy. As a result, the rate at which hydrogen is fusing into helium in a shell around the core increases again. The star gets larger again. The outer layers, this time, continue to drift outward until they leave the star. The ions in the gas combine with the electrons to form neutral atoms.

Perhaps the outer layers drift off as a shell of gas. Or perhaps the gas drifts off gradually and a second round of gas comes off at a more rapid pace. This second round of gas plows into the first round, creating a visible

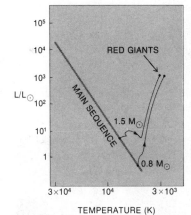

FIGURE 20–1 An H-R diagram, showing the evolutionary tracks of 0.8 and 1.5 solar mass stars, as the stars evolve from the main sequence to become red giants. For each star, the first dot represents the point where hydrogen "burning" starts and the second dot the point where helium "burning" starts. The vertical axis is the luminosity of the star—the total amount of energy it gives off—in units of the sun's luminosity; the horizontal axis is the temperature of the star's surface.

FIGURE 20–2 NGC 2392, a planetary nebula in the constellation Gemini, the Twins. This object is sometimes known as the Eskimo Nebula for obvious reasons.

FIGURE 20–3 M97, the Owl Nebula, a planetary nebula in the constellation Ursa Major.

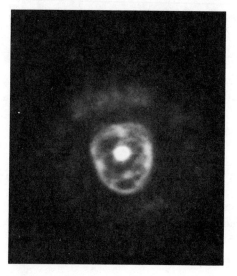

FIGURE   20–2

FIGURE 20–3

shell. Each of these two models has its proponents, and observations are being carried out to discover which is valid.

In any case, we know of a thousand such shells of gas in our galaxy (Figs. 20–2 and 20–3 and Color Plates 49 through 51). Each contains about 20 per cent of the sun's mass. The shells are exceedingly beautiful. In the small telescopes of a hundred years ago, though, they appeared as faint greenish objects, similar in appearance to the planet Uranus. They were thus named *planetary nebulae*. We now know that planetary nebulae appear greenish because the gas in them emits mainly a few strong spectral lines that include greenish ones. Uranus appears green for an entirely different reason (principally the molecule methane). But the name "planetary nebulae" remains.

The best known planetary nebula is the Ring Nebula in the constellation Lyra (Color Plate 49). It is visible in even a medium-sized telescope as a tiny apparent smoke ring in the sky. Only photographs reveal the colors.

This stage in the life of a sun-like star lasts only about 50,000 years. After that time, the nebula spreads out and fades too much to be seen at a distance.

20.3 WHITE DWARFS

Now that we have followed the outer layers, let us return to the core of the star. It has such a high temperature that it appears very bluish. Such stars are detectable at the centers of many planetary nebulae. You can pick them out in Color Plates 49 through 51, though their color doesn't show clearly.

The theory of what happens next was worked out by an Indian university student en route to England half a century ago. The student, S. Chandrasekhar, became one of the most distinguished astronomers in a long career in the United States, and shared the 1983 Nobel Prize in Physics for this early work.

Chandra (as he is known) worked out the idea that a limit exists at 1.4 solar masses, that is, stars with less than 1.4 times the mass of the sun evolve in a different way from stars with greater masses. Below the "Chandrasekhar

SURFACE OF SUN

EARTH

SIRIUS B
(WHITE DWARF)

FIGURE 20–4 The sizes of the white dwarfs are not very different from that of the earth. A white dwarf contains about 300,000 times more mass than does the earth, however.

limit," the remainder of the star contracts further. But when it reaches about the size of the earth, about 100 times smaller in diameter than it had been on the main sequence, a new type of pressure succeeds in counterbalancing gravity so that the contracting stops. This new pressure is the result of processes that can be understood only with quantum mechanics. It results from the resistance of electrons to be packed too closely together. This resistance results in a type of star called a *white dwarf* (Fig. 20–4).

The sun is 1.4 million km (a million miles) across. When essentially all its mass is compressed into a volume 100 times smaller across, which is a million times smaller in volume, the density of matter goes up incredibly. A single teaspoonful of a white dwarf would weigh 5 tons! We cannot experiment with such matter on earth, though such a high density may have been momentarily achieved in a recent terrestrial laboratory experiment.

Because they are so small, white dwarfs are so faint that they are hard to detect. Only a few single ones are known. We find most of them as members in binary systems. Even the brightest star Sirius, the Dog Star, has a white-dwarf companion, which is named Sirius B and sometimes called "the Pup" (Fig. 20–5).

White-dwarf stars have all the energy they will ever have. Over billions of years, they will gradually radiate their energy, gradually cooling off until they can no longer be seen.

20.4 NOVAE

For millennia, new stars have occasionally become visible in the sky. Some of them turn out, by recent theory, to be the result of an interaction of a large star with a white dwarf. In this section, we will discuss the role of white dwarfs in some of the apparently new stars, *novae* (singular: *nova*). In the next chapter, we will see how white dwarfs may also be contributors to even more luminous objects now known as supernovae.

A nova is newly visible, but is not really new. It represents a star system's brightening by 10 magnitudes or so, a factor of about 10,000 (Fig. 20–6). It

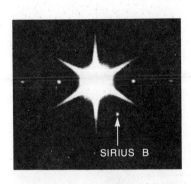

FIGURE 20–5 Sirius A appears as a bright overexposed image with six spikes caused in the telescope system; fainter additional images of Sirius A appear at regular spacing to both sides. Sirius B appears to the lower right of the brightest image of Sirius A; its additional images are too faint to see. Sirius A and B differ by 10 magnitudes, a factor of 10,000, and are very close together, making them very difficult to photograph together. This led to the use of the special technique in which much of the overexposed light from Sirius A is caused to fall in the six spikes so that Sirius B can be seen between the spikes.

FIGURE 20–6 These photographs are part of a unique series of observations covering the eruption of Nova Cygni 1975 during the period of its brightening. A Los Angeles amateur astronomer, Ben Mayer, was repeatedly photographing this area of the sky at the crucial times to search for meteors. When he heard of the nova, he retrieved his meteor-less film from the waste basket. Never before had a nova's brightening been so well observed. On August 28 *(left)*, no star was visible at the arrow. By August 30 *(right)*, the nova had reached 2nd magnitude, almost the brightness of Deneb, which is seen at the right.

Mayer has now organized an international amateur effort, Problicom (**Pro**jection **Bli**nk **Com**parator), in which each participant photographs a region of sky every few weeks and checks over the film using a simple combination of two ordinary slide projectors. Volunteers are welcome.

August 28, 1975   11:30 U.T.   August 30, 1975   6:45 U.T.

FIGURE 20–7 Nova Herculis 1934 (also known as DQ Herculis), showing its rapid fading from 3rd magnitude on March 10, 1935, to below 12th magnitude on May 6, 1935.

A

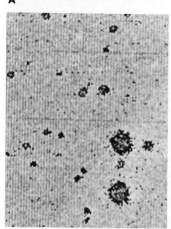

B

may remain bright for only a few days or weeks, and then fades over the years (Fig. 20–7). The ejected gas may eventually become visible (Fig. 20–8).

By the current theory, novae occur when one star in a binary system has evolved to the white-dwarf stage, and the other component is a red giant or almost so. Since the outer layers of the red giant are not strongly held in by the star's gravity, material from them can be pulled off. This matter surrounds the white dwarf. Whenever some of that material falls down to the white dwarf's surface, it may heat up enough to begin nuclear fusion there. The process involves only 1/10,000 or so of a solar mass, so can happen many times. We do indeed see some novae repeat their outbursts.

20.5 OBSERVING WITH THE IUE SPACECRAFT

The most widely used telescope now available to astronomers is not on a mountain top, but is rather hovering over earth. Known as IUE, the International Ultraviolet Explorer, it has supplied data to hundreds of astronomers in the United States and around the world since its launch in 1978. The part of the spectrum that it studies does not penetrate the atmosphere, so cannot be observed by astronomers on the ground, no matter how high the mountain top.

IUE carries a telescope whose mirror is 45 cm (18 inch) in diameter. It uses a television-type device for a detector, and can survey the spectrum of a star through the entire ultraviolet in an hour or so. It has studied many white dwarfs; their surfaces are so hot that they emit most of their radiation in the ultraviolet. IUE is so sensitive that it can also be used to observe much fainter objects, including galaxies and even quasars.

IUE is in an orbit that carries it around the earth once every 24 hours. Since the earth rotates underneath at the same rate, the spacecraft hovers within view of NASA's Goddard Space Flight Center in Greenbelt, Maryland, all the time—it is in a *synchronous orbit,* like a communications satellite.

American astronomers prepare proposals that describe what they want to observe with IUE and why, and send them periodically to NASA. European astronomers apply to the European Space Agency. For the successful proposers, the observing time follows some months later. Three days or a week of 16-hour shifts, for example, might be available for a given project. Let us say that our proposal to observe the outer atmospheres of stars is accepted. We take our leave from home for a week or so and set out for the Goddard Space Flight Center, which is in a suburb of Washington, D.C.

We arrive at an office building, not at all like a telescope dome. In a small room on an ordinary office corridor, we find the American control room for this fabulous telescope in space.

Before we arrive, we have carefully figured out exactly what objects we want to observe, and indeed have a list of specific objects approved by the IUE team. We meet the Telescope Operator, the person who actually sends commands from the control room. We are really there in an advisory capacity.

Acquiring the star in the field of view might take fifteen minutes. Then an image of a field of view 16 minutes of arc across—about half the diameter of the moon—appears on a television screen before us (Fig. 20–9).

Now we tell the Telescope Operator to take the first exposure. Only the upper tip of IUE's spectral range comes through the earth's atmosphere, so just what is to be found in these spectra is often a tremendous surprise.

FIGURE 20–8 *(A)* In 1951, a shell of gas could be seen to surround Nova Herculis 1934 (DQ Her). Nova shells expand at 100 times the rate of planetary nebulae. *(B)* In 1984, the shell of gas from Nova Cygni 1975 (V1500 Cyg) was detected. It is just above center of this CCD image.

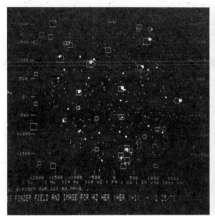  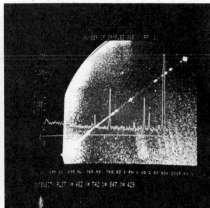

After our hour-long exposure, the data are radioed back from the spacecraft to an antenna at Goddard, and soon appear on our video screen. The spectrum is displayed both on a television screen on the console and also on a large screen overhead (Fig. 20–10 and Color Plate 3). We have the Telescope Operator begin the next exposure, perhaps the other wavelength band for the same star.

As soon as these instructions are radioed to the spacecraft, we can make use of one of the nicest features of IUE—an interactive computer. We can tell the computer to display an enlarged region of the spectrum on the screen.

We can even have the computer display, in a few seconds, a graph of a part of the spectrum (Fig. 20–11). No more do we have to develop photographic plates and then have the densities on the emulsion traced laboriously. Within seconds, we have a spectrum in view, and can even lay a plastic overlay on it to make a few measurements right away. So much of the spectrum of objects is completely unknown in the ultraviolet that important results have regularly been found in this way.

A rest at our motel—our shifts may well run through the night—brings us up to the time to observe again. We see where the telescope is pointing, figure out what to observe first, and start again. A week of this is exhausting but rewarding. We even have preliminary data to take home with us, though computer tapes with final treatments of the data will follow us by mail.

IUE—the first international observatory in space—has been a tremendous success. It has sent back data on stars, planets, nebulae, galaxies, and quasars, and we look forward to continued productivity in the years to come.

FIGURE 20–9 Within a couple of minutes, IUE displays its field of view so that the astronomers can specify which object to observe and to use as a guide star.

FIGURE 20–10 The control room contains video screens to display the data *(left)* and spacecraft housekeeping equipment *(right)*, plus a large projection tv screen *(top)* for a second view of the data.

FIGURE 20–11 Within seconds after the observer's request, the computer can plot on its screen a graph of any part of the spectrum that the astronomer designates.

KEY WORDS

red giant, planetary nebula (nebulae), white dwarf, nova (novae), synchronous orbit

QUESTIONS

1. Why does a red giant appear reddish?
2. Sketch a color-magnitude diagram, label the axes, and point out where red giants are.
3. What are two theories for planetary nebulae?
4. How far in the future might it be before the sun becomes a planetary nebula?

5. What forces balance to make a white dwarf?
6. What is the relation of the nova phenomenon to white dwarfs?
7. Is a nova really a "new star"?
8. How is it helpful for IUE to be in a synchronous orbit?

Chapter 21

Filaments at the north end of the supernova remnant known as the Cygnus Loop.

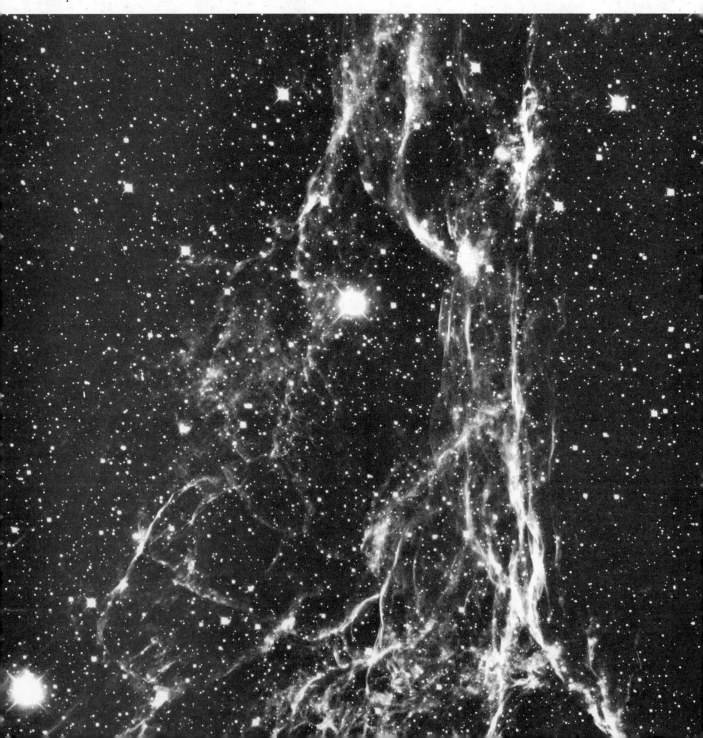

SUPERNOVAE: STELLAR RECYCLING

Although stars with about the mass of the sun gradually puff off faint planetary nebulae, some of them join more massive stars in finally going off with a spectacular bang. In this chapter, we explain these celestial fireworks.

21.1 OBSERVING SUPERNOVAE

Only in the 1920's was it realized that some of the "novae"—apparently new stars—that had been seen in distant galaxies were really much brighter than ordinary novae seen in our own galaxy. These *supernovae* are very different kinds of objects. Whereas novae are small eruptions involving only a tiny fraction of a star's mass, supernovae involve entire stars. The object we see may have brightened by 20 magnitudes—100 million times. A supernova may appear about as bright as the entire galaxy it is in (Fig. 21–1).

Unfortunately, we have seen very few supernovae in our own galaxy. The most recent ones definitely noticed were observed by Kepler in 1604 and Tycho in 1572. Since studies in other galaxies show that supernovae erupt every 30 years or so on the average, we are due. Maybe the light from a nearby supernova will reach us tonight.

Photography of the sky has revealed some two dozen regions of gas that are *supernova remnants,* the gas spread out by the explosion of a supernova (Fig. 21–2 and Color Plates 54 through 55). Dozens more have been ob-

FIGURE 21–1 The central part of the galaxy NGC 5253, photographed in 1959 and in 1972. The supernova that appeared in 1972 was nearly as bright as the rest of the galaxy.

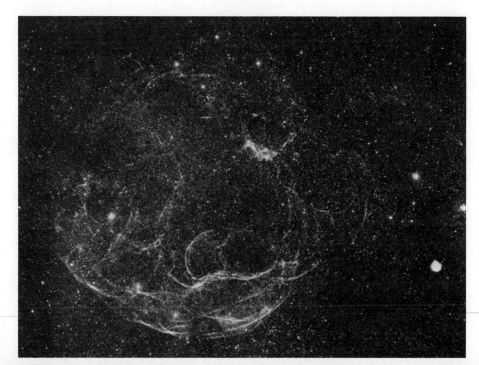

FIGURE 21–2 S147, the remnant of a supernova explosion in our galaxy. The long delicate filaments shown cover an area of 3° × 3°, about 40 times the area of the moon.

FIGURE 21–3 Views of the supernova that erupted in 1979 in the galaxy M100. Less spiral structure appears in the later view, chiefly because of a different exposure time, yet the supernova appears prominently at lower left. At 12th magnitude (apparent), it was the brightest supernova since 1970.

served in the radio part of the spectrum, and still others have more recently been observed from x-ray satellites (Color Plate 53). The supernova shown there, in the constellation Cassiopeia, was not widely noticed when it went off, though there is a possibility that it was plotted on one star map of 1680.

The most studied supernova remnant is the Crab Nebula (Color Plate 54) in the constellation Taurus. The explosion was noticed widely in China, Japan, and Korea in A.D. 1054; there is still debate as to why Europeans did not see it. If we compare photographs of the Crab taken decades apart, we can measure the speed at which its filaments are expanding. Tracing them back shows that they were together at about the time the bright "guest star" was seen in the sky by the Oriental observers, confirming the identification. The rapid speed of expansion—thousands of kilometers per second—also confirms that the Crab Nebula comes from an explosive event. The nebula's distance from us is about 1/5 the distance between the sun and the center of our galaxy; perhaps dust in our galaxy has masked most of the more recent supernova explosions.

Studies of other galaxies have given us many supernovae to observe (Fig. 21–3). We can distinguish two major classes of supernovae from (a) the speed with which the light level declines, and (b) the presence or absence of hydrogen spectral lines. They are thought to originate in different ways, as we shall now discuss.

21.1a Supernovae from Lightweight Stars

Some supernovae have no hydrogen in their spectra. These "Type I supernovae" are currently thought to be a further stage of evolution of a lightweight star, a star with about the mass of the sun. When a white dwarf is in a binary system, mass from the second object sometimes falls onto the white dwarf. If enough extra mass is added to put the white dwarf over the 1.4-solar-mass Chandrasekhar limit, the white dwarf can explode in thermonuclear fusion. This incineration makes the supernova that can be seen. Later light from the supernova comes from radioactive decay of elements that are formed in the incineration; theoretical models can now account fairly well for the spectrum and the amount of light.

NASA's Gamma Ray Observatory, hoped for in the 1990's, should be able to test this idea by detecting gamma rays emitted as the radioactive elements are formed. The absence of hydrogen in the spectrum is explained by the idea that the white dwarf could have lost its outer atmosphere before it is incinerated.

21.1b Supernovae from Heavyweight Stars

Some supernovae show many hydrogen lines in their spectra. These "Type II supernovae" are thought to be the death of massive stars, stars with more than 8 times the mass of the sun. (The fate of stars intermediate in mass between about 4 and 8 solar masses is not now understood.) They don't get quite as bright as Type I supernovae. The fact that they appear only in the spiral arms of galaxies, where massive stars are uniquely found, fits with the idea that they come from massive stars. Massive stars are short lived, so remain near the spiral arms where they are born.

A massive star runs through its main-sequence lifetime much more rapidly than does a lightweight star. A star of 15 solar masses may live on the main sequence for only 10 million years, a thousand times shorter than the sun's main-sequence stage. As the massive star's outer layers expand to make it a red giant, the core contracts and reaches 100 million kelvins. The triple-alpha process operates at this extremely high temperature. It burns helium into carbon steadily, unlike the flash of helium burning in stars like the sun.

By the time helium burning ends, the outer layers have expanded so that the star is much bigger and brighter than even a red giant. Such a *red supergiant* is a million times brighter than the sun. Betelgeuse, in Orion's shoulder (Fig. 21–4) in the winter sky, is the best-known example and is noticeably reddish to the naked eye.

A supergiant contains so much mass that when all the helium is fused into carbon, gravity makes the star contract further until it is hot enough for the carbon to fuse into heavier elements. Eventually, even iron builds up. The iron core is surrounded by layers of elements of different mass, with the lightest toward the periphery. The new models are calculated with the aid of powerful computers.

While other elements give off energy when they either fuse or break up into different elements, iron takes up energy when it does so. Thus when

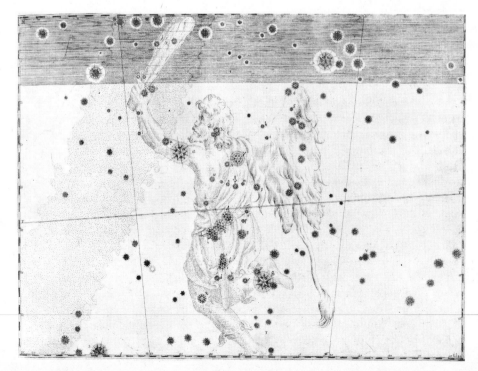

FIGURE 21–4 The red supergiant Betelgeuse is the star labelled α in the shoulder of Orion, the Hunter, shown here in Johann Bayer's *Uranometria,* first published in 1603. Betelgeuse, visible in the winter sky, appears reddish to the eye.

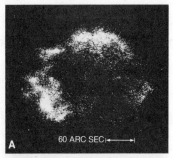

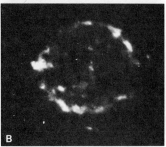

60 ARC SEC

A

B

FIGURE 21–5 The Cassiopeia A supernova remnant. *(A)* An x-ray image made with the Einstein Observatory. *(B)* A radio image made with the Very Large Array. The source is a relatively recent supernova remnant, but the supernova itself was not observed. We can date it to the late 17th century by studying its expansion. Both images show a bright shell of emission, presumably the supernova matter encountering the interstellar medium. Analysis of the x-ray spectrum shows that there are higher proportions of some of the heavy elements, showing that heavy elements are indeed formed in supernovae.

FIGURE 21–6 The Long-Duration-Exposure Facility, LDEF, was launched by a space shuttle in 1984, carrying plastic aloft to study cosmic rays.

the core of the star is iron, the star is set for disaster. No new energy is released to make enough pressure to hold up the star against the force of gravity pulling in. Within seconds, the star collapses. Though the details are still uncertain, the core apparently becomes so dense that it bounces outwards, since nuclear processes prevent it from collapsing too far. Shock waves—like sonic booms—that result cause heavy elements to form and then throw off the outer layers. The many neutrinos that are formed may also help blow off the outer layers. Only the core may be left behind.

21.1c Supernovae and Us

The heavy elements that are formed and thrown out by both Type I and Type II supernovae are necessary for life. Supernovae are the only known source of such elements. These heavy elements are spread through space and are incorporated in stars that form later on. The sun is such a second-generation star. So we humans, who depend on heavy elements for our existence, are here because of supernovae and this process of recycling of material.

No supernova has been detected in our galaxy since the invention of the telescope. Astronomers would dearly love one to study. It might appear as bright as the moon in the sky for months, and be visible night and day. But we don't want one too close, or it could blow off our protective atmosphere.

A few searches, some automated, are newly set up to look for supernovae in other galaxies. The hope is to find them early enough—even before they have reached maximum light—so that their beginning stages can be studied. The first success came in 1983, when a Type II supernova was found before it reached its maximum light.

Studies of supernovae in different parts of the spectrum (Fig. 21–5) are giving us lots of new information.

21.2 COSMIC RAYS

So far, our study of the universe in this book has relied on information that we get by observing radiation—including not only light but also gamma-rays, x-rays, ultraviolet, infrared, and radio waves. But we also receive a few high-energy particles from space. These *cosmic rays* (misnamed historically; we now know that they aren't rays at all) are nuclei of atoms moving at tremendous velocities. Some of the weaker cosmic rays come from the sun while other cosmic rays come from farther away. Cosmic rays provide about 1/5 of the radiation environment of earth's surface and of the people on it. (Almost all the radiation we are exposed to comes from cosmic rays, from naturally occurring radioactive elements in the earth and in our bodies, and from medical x-rays.)

For a long time, scientists have debated the origin of the non-solar "primary cosmic rays," the ones that actually hit the earth's atmosphere as opposed to cosmic rays that hit the earth's surface. Because cosmic rays are charged particles—mostly protons and also some nuclei of atoms heavier than hydrogen—our galaxy's magnetic field bends them. Thus we cannot trace back the paths of cosmic rays we detect to find out their origin. It seems that most middle-energy cosmic rays were accelerated to their high velocities in supernova explosions. Evidence released in 1985 indicates that most of the cosmic rays in our galaxy come from a single source—Cygnus X-3. This object was a discovery of space research. Its name means that it was the third x-ray source to be detected in the constellation Cygnus. Current evidence indicates that Cygnus X-3 is a neutron star (Chapter 22) in orbit with a larger companion. The interaction between the two gives the protons millions of times more energy than we can give protons with even the largest "accelerators" (atom smashers) in laboratories on earth. Some rare cosmic rays of hundreds of times higher energy than Cygnus X-3 may come from sources outside our galaxy.

Our atmosphere filters out most of the primary cosmic rays. When they hit the earth's atmosphere, the collisions with air molecules generate "secondary cosmic rays." Primary cosmic rays can be captured with high-altitude balloons or satellites (Fig. 21–6). Stacks of suitable plastics (the observations were formerly made with thick photographic emulsions) show the damaging effects of cosmic rays passing through them. Scientists are now worried about cosmic rays damaging computer chips vital for navigation in airplanes as well as spacecraft, and are building in redundancy to the chips for safety.

When primary cosmic rays hit the earth's atmosphere, they cause flashes of light that can be studied electronically with telescopes on earth. A project for observing secondary cosmic rays by studying light they generate as they plow through a large volume of clear seawater is being planned (Fig. 21–7). In 1985, a space shuttle took aloft Spacelab 2, which carried an electronic cosmic-ray detector. Nuclei from lithium to nickel, when they passed through the detector with high energy, generated bursts of light or x-rays that could be detected to reveal the mass, direction, and velocity of each particle.

FIGURE 21–7 A University of Hawaii professor holds one of the giant photomultipliers being deployed in tests of Project DU-MAND in the clear seawater off Hawaii. Cosmic rays moving through the water should give off pulses of light, which the giant photomultipliers will detect.

KEY WORDS

supernovae, supernova remnants, red supergiant, cosmic rays

QUESTIONS

1. Distinguish observationally between novae and supernovae.

2. Distinguish between what is going on in novae and supernovae.

3. What are two ways that we distinguish observationally between the two types of supernovae?

4. Since there have been so few supernovae in our galaxy, how can we tell that a supernova goes off in a galaxy about every 30 years?

5. Why do we think that the Crab Nebula is a supernova remnant?

6. Why does iron help a supernova collapse?

7. For what elements were supernovae rather than only normal stars vital?

8. What is the difference between cosmic rays and x-rays?

Chapter 22

The Crab Nebula, the remnant of a supernova explosion that became visible on earth in A.D. 1054. The pulsar, the first pulsar detected to be blinking on and off in the visible part of the spectrum, is marked with an arrow. In the long exposure necessary to take this photograph, the star turns on and off so many times that it appears to be an ordinary star. Only after its radio pulsation had been detected was it observed to be blinking in optical light. It had long been suspected, however, of being the leftover core of the supernova because of its unusual spectrum, which has only a continuum and no spectral lines.

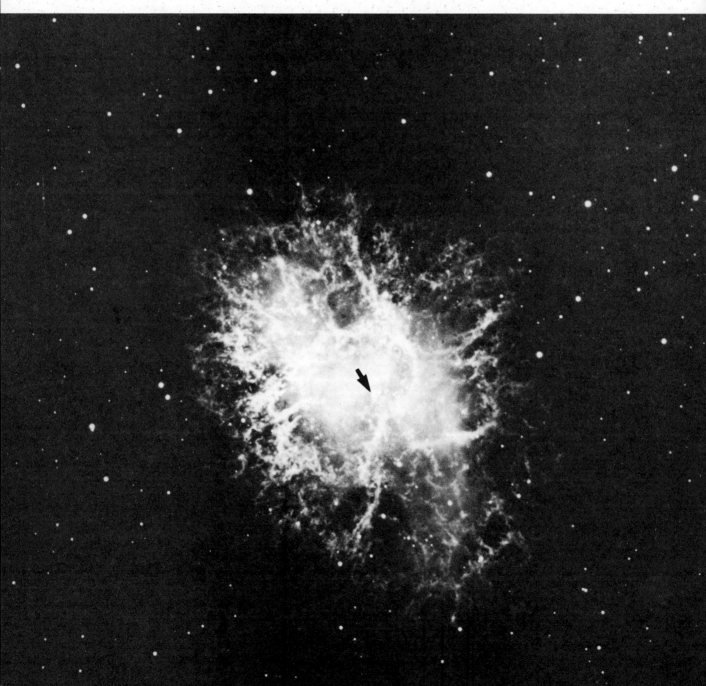

PULSARS:
STELLAR BEACONS

We have discussed the fate of the outer layers of a massive star that explodes as one type of supernova. But what about the core? In this chapter, we will discuss cores that wind up as superdense stars. In the next chapter, we will discuss what happens when the core is too massive to ever stop collapsing.

22.1 NEUTRON STARS

As iron fills the core of a supergiant, the temperature becomes so high that the iron begins to break down into smaller units like helium nuclei. This breakdown soaks up energy. The core can no longer counterbalance gravity, and it collapses.

The core's density becomes so high that electrons are squeezed into the nuclei. They react with the protons there to produce neutrons and neutrinos. The neutrinos escape, perhaps helping to blow off the supernova remnants. A gas composed mainly of neutrons remains. If between a few tenths of a solar mass and about two solar masses are left in the remaining core, it can reach a new stable stage.

When this remaining core is sufficiently compressed, the neutrons resist being further compressed, as we can explain using laws of quantum mechanics. A pressure is created, which counterbalances the inward force of gravity. The star is now basically composed of neutrons, and is so dense that it is like a single, giant nucleus. We call it a *neutron star*. It is only about 20 kilometers or so across (Fig. 22–1). A teaspoonful could weigh a billion tons.

As an object contracts, its magnetic field is compressed. As the magnetic-field lines come together, the field gets stronger. A neutron star is so much smaller than the sun that its field should be over a million times stronger.

When neutron stars were first discussed theoretically in the 1930's, the chances of observing one seemed hopeless. We presently have signs of them in several independent and surprising ways, as we now discuss.

22.2 THE DISCOVERY OF PULSARS

Just as the light from stars twinkles in the sky because the stars are point objects, point radio sources (radio sources that are so small or so far away that they have no length or breadth) fluctuate in brightness as well. In 1967, a special radio telescope was built to study this radio twinkling; previously, radio astronomers had mostly ignored and blurred out the effect, to study the objects themselves.

Jocelyn Bell (now Jocelyn Burnell), in 1967, was a graduate student working on Professor Antony Hewish's special radio telescope (Fig. 22–2). As the sky swept over the telescope, which pointed in a fixed direction, she

FIGURE 22–1 A neutron star may be only 20 kilometers in diameter (10 km in radius), the size of a city, even though it may contain a solar mass or more. A neutron star might have a solid, crystalline crust about a hundred meters thick. Above these outer layers, its atmosphere probably takes up only another few centimeters. Since the crust is crystalline, there may be irregular structures like mountains, which would poke up only a few centimeters through the atmosphere.

FIGURE 22–2 The radio telescope—actually a field of aerials—with which pulsars were discovered at Cambridge, England. The total collecting area was large, so that it could detect faint sources, and the electronics was set so that rapid variations could be observed.

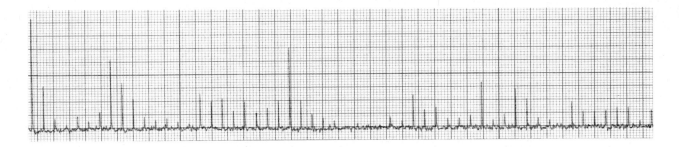

FIGURE 22–3 This pulsar, PSR 0329+54, has a period of 0.7145 second. The horizontal axis is time and the vertical axis is intensity of radio radiation at a certain wavelength.

noticed that the signal occasionally wavered a lot in the middle of the night, when radio twinkling was usually low. Her observations eventually showed that the position of the source of the signals remained fixed with respect to the stars rather than constant in terrestrial time. This showed her that the phenomenon was celestial rather than terrestrial or solar.

Bell and Hewish found that the signal, when spread out, was a set of regularly spaced pulses, with one pulse every 1.3373011 seconds (Fig. 22–3). The source was briefly called LGM, for "Little Green Men," for such a signal might come from an extraterrestrial civilization. But soon Bell located three other sources, pulsing with regular periods of 0.253065, 1.187911, and 1.2737635 seconds, respectively. Though they could be LGM2, LGM3, and LGM4, it seemed unlikely that extraterrestrials would have put out four such beacons at widely spaced locations in our galaxy, or beacons that so wastefully radioed energy at so many frequencies simultaneously. The objects were named *pulsars* and announced to an astonished world. It was immediately apparent that it was an important discovery, but what were they?

22.3 WHAT ARE PULSARS?

Other observatories set to work searching for pulsars, and dozens were found. They were all characterized by very regular periods, with the pulse itself taking up only a small fraction of a period. When the positions of all the known pulsars are plotted on a celestial map (Fig. 22–4), they are concentrated along the plane of our galaxy. Thus they must be in our galaxy;

FIGURE 22–4 The distribution of most of the 369 known pulsars on a projection that maps the entire sky, with the plane of the Milky Way along the zero degree line of latitude on the map. The concentration of pulsars near 60° galactic longitude on this map merely represents the fact that this section of the sky has been especially carefully searched for pulsars because it is the area of the Milky Way best visible from the Arecibo Observatory. We can extrapolate that there are at least 100,000 pulsars in our galaxy.

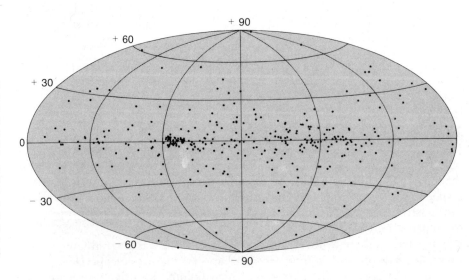

if they were located outside our galaxy we would see them distributed uniformly around the sky or even partly obscured near our galaxy's plane where the Milky Way might obscure something behind it.

The question of what a pulsar is can be divided into two parts. First, we want to know why the pulses are so regular, that is, what the "clock" is. Second, we want to know where the energy comes from.

From the fact that the pulses are so short, we can deduce that pulsars are very small. If the sun, for example, were to wink out all at once, we would see its side nearest to us disappear a few seconds before its far side disappeared, since the sun is a few light seconds across. So we knew that pulsars were much smaller than the sun. That left only white dwarfs and neutron stars as possibilities.

We can get pulses from a star in two ways: if the star oscillates or if it rotates. (The only other possibility—collapsed stars orbiting each other—would give off too much energy to match the observations.) The theory worked out for ordinary variable stars had shown that the speed with which a star oscillates depends on its density. Ordinary stars would oscillate much too slowly to be pulsars, and even white dwarfs would oscillate somewhat too slowly. Further, neutron stars would oscillate too rapidly to be pulsars. So oscillations were excluded.

That left only rotation as a possibility. And it can be calculated that a white dwarf is too large to rotate rapidly enough to cause pulsations as rapidly as occur in a pulsar. So the only remaining possibility is the rotation of a neutron star. We have solved the problem by the process of elimination. There is agreement on this *lighthouse model* for pulsars (Fig. 22–5). Just as a

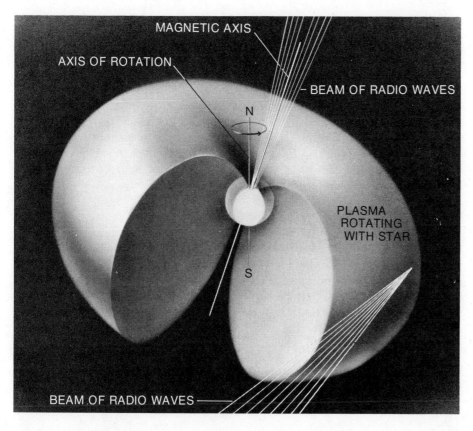

FIGURE 22–5 In the lighthouse model for pulsars, which is now commonly accepted, a beam of radiation flashes by us once each pulsar period, just as a lighthouse beam appears to flash by a ship at sea. It is believed that the generation of a pulsar beam is related to the neutron star's magnetic axis being aligned in a different direction than the neutron star's axis of rotation. The mechanism by which the beam is generated is not currently understood. Two possible variations of the lighthouse theory are shown here. The beam may be emitted along the magnetic axis, as shown in the top half of the figure. Alternatively, the beam could be generated on the surface of a doughnut of magnetic field (shown cut away) that surrounds and rotates with the neutron star, as shown in the bottom half of the figure.

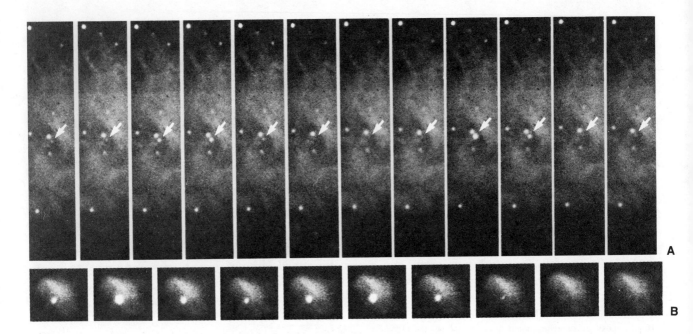

FIGURE 22–6 *(A)* A sequence of pictures taken by adding light from different phases of the pulsar's period shows the pulsar apparently blinking on and off. Other stars in the field remain constant in intensity. Observations like this one clinched the identification of this star with the pulsar. *(B)* High-resolution x-ray images of the Crab Nebula and its pulsar, displayed in 10 time intervals covering the 33-millisecond pulse period. A total of 10 hours of observation is included. The main pulse occurs in the second interval, and an in-between pulse called the "interpulse" falls in the sixth.

lighthouse seems to flash light at you every time its beam points toward you, a pulsar is a rotating neutron star.

What about the way in which energy is generated? There is much less agreement about that, and the matter remains unsettled. Remember that the magnetic field of a neutron star is extremely high. This can lead to a powerful beam of radio waves. If the magnetic axis is tilted with respect to the axis of rotation (which is also true for the earth, whose magnetic north pole is in Hudson Bay, Canada), the beam from stars oriented in certain ways will flash by us at regular intervals. We wouldn't see other neutron stars, if their beams were oriented in other directions.

22.4 THE CRAB, PULSARS, AND SUPERNOVAE

Several months after the first pulsars had been discovered, strong bursts of radio energy were found to be coming from the direction of the Crab Nebula. Observers detected that the Crab pulsed 30 times a second, almost ten times more rapidly than the fastest other known pulsar. This rapid pulsation endorsed the exclusion of white dwarfs from the list of possible explanations.

The discovery of a pulsar in the Crab Nebula made the theory that pulsars were neutron stars look more plausible, since neutron stars **should** exist in supernova remnants like this one. And the case was clinched when it was discovered that the clock in the Crab pulsar was not precise—it was slowing down slightly. The energy given off as the pulsar slowed down was precisely the amount of energy needed to keep the Crab Nebula shining. The source of the Crab Nebula's energy had been discovered!

Astronomers soon found, to their surprise, that a star in the center of the Crab Nebula could actually be seen apparently to turn on and off 30 times a second (Fig. 22–6). Actually, the star only appears "on" when its beamed light is pointing toward us as it sweeps around. Long photographic exposures had always hidden this fact, though the star had long been thought to be the remaining core because of its spectrum, which oddly doesn't show any spectral lines. Subsequently, similar observations of the star's blinking on and off in x-rays were also found.

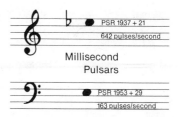

FIGURE 22–7 Two of the millisecond pulsars have periods fast enough to hear as musical notes, when we amplify the radio waves they give off.

22.5 SLOWING PULSARS AND THE FAST PULSAR

The Crab is one of the most rapidly pulsing pulsars, and is slowing by the greatest amount. But most other pulsars have also been found to be slowing gradually. The theory had been that the younger the pulsar, the faster it was spinning and the faster it was slowing down. After all, the Crab came from a supernova explosion only 900 years ago.

So scientists were surprised to find, in 1982, a pulsar spinning 20 times faster—642 times per second. Even a neutron star rotating at that speed would be on the verge of being torn apart. And this pulsar is hardly slowing down at all; it may be useful as a long-term time standard to test even the atomic clocks that are presently the best available to scientists. The object, which is in a binary system, is thought to be old—over a million years old— because of its gradual slowdown rate. Possibly its rotation rate has been speeded up in an interaction with its companion.

This pulsar's period is 1.4 milliseconds (0.0014 second), so it is known as the "millisecond pulsar." A second millisecond pulsar has since been discovered. Each pulses rapidly enough to sound like a note in the middle of a piano keyboard (Fig. 22–7).

22.6 THE BINARY PULSAR AND GRAVITY WAVES

We know of only a few pulsars in binary systems. One—the "binary pulsar"—has an elliptical orbit that can be traced out by studying small differences in the time of arrival of the pulses. The pulses come a little less often (3 seconds) when the pulsar is moving away from us in its orbit, and a little more often when the pulsar is moving toward us.

Einstein's general theory of relativity explained the slight change over decades in the orientation of Mercury's orbit around the sun. The gravity in the binary-pulsar system is much stronger, and the effect is much more pronounced. Calculations show that the orientation of the pulsar's orbit should change by 4° per year (Fig. 22–8), which is verified precisely by measurements.

Another prediction of Einstein's general theory is that *gravitational waves,* caused by fluctuations of the positions of masses, should travel through space. The process would be similar to the way that radiation, caused by fluctuations in electricity and magnetism, travels through space. But gravity waves have never been detected directly. The motion of the binary pulsar in its orbit is slowing down by precisely the rate that would be expected if the system were giving off gravitational waves. So scientists consider that the existence of gravitational waves has been verified in this way.

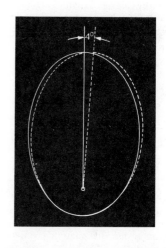

FIGURE 22–8 The orbit of the binary pulsar PSR 1913 + 16 with respect to its companion swings around by 4° per year, providing a strong endorsement of Einstein's general theory of relativity.

A   **B**

FIGURE 22–9 *(A)* Joseph Weber and one of his original gravitational-wave detectors from the 1960's: a large aluminum cylinder weighing 4 tons, delicately suspended so that gravitational waves would set it vibrating 1660 times per second. Weber's efforts have stimulated interest in detecting gravity waves, and even more sensitive detectors of this type are currently in use or under construction at Louisiana State and the Universities of Maryland, Rochester, Rome, and Tokyo. *(B)* MIT's laser interferometer to measure gravity waves. Future larger versions might be sensitive enough to detect gravity waves from supernovae in nearby galaxies. Scientists at several institutions here and abroad are developing other laser detectors.

Several experiments have been carried out and are under way (Fig. 22–9) to detect gravitational waves directly.

22.7 X-RAY BINARIES

Neutron stars are now routinely studied in a way other than their existence as pulsars. Many neutron stars in binary systems interact with their companions. As gas from the companion is funneled toward the neutron star's poles by the strong magnetic field, the gas heats up and gives off x-rays. X-ray telescopes in orbit detect such pulses of x-rays. But in these binary systems, unlike the case for normal pulsars, the pulse rate usually speeds up.

FIGURE 22–10 A model of SS433 in which the radiation emanates from two narrow beams of matter that are given off by the accretion disk that surrounds a neutron star or a black hole *(right center)*. Material is being fed into the accretion disk from the companion star. The velocity of gas in the beams is 80,000 km/sec in each direction, though we see a somewhat lower velocity because we do not see the jets head on.

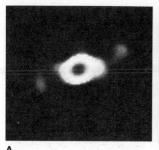

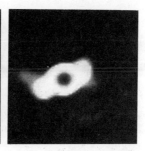

A

B

One of the oddest x-ray binaries is known as SS433, from its number in a catalogue. From measurements of Doppler shifts, we detect gas coming out of this x-ray binary at about 25 per cent of the speed of light, far greater than any other velocity ever measured in our galaxy. The most widely accepted model (Fig. 22–10) considers that SS433 is a neutron star surrounded by a disk of matter it has taken up from a companion star. Our measurements of the Doppler shifts in optical light show us light coming toward us from one jet and going away from us from the other jet at the same time. So the star appears to be coming and going at the same time. The disk would wobble like a top (a precession, similar to the one we discussed in Chapter 3). As it wobbles, the velocities decrease and increase again, as we see the jets at different angles. We have even detected the jets in radio waves and x-rays (Fig. 22–11).

FIGURE 22–11 *(A)* SS433 observed at a radio wavelength of 6 cm with the Very Large Array (VLA). SS433 is in the center of a supernova remnant. *(B)* An x-ray image of SS433 from the Einstein Observatory. Two jets extend to the sides of the bright central image. SS433 is not an especially strong x-ray source; some of the other binary x-ray sources are 100 times stronger.

KEY WORDS

neutron stars, pulsars, lighthouse model, gravitational waves

QUESTIONS

1. What keeps a neutron star from collapsing?

2. What force tends to collapse the remaining core of a supergiant after the supernova explosion?

3. Compare the sun, a white dwarf, and a neutron star in size.

4. Sketch the sun, a white dwarf, and a neutron star to scale.

5. In what part of the spectrum do all pulsars give off energy that we study?

6. How do we know that pulsars are in our galaxy?

7. Why do we think that the lighthouse model explains pulsars?

8. How did studies of the Crab Nebula pin down the explanation of pulsars?

9. How has the binary pulsar been especially useful?

10. How do the theories of x-ray pulsars differ from those that explain normal pulsars?

Chapter 23

The overexposed dark object in the center of this negative print is the blue supergiant star HDE 226868, which is thought to be the companion of the first black hole to be discovered, Cygnus X-1.

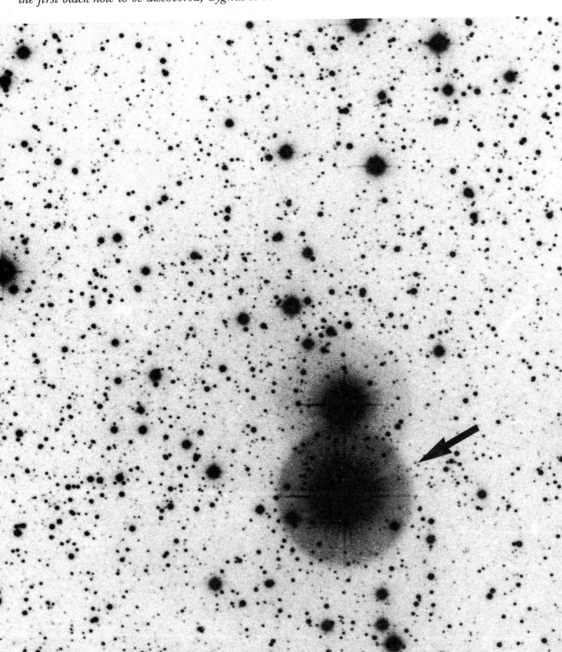

Color Plate 61 (top): M16, the Eagle Nebula in the constellation Serpens. Hydrogen radiation makes it appear red. The bright stars at the upper right are hot and young, and are part of a galactic cluster. The small, dark regions may be protostars. (Palomar Observatory, California Institute of Technology photo with the 5-m telescope)

Color Plate 62 (bottom): M17, the Omega Nebula in Sagittarius. (Palomar Observatory, California Institute of Technology photo with the 1.2-m Schmidt camera)

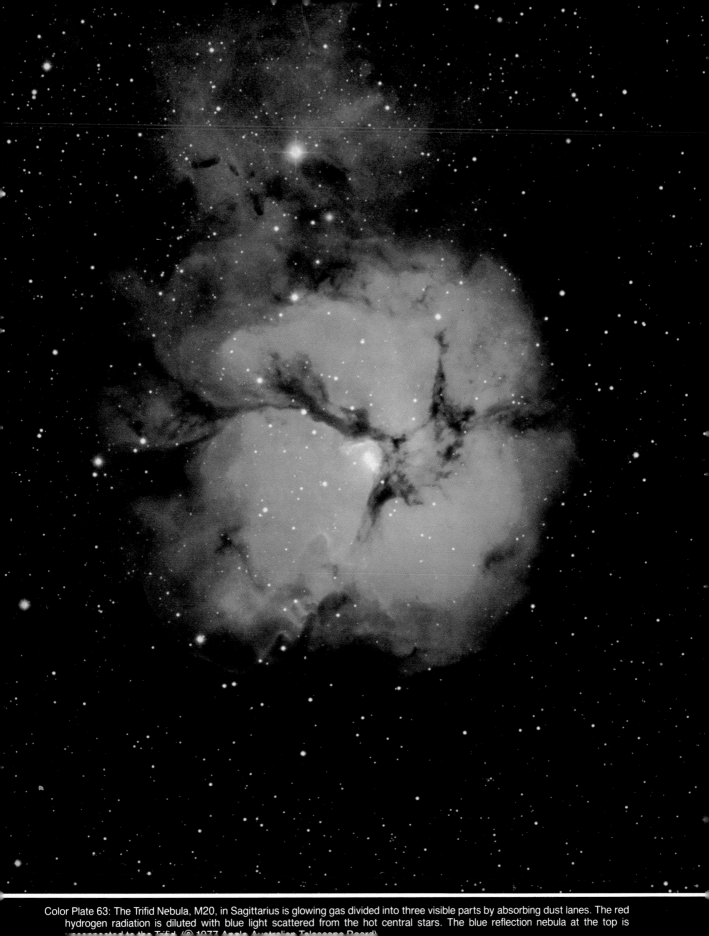

Color Plate 63: The Trifid Nebula, M20, in Sagittarius is glowing gas divided into three visible parts by absorbing dust lanes. The red hydrogen radiation is diluted with blue light scattered from the hot central stars. The blue reflection nebula at the top is

Color Plate 64 (top): The Lagoon Nebula, M8, in Sagittarius. Red hydrogen light is emitted by gas that is excited by the radiation of very hot stars buried within the nebula; dark filaments of material within the cloud emit strong infrared radiation. Also, several peculiar variable stars in this nebula occasionally flare up. The Lagoon Nebula is about 60 light years across and is located about 6500 light years away from us. (Kitt Peak National Observatory photo with the 4-m telescope)

Color Plate 65 (bottom): The Rosette Nebula, NGC2237–39, is an emission nebula in Monoceros. It appears twice the diameter of the moon. We see the reddish emission typical of ionized hydrogen. The bright central stars have blown out a cavity. (© 1984 Royal Observatory, Edinburgh)

BLACK HOLES: COSMIC VACUUM CLEANERS

The strange forces of electron and neutron pressure support dying light-weight and some heavyweight stars against gravity. The strangest case of all occurs at the death of the most massive stars, which contained much more than 8 and up to about 60 solar masses when they were on the main sequence. After they undergo supernova explosions, some may retain cores of over 2 or 3 solar masses. Nothing in the universe is strong enough to hold up the remaining mass against the force of gravity. The remaining mass collapses, and continues to collapse forever. The result is a black hole, in which the matter disappears from contact with the rest of the universe. Later, we shall discuss the formation of black holes in processes other than those that result from the collapse of a star.

23.1 THE FORMATION OF A STELLAR BLACK HOLE

Astronomers had long assumed that the most massive stars would somehow lose enough mass to wind up as white dwarfs. When the discovery of pulsars ended this prejudice, it seemed more reasonable that black holes could exist. If more than 2 or 3 times the mass of the sun—two or three "solar masses," we say—remain after the supernova explosion, the star collapses through the neutron star stage. We know of no force that can stop the collapse.

We may then ask what happens to a 5- or 10- or 50-solar-mass star as it collapses, if it retains more than 2 or 3 solar masses. It must keep collapsing, getting denser and denser. We have seen that Einstein's general theory of relativity predicts that a strong gravitational field will redshift and appear to bend radiation (Section 18.6).

As the mass contracts and the star's surface gravity increases, radiation is continuously redshifted more and more, and radiation leaving the star other than perpendicularly to the surface is bent more and more. Eventually, when the mass has been compressed to a certain size, radiation from the star can no longer escape into space. The star has withdrawn from our observable universe, in that we can no longer receive radiation from it. We say that the star has become a *black hole*.

Why do we call it a black hole? We think of a black surface as a surface that reflects none of the light that hits it. Similarly, any radiation that hits a black hole continues into its interior and is not reflected. In this sense, the object is perfectly black. (A "black" piece of paper on earth, in contrast, may radiate in the infrared, and is not truly black.)

23.2 THE PHOTON SPHERE

Let us consider what happens to radiation emitted by the surface of a star as it contracts. Although what we will discuss applies to radiation of all wavelengths, let us simply visualize standing on the surface of the collapsing star while holding a flashlight.

FIGURE 23–1 As the star contracts, a light beam emitted other than radially outward will be bent.

On the surface of a supergiant star, if we shine the beam at any angle, it seems to go straight out into space. As the star collapses, two effects begin to occur. (We will ignore the outer layers, which are unimportant here.) Although we on the surface of the star cannot notice the effects ourselves, a friend on a planet revolving around the star could detect them and radio back information to us about them. For one thing, our friend could see that our flashlight beam is redshifted. Second, our flashlight beam would be bent by the gravitational field of the star (Fig. 23–1). If we shine the beam straight up, it would continue to go straight up. But the further we shine it away from the vertical, the further it would be bent even farther away from the vertical. When the star reaches a certain size, a horizontal beam of light would not escape (Fig. 23–2).

From this time on, only if the flashlight is pointed within a certain angle of the vertical does the light continue outward. This angle forms a cone, with its apex at the flashlight, and is called the *exit cone* (Fig. 23–3). As the star grows smaller yet, we find that the flashlight has to be pointed more directly upward in order for its light to escape. The exit cone grows smaller as the star shrinks.

When we shine our flashlight in a direction outside the exit cone, the light is bent sufficiently that it falls back to the surface of the star. When we shine our flashlight exactly along the side of the exit cone, the light goes into orbit around the star, neither escaping nor falling onto the surface (Fig. 23–3).

The sphere around the star in which light can orbit is called the *photon sphere*. Its size is calculated theoretically to be 4.5 km for each solar mass present. It is thus 13.5 km in radius for a star of 3 solar masses, for example.

As the star continues to contract, theory shows that the exit cone gets narrower and narrower. Light emitted within the exit cone still escapes. The photon sphere remains at the same height even though the matter inside it has contracted further, since the total amount of matter within has not changed.

23.3 THE EVENT HORIZON

We might think that the exit cone would simply continue to get narrower as the star shrinks. But the general theory of relativity predicts that the cone vanishes when the star contracts beyond a certain size. Even light travelling straight up no longer can escape into space, as was worked out by Karl Schwarzschild in solving Einstein's equations in 1916. The radius of the star at this time is called the *gravitational radius*. The imaginary surface at that radius is called the *event horizon* (Fig. 23–4). Its radius is exactly 2/3 times that of the photon sphere, 3 km for each solar mass.

We can visualize the event horizon in another way, by considering a classical picture based on the Newtonian theory of gravitation. The picture is essentially that conceived in 1796 by Laplace, the French astronomer and mathematician. You must have a certain velocity, called the *escape velocity,* to escape from the gravitational pull of another body. For example, we have to launch rockets at 11 km/sec (40,000 km/hr) in order for them to escape from the earth's gravity. For a more massive body of the same size, the escape velocity would be higher. Now imagine that this body contracts. We are drawn closer to the center of the mass. As this happens, the escape velocity rises. When all the mass of the body is within its gravitational radius, the escape velocity becomes equal to the speed of light. Thus even light cannot

FIGURE 23–2 Light can be bent so that it falls back onto the star.

escape. If we begin to apply the special theory of relativity, which explains motion at very high speeds, we might then reason that since nothing can go faster than the speed of light, nothing can escape. Now let us return to the picture according to the general theory of relativity, which explains gravity and the effects caused by large masses.

The size of the gravitational radius depends linearly on the amount of mass that is collapsing. A star of 3 solar masses, for example, would have a gravitational radius of 9 km. A star of 6 solar masses would have a gravitational radius of 18 km. One can calculate the gravitational radii for less massive stars as well, although the less massive stars would be held up in the white dwarf or neutron-star stages and not collapse to their gravitational radii. The sun's gravitational radius is 3 km. The gravitational radius for the earth is only 9 mm; that is, the earth would have to be compressed to a sphere only 9 mm in radius in order to form an event horizon and be a black hole.

Anyone or anything on the surface of a star as it passed its event horizon would not be able to survive. An observer would be torn apart by the tremendous difference in gravity between head and foot. (This is called a tidal force, since this kind of difference in gravity also causes the tides on earth.) If the tidal force could be ignored, though, the observer on the surface of the star would not notice anything particularly wrong as the star passed its event horizon. But the observer's flashlight signal would never get out.

Once the star passes inside its event horizon, it continues to contract. Nothing can ever stop its contraction. In fact, the mathematical theory predicts that it will contract to zero radius—it will reach a *singularity*.

Even though the mass that causes the black hole has contracted further, the event horizon doesn't change. It remains at the same radius forever, as long as the amount of mass inside doesn't change.

23.4 ROTATING BLACK HOLES

Once matter is inside a black hole, it loses its identity in the sense that from outside a black hole, all we can tell is the mass of the black hole, the rate at which it is spinning, and what total electric charge it has. These three quantities are sufficient to completely describe the black hole. Thus, in a sense,

FIGURE 23–3 When the star has contracted enough (the inner sphere), only light emitted within the exit cone escapes. Light emitted on the edge of the exit cone goes into the photon sphere. The further the star contracts within the photon sphere, the narrower the exit cone becomes.

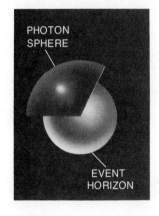

FIGURE 23–4 When the star becomes smaller than its gravitational radius, we can no longer observe it. We say that it has passed its *event horizon*, by analogy to the statement that we cannot see a thing on earth once it has passed our horizon.

FIGURE 23–5 This drawing by Charles Addams is reprinted with permission of The New Yorker Magazine, Inc., © 1974.

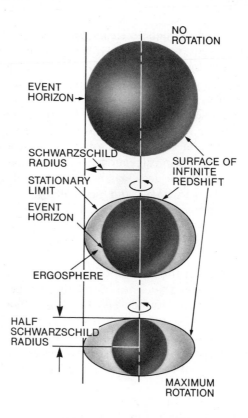

NO ROTATION

EVENT HORIZON→

SCHWARZSCHILD RADIUS

STATIONARY LIMIT

EVENT HORIZON

SURFACE OF INFINITE REDSHIFT

ERGOSPHERE

HALF SCHWARZSCHILD RADIUS

MAXIMUM ROTATION

black holes are simple objects to describe physically, because we only have to know three numbers to characterize each one. The theorem that describes the simplicity of black holes is often colloquially stated by astronomers active in the field as "a black hole has no hair."

The theoretical calculations about black holes we have just discussed are based on the assumption that black holes do not rotate. But this assumption is only a convenience; we think, in fact, that the rotation of a black hole is one of its important properties. It was not until 1963 that Roy P. Kerr solved Einstein's equations for a situation that was later interpreted in terms of the notion that a black hole is rotating. In this more general case, an additional special boundary—the *stationary limit*—appears, with somewhat different properties from the original event horizon. Within the stationary limit, no particles can remain at rest even though they are outside the event horizon.

The equator of a stationary limit of a **rotating** black hole has the same diameter as the event horizon of a **non-rotating** black hole of the same mass. But a rotating black hole's stationary limit is squashed. The event horizon touches the stationary limit at the poles. Since the event horizon remains a sphere, it is smaller than the event horizon of a non-rotating black hole (Fig. 23–6).

The region between the stationary limit and the event horizon is the *ergosphere*. In principle, we can get energy and matter out of the ergosphere.

A black hole can rotate up to the speed at which a point on the event horizon's equator is travelling at the speed of light. The event horizon's radius is then half the gravitational radius. If a black hole rotated faster than this, its event horizon would vanish. Unlike the case of a non-rotating black hole, for which the singularity is always unreachably hidden within the event horizon, in this case distant observers could receive signals from the singularity. Such a point would be called a *naked singularity* and, if one exists, we might have no warning—no photon sphere or orbiting matter, for example—before we ran into it. Most theoreticians assume the existence of a law of "cosmic censorship," which requires all singularities to be "clothed" in event horizons, that is, not naked. Since so much energy might erupt from a naked singularity, we can conclude that there are none in our universe from the fact that we do not find signs of them.

23.5 DETECTING A BLACK HOLE

A star collapsing to be a black hole would blink out in a fraction of a second, so the odds are unfavorable that we would actually see a collapsing star as it went through the event horizon. And a black hole is too small to see directly.

But all hope is not lost for detecting a black hole. Though the black hole disappears, it leaves its gravity behind. It is a bit like the Cheshire Cat from *Alice in Wonderland,* which fades away, leaving only its grin behind (Fig. 23–7).

Like all objects in the universe, the black hole attracts matter, and the matter accelerates toward it. Some of the matter will be pulled directly into the black hole, never to be seen again. But other matter will go into orbit around the black hole, and will orbit at a high velocity. This added matter forms an *accretion disk* (Fig. 23–8); "accretion" is growth in size by the gradual addition of matter.

It seems likely that the gas in orbit will be heated to a very high temperature by friction. The gas will radiate strongly in the x-ray region of the spectrum, giving off bursts of radiation sporadically as hot spots rotate. The inner 200 km should reach hundreds of millions of kelvins. Thus, though we cannot observe the black hole itself, we can hope to observe x-rays from the gas surrounding it.

In fact, a large number of x-ray sources have been detected from satellites. Some may be black holes. How can we tell?

It is not enough to find an x-ray source that gives off sporadic pulses, for other mechanisms besides matter revolving around a black hole can lead to such pulses. We also have to show that a collapsed star of greater than 3

looked goodnatured, she thought: still it had *very* long claws and a great many teeth, so she felt it ought to be treated with respect.

"Cheshire Puss," she began, rather timidly, as she did not at all know whether it would like the

FIGURE 23–7 Lewis Carroll's Cheshire Cat, from *Alice in Wonderland,* shown here in John Tenniel's drawing, is analogous to a black hole in that it left its grin behind when it disappeared, while a black hole leaves its gravity behind when its mass disappears. Alice thought that the Cheshire Cat's persisting grin was "the most curious thing I ever saw in my life!" We might say the same about the black hole and its persisting gravity.

FIGURE 23–8 *(A)* The optical appearance of a rotating black hole surrounded by a thin accretion disk. The drawing shows, on the basis of computer calculations, how curves of equal intensity are affected by the mass of the black hole. The observer is located 10° above the plane of the accretion disk. The asymmetry in appearance is caused by the rotation of the black hole. *(B)* Simulation on a supercomputer allowed study of the matter closer to the event horizon than previously possible. A splashing of material was discovered.

A

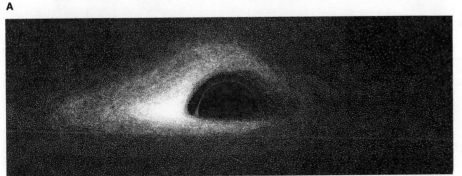

B

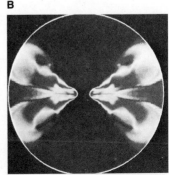

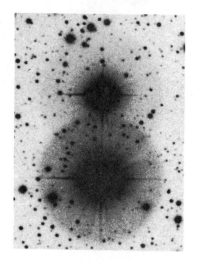

FIGURE 23–9 A section of the image reproduced in the photograph opening this chapter of the blue supergiant star HDE 226868. The black hole Cygnus X-1 that is thought to be orbiting the supergiant star is not visible. The image of the supergiant appears large because it is overexposed on the film; the star is too small and too far away to be seen from earth as other than a point.

solar masses is present. We can determine masses only for certain binary stars. When we search the position of the x-ray sources, we look for a spectroscopic binary (that is, a star whose spectrum shows a Doppler shift that indicates the presence of an invisible companion). Then, if we can show that the companion is too faint to be a normal, main-sequence star, it must be a collapsed star. If, further, the mass of the unobservable companion is greater than 3 solar masses, it must be a black hole.

The most persuasive case is named Cygnus X-1, the first x-ray source to be discovered in the constellation Cygnus. A 9th magnitude star called HDE 226868 has been found at its location (Fig. 23–9). This star has the spectrum of a blue supergiant, and thus has a mass of about 15 times that of the sun. Its radial velocity varies with a period of 5.6 days, indicating that the supergiant and the invisible companion are orbiting each other with that period. From the orbit, it is deduced that the invisible companion must certainly have a mass greater than 4 solar masses; the best estimate is 8 solar masses. This mass makes it likely that it is a black hole (Fig. 23–10).

Another promising candidate is LMC X-3 (the third x-ray source to be found in the Large Magellanic Cloud, a small galaxy that is a satellite of the Milky Way Galaxy in which we live). Since we know the distance to the Large Magellanic Cloud and thus to anything in it much more accurately than we know the distance to Cygnus X-1, our calculations for the invisible object's mass may be more accurate. Again, the invisible object seems to have about 8 solar masses of matter, and so is probably a black hole.

23.6 NON-STELLAR BLACK HOLES

We have discussed how black holes can form by the collapse of massive stars. But theoretically a black hole should result if a mass of any amount is sufficiently compressed. No object containing less than 2 or 3 solar masses will contract sufficiently under the force of its own gravity in the course of stellar evolution. But the density of matter was so high at the time of the origin of the universe (see Chapter 28) that smaller masses may have been sufficiently compressed to form "mini black holes."

Stephen Hawking, an English astrophysicist, has suggested their existence. Mini black holes the size of pinheads would have masses equivalent to those of asteroids. There is no observational evidence for a mini hole, but they are theoretically plausible. Hawking has deduced that small black holes can seem to emit energy in the form of elementary particles (neutrinos and

FIGURE 23–10 An artist's conception of the disk of swirling gas that would develop around a black hole like Cygnus X-1 *(right)* as its gravity pulled matter off the companion supergiant *(left)*. The x-radiation would arise in the disk.

so forth). The mini holes would thus evaporate and disappear. This may seem to be a contradiction to the concept that mass can't escape from a black hole. But when we consider effects of quantum mechanics, the simple picture of a black hole that we have discussed up to this point is not sufficient. Hawking suggests that a black hole so affects space near it that a pair of particles—a nuclear particle and its antiparticle—can form simultaneously. The antiparticle disappears into the black hole, and the remaining particle reaches us. Photons, which are their own antiparticles, appear too.

On the other extreme of mass, we can consider what a black hole would be like if it contained thousands or millions of times the mass of the sun. The more mass involved, the lower the density needed for a black hole to form. For a very massive black hole, one containing hundreds of millions or billions of solar masses, the density would be so low when the event horizon formed that it would be close to the density of water.

Thus if we were travelling through the universe in a spaceship, we couldn't count on detecting a black hole by noticing a volume of high density. We could pass through the event horizon of a high-mass black hole without even noticing. We would never be able to get out, but it would be hours before we would notice that we were being drawn into the center at an accelerating rate.

Where could such a supermassive black hole be located? The center of our galaxy may contain a black hole of a million solar masses. Though we would not observe radiation from the black hole itself, the gamma-rays, x-rays, and infrared radiation we detect would be coming from the gas surrounding the black hole. Active galaxies and quasars are other probable locations for massive black holes. Black holes may be widespread and very important in our universe.

KEY WORDS

black hole, exit cone, photon sphere, gravitational radius, event horizon, escape velocity, singularity, stationary limit, ergosphere, naked singularity, accretion disk

QUESTIONS

1. Why doesn't the pressure from electrons or neutrons prevent a star from becoming a black hole?

2. (a) Is light acting more like a particle or more like a wave when it is bent by gravity? (b) Explain the bending of light as a property of a warping of space (Section 18.6).

3. What is the gravitational radius for a 10 solar mass star?

4. What is your gravitational radius?

5. What are radii of the photon sphere and the event horizon of a non-rotating black hole of 18 solar masses?

6. What is the relation in size of the photon sphere and the event horizon? If you were an astronaut in space, could you escape from within the photon sphere of a rotating black hole? From within its ergosphere? From within its event horizon?

7. (a) How could the mass of a black hole that results from a collapsed star increase? (b) How could mini black holes, if they exist, lose mass?

8. Would we always notice when we reached a black hole by its high density? Explain.

9. Could we detect a black hole that was not part of a binary system?

10. Under what circumstances does the presence of an x-ray source associated with a spectroscopic binary suggest to astronomers the presence of a black hole?

PART V   Our Milky Way

The Origin of the Milky Way *by Tintoretto, circa 1578. (The National Gallery, London)*

On the clearest nights, when we are far from city lights, we can see a hazy band of light stretching across the sky. This band is the *Milky Way*—the dust, gas, and stars that make up the galaxy in which our sun is located.

Don't be confused by the terminology: the Milky Way itself is the band of light that we can see from the earth, and the Milky Way Galaxy is the whole galaxy in which we live. Like other galaxies, our Milky Way Galaxy is composed of perhaps a trillion stars plus many different types of gas, dust, planets, and so on. In the directions we see the Milky Way in the sky we are looking through the disk of matter that forms a major part of our Milky Way Galaxy.

The Milky Way appears very irregular when we see it stretched across the sky—there are spurs of luminous material that stick out in one direction or another, and there are dark lanes or patches in which nothing can be seen. This patchiness is due to the splotchy distribution of gas, dust, and stars.

Here on earth, we are inside our galaxy together with all of the matter we see as the Milky Way. Because of our position, we see a lot of our own galaxy's matter when we look along the plane of our galaxy. On the other hand, when we look "upward" or "downward" out of this plane, our view is not obscured by matter, and we can see past the confines of our galaxy.

The gas in our galaxy is more or less transparent to visible light, but the small solid particles that we call "dust" are opaque. So the distance we can see through our galaxy depends mainly on the amount of dust that is present. This is not surprising: we can't always see across a smoke-filled room. Similarly, the dust between the stars in our galaxy dims the starlight by absorbing it or by scattering it in different directions.

The dust in the plane of our galaxy prevents us from seeing very far toward its center. With visible light, we can see only one tenth of the way in. Because of widespread dust, we can see just about the same distance in any direction we look in the plane of the Milky Way. These direct optical observations fooled scientists at the turn of the century into thinking that the earth was near the center of the universe.

We shall see in the next chapter how the American astronomer Harlow Shapley in the 1920's realized that our sun was not in the center of the Milky Way. This fundamental idea took humanity one step further away from thinking that we were at the center of the universe. Copernicus, in 1543, had already made the first step in removing the earth from the center of the universe.

In recent years astronomers have been able to use wavelengths other than optical ones to study the Milky Way Galaxy. In the 1950's and 60's especially, radio astronomy gave us a new picture of our galaxy. In the 1980's, we have benefited from infrared observations, most recently from the IRAS spacecraft. Infrared and radio radiation can pass through the galaxy's dust and allow us to see our galactic center and beyond.

In this part of the book, we shall first discuss the types of objects that we find in the Milky Way Galaxy. Chapter 24 considers the general structure of the galaxy, and its major parts. In Chapter 25, we describe the matter between the stars, and what studying this matter has told us about how stars form.

Chapter 24

The Milky Way in Sagittarius. The Great Rift across the center may be an overlap of many dark nebulae that shield "giant molecular clouds," huge regions where hydrogen and other molecules have formed.

THE MILKY WAY: OUR HOME IN THE UNIVERSE

We have now described the stars, which are important parts of any galaxy, and how they are born, live, and die. In this chapter, we describe the gas and dust that accompany the stars. Clouds of this gas and dust are called nebulae (singular: nebula). We also discuss the overall structure of the Milky Way Galaxy and how, from our location inside it, we detect this structure.

24.1 NEBULAE

A *nebula* is a cloud of gas and dust that we see in visible light. When we see the gas actually glowing in the visible part of the spectrum, we call it an *emission nebula*. Sometimes we see a cloud of dust that obscures our vision in some direction in the sky. When we see the dust appear as a dark silhouette, we call it a *dark nebula* (or, sometimes, an *absorption nebula*, since it absorbs light from stars behind it). The photo of the Milky Way on the facing page shows both emission nebulae (some of the brightest parts) and dark nebulae (some of the dark parts where relatively few stars can be seen). In the heart of the Milky Way, shown in the picture, we also see the great *star clouds* of the galactic center, the bright regions where the stars are too close together to tell them apart.

The clouds of dust in Figure 24–1 or surrounding some of the stars in the Pleiades (Color Plate 48) are examples of *reflection nebulae*—they merely reflect the starlight toward us without emitting visible radiation of their own. Reflection nebulae usually look bluish because dust reflects blue light more efficiently than it does red light. (Similar scattering of sunlight in the earth's atmosphere makes the sky blue.) Whereas an emission nebula has its own spectrum, as does a neon sign on earth, a reflection nebula shows the spectral lines of the star or stars whose light is being reflected.

The Great Nebula in Orion (Color Plate 56) is an emission nebula. In the winter sky, we can readily observe it through even a small telescope, but only with long photographic exposures or large telescopes can we study its structure in detail. In the Orion Nebula we think stars are being born this very minute.

The North America Nebula (Color Plate 60), gas and dust that has a shape in the sky similar to the shape of North America on the earth's surface, is an example of an object that is simultaneously an emission and an absorption nebula. The reddish emission comes from glowing gas spread across the sky. Obscuring dust causes the boundary we see for the North America Nebula. We can tell this because we see fewer stars there.

The Horsehead Nebula (Fig. 24–2 and Color Plate 59) is another example of an object that is both an emission and an absorption nebula simultaneously. A bit of absorbing dust intrudes onto emitting gas, outlining the shape of a horse's head. We can see in the pictures that the horsehead is a continuation of a dark area in which very few stars are visible.

Nebula *is Latin for fog or mist. The plural is usually* **nebulae** *rather than "nebulas."*

FIGURE 24–1 A reflection nebula, NGC 7129, in Cepheus.

FIGURE 24–2 The Horsehead
Nebula in Orion, which descends
from the left-most star in Orion's
belt.

We have already discussed some of the most beautiful nebulae in the
sky, composed of gas thrown off in the late stages of stellar evolution. They
include planetary nebulae and supernova remnants.

24.2 THE PARTS OF OUR GALAXY

It was not until the 1920's that the American astronomer Harlow Shapley
realized that we were not in the center of the galaxy. He was studying the
distribution of globular clusters and noticed that they were all in the same
general area of the sky as seen from the earth. They mostly appear above
or below the galactic plane and thus are not obscured by the dust. When he
plotted their distances and directions, he noticed that they formed a spherical halo around a point thousands of light years away from us (Fig. 24–3).
Shapley's touch of genius was to realize that this point must be the center of
the galaxy.

The picture that we have of our own galaxy has changed in the last few
years. Let us now discuss our current view:

1. *The nuclear bulge:* Our galaxy has the general shape of a pancake with a
bulge at its center that contains millions of stars. This *nuclear bulge* is about
16,000 light years in radius, with the galactic *nucleus* at its midst. The nucleus
itself is only about 10 light years across.

2. *The disk:* The part of the pancake outside the bulge is called the galactic
disk. It extends 50,000 light years or so out from the center of the galaxy.
The sun is located about half-way out. The disk is very thin, 2 per cent of
its width, like a phonograph record. It contains all the young stars and interstellar gas and dust; stars of other ages pass through it. The disk is
slightly warped at its ends, perhaps by interaction with our satellite galaxies,
the Magellanic Clouds. Our galaxy looks a bit like a hat with a turned-down
brim.

It is very difficult for us to tell how the material is arranged in our
galaxy's disk, just as it would be difficult to tell how the streets of a city were
laid out if we could only stand still on one street corner without moving.
Still, other galaxies have similar properties to our own, and their disks are

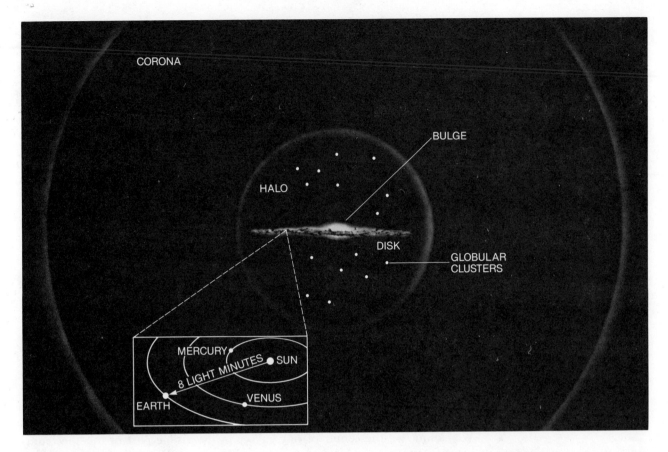

filled with great *spiral arms,* regions of dust, gas, and stars in the shape of a pinwheel (see, for example, Color Plates 66, 68, 69, and 71). So we assume the disk of our galaxy has spiral arms too, though the direct evidence (Section 24.5) is ambiguous.

3. *The halo:* Older stars (including the globular clusters) and interstellar matter form a galactic *halo* around the disk. This halo is at least as large across as the disk, perhaps 65,000 light years in radius. It extends far above and below the plane of our galaxy. Spectra from the International Ultraviolet Explorer spacecraft show that gas in the halo is hot—100,000 K. The gas in the halo contains only about 2 per cent of the mass of the gas in the disk.

4. *The galactic corona:* We shall see in the next chapter how studies of the rotation of material in the outer parts of galaxies tell us how much mass is present. These studies have told us of the existence of a lot of mass we had previously overlooked because we couldn't see it. This mass extends outward some 200,000 or 300,000 light years. Believe it or not, this *galactic corona* apparently contains 5 or 10 times as much mass as the nucleus, disk, and halo together. And it makes our galaxy 3 or 5 times larger across than we had thought. (The subject is so new that the vocabulary is not settled; "galactic corona" and "outer halo" are among the names for this material.)

If the material in the galactic corona was of an ordinary type, we would have seen it directly, if not in visible light than in radio waves, x-rays, infrared, etc. But we don't see it at all! We only detect its gravitational properties. What is the galactic corona made of? We just don't know. A tremendous number of very faint stars is a possibility, though it seems unlikely, extrapolating from the numbers of the faintest stars that we can study. A

FIGURE 24–3 The current model of our galaxy. Absorption by interstellar matter blocks our view of almost all objects between the dotted lines. The drawing shows the *nuclear bulge* surrounded by the *disk,* which contains the spiral arms. The globular clusters are part of the *halo,* which extends above and below the disk. Extending even farther is the *galactic corona.*

FIGURE 24–4 An IRAS view of 45° across the sky, showing the Milky Way. The warm dust that IRAS images is close to the plane of the galaxy. The bright spots are regions where stars are forming; we see dust heated by the stars. The bright region at the core is about 25 K. Infrared cirrus seems to stream from the plane.

large number of small black holes has been suggested. And another possibility is a huge number of neutrinos, if neutrinos have mass.

If it is unsatisfactory to you that 95 per cent of our galaxy's mass is in some unknown form, you may feel better by knowing that astronomers find the situation unsatisfactory too. But all we can do is go out and do our research, and try to find out more. We just don't know the answer . . . yet.

24.3 THE CENTER OF OUR GALAXY AND INFRARED STUDIES

We cannot see the center of our galaxy in the visible part of the spectrum because our view is blocked by interstellar dust. But radio waves and infrared penetrate the dust. IRAS, the NASA/Netherlands/UK Infrared Astronomical Satellite, with its 0.6-m telescope, mapped the sky at much longer wavelengths in 1983. With its detectors and telescope cooled to only 2 K by liquid helium to eliminate the contribution from their own heat, IRAS was at least 1000 times more sensitive than past infrared telescopes. It discovered hundreds of thousands of new sources—so much data that it will take years to interpret.

Many of the infrared sources are cool stars but many others do not coincide in space with known optical sources. Some may be galaxies too distant to see optically. The map of the radio sky doesn't look like the map of the optical sky, and the map of the infrared sky doesn't resemble either of the others. Most of the identified infrared-emitting objects, however, unlike the radio objects, are in our galaxy.

IRAS mapped the entire galaxy; its view of the disk penetrated to the galactic center (Fig. 24–4). Another of IRAS's discoveries was that the sky is covered with infrared-emitting material, probably outside our solar system but in our galaxy. Since its shape resembles terrestrial cirrus clouds, the material is being called "infrared cirrus." It is shown in the Color Essay on IRAS toward the beginning of this book.

In the early 1990's, NASA plans to carry aloft an 0.85-m infrared telescope—the Space Infrared Telescope Facility (SIRTF). Its infrared detectors, developed more recently than those of IRAS, are 100 times more sensitive. The relation between IRAS and SIRTF should be similar to the relation shared by two of the telescopes at Palomar: the wide-field 1.2-m Schmidt, and the narrow-field 5-m reflector. Also for the early 1990's, the

European Space Agency is planning the Infrared Space Observatory with an 0.6-m telescope.

24.3a The Galactic Nucleus

One of the brightest infrared sources in our sky is the center of our galaxy. It is only about 10 light years across. This makes it a very small source for the prodigious amount of energy it emits: as much energy as if there were 80 million suns radiating. It is also a radio source and a strong and variable x-ray source.

In the very center of the nucleus, radio and infrared astronomers have discovered an extremely narrow source. It is only about 10 astronomical units across, smaller than the orbit of Jupiter around the sun. It is giving off a great deal of energy (though not as much as the nuclei of some distant galaxies). The leading theory is that a high-mass black hole millions of times the mass of the sun is present. This theory explains particularly well why the radio and infrared source at the galactic center is so small. Interstellar gas and dust spiralling in toward the black hole would heat up and give off the large amount of energy that we detect. Much observational and theoretical work remains to be done here.

The high-resolution radio maps of our galactic center, now made with the Very Large Array, show a small bright spot that could well be the central giant black hole (Fig. 24–5).

24.4 HIGH-ENERGY SOURCES IN OUR GALAXY

The study of our galaxy provides us with a wide range of types of sources. Many of these have been known for many years from optical studies (Fig. 24–6). We have just seen how the infrared sky looks quite different. The radio sky provides still a different picture. Technological advances have enabled us to study sources in our galaxy in the x-ray and gamma-ray region of the spectrum as well.

The first reasonable map of the x-ray sky was made with the Uhuru satellite, which observed hundreds of x-ray sources starting in 1970. Most of our current knowledge of x-ray sources came from the U.S. **High-Energy**

FIGURE 24–5 *(A)* The nucleus of our galaxy at the center of Sagittarius A, observed at the radio wavelength of 6 cm with the VLA. The brightest spot, at the center of the spiral, is the image of the nucleus, though the point radio source is known to be 100 times smaller than it appears here. The spirals may be the matter of stars being drawn into the giant black hole in the nucleus. *(B)* In a field of view 10 times larger across, also centered near the nucleus, the VLA shows that a vast arc of parallel filaments stretched over 130 light years perpendicularly to the plane of the galaxy. Resembling one type of solar prominence (which are held up by the sun's magnetic field), the structure seems to indicate that a strong magnetic field is present.

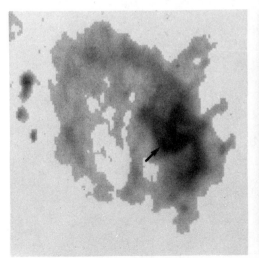

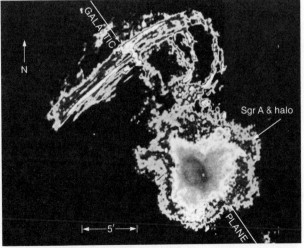

A   **B**

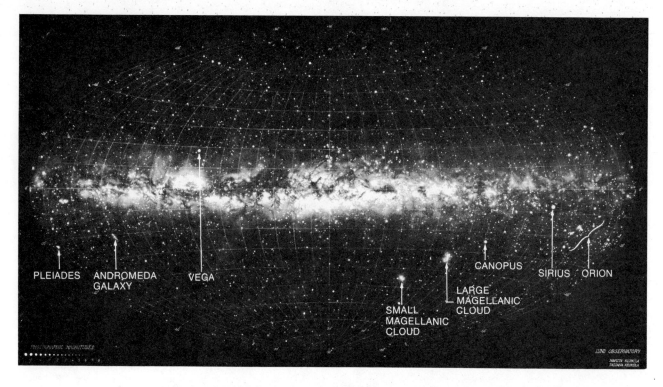

FIGURE 24–6 A drawing of the Milky Way, made under the supervision of Knut Lundmark at the Lund Observatory in Sweden. 7000 stars plus the Milky Way are shown in this panorama, which lies in coordinates such that the Milky Way falls along the equator.

FIGURE 24–7 An x-ray map of the Milky Way showing the objects observed by the first High Energy Astronomy Observatory, drawn in the same "galactic coordinates" as the preceding figure. (Kent S. Wood/Naval Research Laboratory)

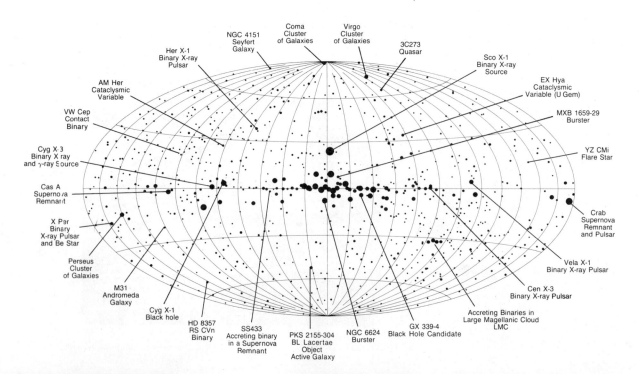

Astronomy Observatories. HEAO-1 mapped 1500 x-ray sources (Fig. 24–7). Scientists using HEAO-1 and HEAO-2 (Einstein) studied many of these sources over extended periods of time.

Among the many discoveries were the x-ray bursts from globular clusters and from other locations in the sky. These *bursters,* which give off bursts of x-rays on a time scale of seconds or hours, turn out to be x-ray binaries, some of which are located in globular clusters.

Another discovery was the bright x-ray source at the galactic center (Fig. 24–8), which fits in with the theory that a giant black hole is present there.

Although an occasional strange object, like the Crab Pulsar, can be detected at every wavelength of the spectrum, astronomers know of very few gamma-ray objects. The European COS-B spacecraft has been mapping the sky in gamma rays since 1975. It has discovered 25 sources emitting gamma rays of relatively high energy (Fig. 24–9). Only four of these sources correspond to well-known objects: the Crab and Vela pulsars, a nearby quasar (3C 273), and a complex of interstellar clouds. With a single exception, the other sources lie close to the plane of our galaxy. Extrapolating from the observations, several hundred such sources may exist near the plane of our galaxy, but we do not know what they are. U.S. satellites in the Vela series, whose prime purpose is to detect nuclear explosions, and HEAO-3 have also made gamma-ray observations.

A powerful gamma-ray burst was observed on March 5, 1979. The initial burst, only 1/5000 second in duration, was followed for a few minutes by pulses with an 8-second period. The periodicity may indicate the presence of a neutron star. During the burst, the object gave off energy at a rate greater than that of the entire Milky Way Galaxy. Other gamma-ray bursts, most from unknown directions, are also being studied. We await the ability to pinpoint the direction to gamma-ray bursts, for which we may have to wait until the Gamma Ray Observatory is launched.

Studies of electromagnetic radiation like x-rays and gamma rays and of rapidly moving cosmic-ray particles are part of the new field of *high-energy astrophysics.* With the termination of the HEAO program, American x-ray observing has fallen on hard times. The European Exosat now in space, the German Rosat expected in a few years, and small Japanese spacecraft are taking up the slack until the Advanced X-Ray Astrophysics Facility (AXAF) is launched in the early 1990's.

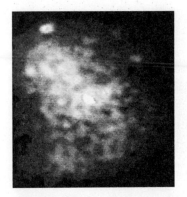

FIGURE 24–8 This x-ray image shows a bright point at the galactic center. (Einstein Observatory image by Michael Watson, Paul Hertz, and colleagues, Harvard-Smithsonian Center for Astrophysics)

24.5 THE SPIRAL STRUCTURE OF THE GALAXY

It is always difficult to tell the shape of a system from a position inside it. Think, for example, of being somewhere inside a maze of tall hedges. We would find it difficult to trace out the pattern. If we could fly overhead in a helicopter, though, the pattern would become very easy to see. Similarly, we have difficulty tracing out the spiral pattern in our own galaxy, even though the pattern would presumably be apparent from outside the galaxy. Still, by

FIGURE 24–9 This gamma-ray map made by the European COS-B spacecraft shows the plane of the Milky Way and several individual gamma-ray sources. The map includes the region within 25° of the plane of the Milky Way.

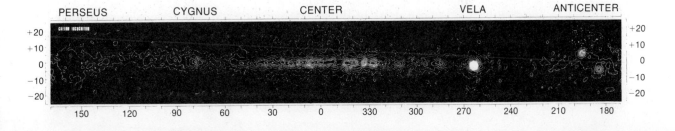

| PERSEUS | | CYGNUS | | CENTER | | VELA | | ANTICENTER |

FIGURE 24-10   The positions of young open clusters and H II regions in our own galaxy are projected on a photograph of the spiral galaxy NGC 2432, scaled to the size of our galaxy. The sun is along the line labelled 0°.

noting the distances and directions to objects of various types, we can tell about the Milky Way's spiral structure.

Open clusters are good objects to use for this purpose, for they are always located in spiral arms. We think that spiral arms are regions where young stars are found. Some of the young stars are the O and B stars; their lives are so short we know they can't be old. But since our methods of determining the distances to O and B stars from their spectra and colors are uncertain to 10 per cent, they give a fuzzy picture of the distant parts of our galaxy.

Other signs of young stars are the presence of emission nebulae, regions of ionized hydrogen also known as H II regions (pronounced "H two" regions). We know from studies of other galaxies that H II regions are preferentially located in spiral arms. In studying the locations of the H II regions, we are really again studying the locations of the O stars and the hotter of the B stars, since it is ultraviolet radiation from these hot stars that provides the energy for the H II regions to glow.

When the positions of the open clusters, the O and B stars, and the H II regions of known distances are studied (by plotting their distances and directions as seen from earth), they appear to trace out bits of three spiral arms (Fig. 24-10). Interstellar dust prevents us from seeing parts of our galaxy farther away from the sun. Another very valuable method of studying the spiral structure in our own galaxy involves spectral lines of hydrogen and of carbon monoxide in the radio part of the spectrum (Sections 25.2 and 25.3). Radio waves penetrate the interstellar dust, and we are no longer limited to studying the local spiral arms.

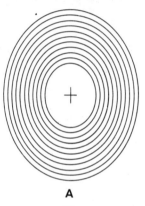

A

24.5a Why Do We Have Spiral Arms?

The sun revolves around the center of our galaxy at a speed of approximately 250 kilometers per second. At this velocity, it would take the sun about 250 million years to travel once around the center, only 2 per cent of the galaxy's lifetime. But stars at different distances from the center of our galaxy revolve around the galaxy's center in different lengths of time. The galaxy does not rotate like a solid disk. Stars closer to the center revolve much more quickly than does the sun. Thus the question arises: Why haven't the arms wound up very tightly?

The leading current solution to this conundrum says, in effect, that the spiral arms we now see are not the same spiral arms that were previously visible. The spiral-arm pattern would be caused by a spiral *density wave*, a wave of increased density that moves through the stars and gas in the galaxy. This density wave is a wave of compression, not of matter being transported. It rotates more slowly than the actual material and causes the density of material to build up as it passes. Stars form at those locations, and give the optical illusion of a spiral (Fig. 24-11).

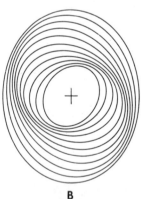

B

We can think of the analogy of a crew of workers painting a white line down the center of a busy highway. A bottleneck occurs at the location of the painters. Observers in an airplane would see an increase in the number

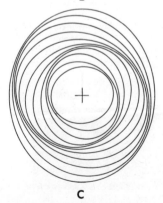

C

FIGURE 24-11   Each part of the figure includes the same set of ellipses; the only difference is the relative alignment of their axes. Consider that the axes are rotating slowly and at different rates. The compression of their orbits takes a spiral form, even though no actual spiral exists. The spiral structure of a galaxy may arise from an analogous effect.

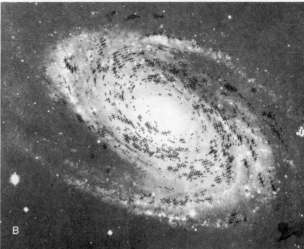

of cars at that place. As the line painters continued slowly down the road, we would seem to see the bottleneck move slowly down the road. We would see the bottleneck move along even if our vision were not clear enough to see the individual cars, which could still speed down the highway, slow down briefly as they cross the region of the bottleneck, and then resume their high speed.

Similarly, we might be viewing only some galactic bottleneck at the spiral arms. The new stars heat the interstellar gas so that it becomes visible. In fact, we do see young, hot stars and glowing gas outlining the spiral arms, which are checks of this prediction of the density-wave theory.

An alternative, very different theory says that stars are produced by a chain reaction and are then spread out into spiral arms by the different speed of rotation of the galaxy at each radius. The chain reaction begins when high-mass stars become supernovae. The expanding shells from the supernovae trigger the formation of stars in nearby regions. Some of these new stars become massive stars, which become supernovae, and so on. Computer modelling gives values that seem to agree with the observed features of spiral galaxies (Fig. 24–12).

FIGURE 24–12 The model galaxies—plotted as +'s—superimposed upon photographs of the galaxies. The models were computed using observed values of rotational velocities for each distance from the center. *(A)* M101 (that is, the 101st object in Messier's Catalogue, Appendix 8). *(B)* M81, including the effect of projecting the model at a 58° angle. (Courtesy of Humberto Gerola and Philip E. Seiden, IBM Watson Research Center)

KEY WORDS

nebula, emission nebula, dark nebula, absorption nebula, reflection nebula, star cloud, nuclear bulge, nucleus, disk, spiral arms, halo, galactic corona, bursters, high-energy astrophysics, density wave

QUESTIONS

1. Why do we think our galaxy is a spiral?

2. How would the Milky Way appear if the sun were closer to the edge of the galaxy?

3. Compare (a) absorption (dark) nebulae, (b) reflection nebulae, and (c) emission nebulae.

4. How can something be both an emission and an absorption nebula? Explain and give an example.

5. If you see a red nebula surrounding a blue star, is it an emission or a reflection nebula? Explain.

6. How do we know that the galactic corona isn't made of ordinary stars like the sun?

7. Why may some infrared observations be made from mountain observatories while all x-ray observations must be made from space?

8. Describe infrared and radio results about the center of our galaxy.

9. What are three tracers that we use for the spiral structure of our galaxy? What are two reasons why we expect them to trace spiral structure?

10. Discuss how observations from space have added to our knowledge of our galaxy.

Chapter 25

The Tarantula Nebula, in the nearby galaxy known as the Large Magel-lanic Cloud, contains many regions of gas and dust in which stars are now forming.

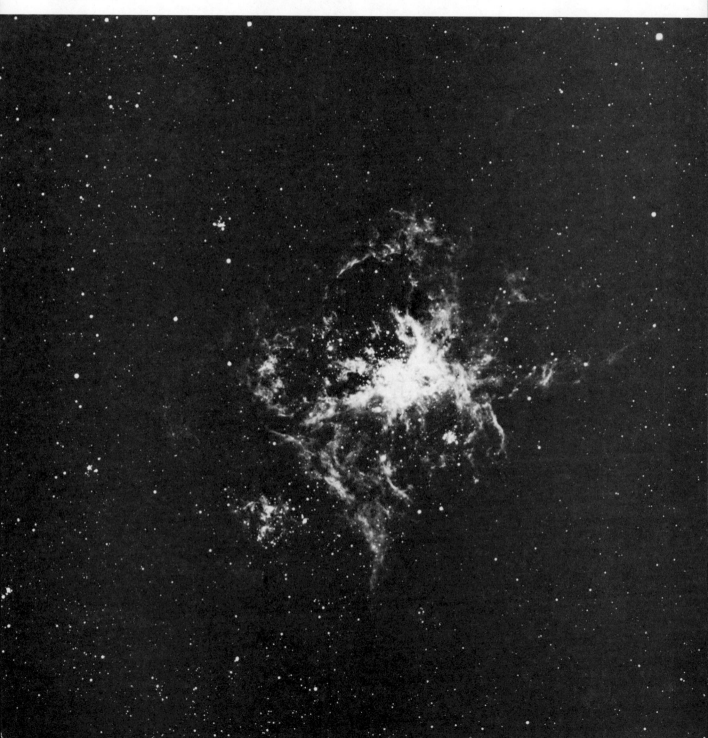

BETWEEN THE STARS: STELLAR NURSERIES

The gas and the dust between the stars is known as the *interstellar medium.* The nebulae represent regions of the interstellar medium in which the density of gas and dust is higher than average.

In this chapter we shall see that the studies of the interstellar medium are vital for understanding the structure of the galaxy and how stars are formed. We shall also discuss the recent realization that units of the interstellar medium called "giant molecular clouds" are basic building blocks of our galaxy.

25.1 MATTER BETWEEN THE STARS

For many purposes, we may consider interstellar space as being filled with hydrogen at an average density of about 1 atom per cubic centimeter. (Individual regions may have densities departing greatly from this average.) Regions of higher density in which the atoms of hydrogen are predominantly neutral are called *H I regions* (pronounced "H one regions"; the Roman numeral "I" refers to the first, or basic, state). Where the density of an H I region is high enough, pairs of hydrogen atoms combine to form molecules (H_2). The densest part of the gas associated with the Orion Nebula might have a million or more hydrogen molecules per cubic centimeter. So hydrogen molecules (H_2) are often found in H I regions.

As we saw in Section 24.5, a region of ionized hydrogen, with one electron missing, is known as an *H II region* (from "H two"; the second state—neutral is the first state and once ionized is the second). Since hydrogen, which makes up the overwhelming proportion of interstellar gas, contains only one proton and one electron, a gas of ionized hydrogen contains individual protons and electrons. Wherever a hot star provides enough energy to ionize hydrogen, an H II region (Fig. 25–1) results. Emission nebulae are such H II regions. They glow because the gas is heated. Emission lines appear.

Studying the optical and radio spectra of H II regions and planetary nebulae tells us the abundances of several of the chemical elements (especially helium, nitrogen, and oxygen). How these abundances vary from place to place in our galaxy and in other galaxies helps us choose between models of element formation and of galaxy evolution.

Tiny grains of solid particles are given off by the outer layers of red giants. They spread through interstellar space, and dim the light from distant stars. The dust never gets very hot, so most of its radiation is in the infrared. The radiation from dust scattered among the stars is too faint to detect, but the radiation coming from clouds of dust surrounding newly formed stars has been observed from the ground and from the IRAS spacecraft. IRAS found infrared radiation from so many stars forming in our galaxy that we now think that about one star forms in our galaxy each year.

Even the nebulae with the highest densities are still many orders of magnitude less dense than the best vacuums that can be made in laboratories on earth, so that the nebulae provide scientists with a way of studying the basic properties of gases under conditions that are unobtainable on earth.

235

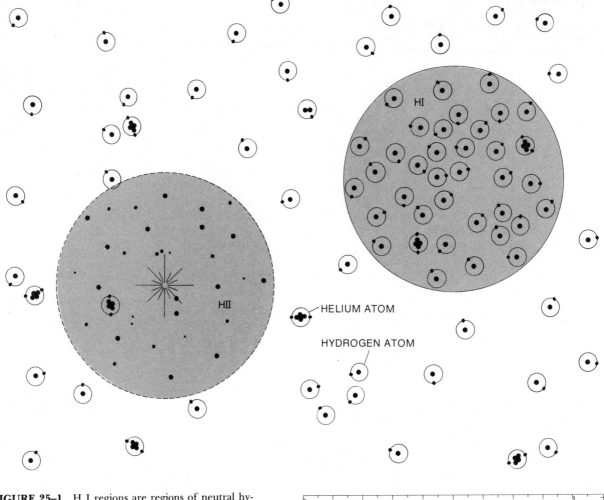

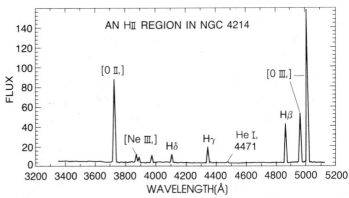

FIGURE 25–1 H I regions are regions of neutral hydrogen of higher density than average, and H II regions are regions of ionized hydrogen. The protons and electrons that result from the ionization of hydrogen by ultraviolet radiation from a hot star, and the neutral hydrogen atoms, are shown schematically. The outlines of the regions are shown for illustrative purposes only; in space, of course, the regions are not outlined, though the edge of the ionized region is abrupt. The larger dots represent protons or neutrons and the smaller dots represent electrons. The star that provides the energy for the H II region is shown. Though all hydrogen is ionized in an H II region, only some of the helium is ionized; heavier elements have often lost 2 or 3 electrons. *(Bottom)* The emission lines in an H II region including some members of hydrogen's Balmer series, helium, and "forbidden lines" of ionized oxygen and neon (enclosed in square brackets to show that they are transitions that happen relatively rarely).

Similarly, since the interstellar gas is "invisible" in the visible part of the spectrum (except at the wavelengths of certain weak spectral lines), special techniques are needed to observe the gas in addition to observing the dust. Radio astronomy is the most widely used technique, so we will now discuss its use for mapping our galaxy.

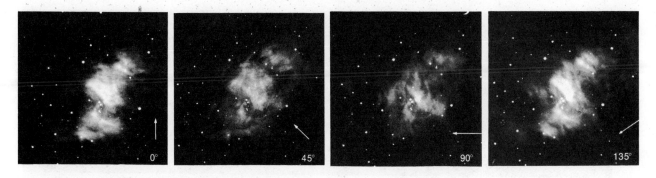

FIGURE 25–2   Polarization, in this case, is a sign that magnetic fields are present. We see here photographs of the Crab Nebula taken through filters that pass visible light polarized at the angles shown with the arrows. A non-polarized source would appear the same when viewed at any angle of polarization. The pictures at different polarization angles look very different from each other. The light and radio waves from the Crab are highly polarized, which implies that strong magnetic fields are present.

Polarization can also occur when radiation bounces off objects. Polarized sunglasses cut glare because they do not pass some of the light polarized by reflection off air molecules or off the ground. Try looking through your polarized sunglasses as you rotate them clockwise.

25.2 RADIO OBSERVATIONS OF OUR GALAXY

The rest of the electromagnetic spectrum carries more information in it than do the few thousand angstroms that we call visible light. At first, all radio astronomy observations were of continuous radiation; no spectral lines were known. One of the major processes that generates such radiation in the radio spectrum involves strong magnetic fields. So we knew that such magnetic fields existed in certain interstellar objects, such as the Crab Nebula (Fig. 25–2).

If a radio spectral line might be discovered, Doppler-shift measurements could be made, and we could tell about motions in our galaxy. What is a radio spectral line? Remember that an optical spectral line corresponds to a wavelength (or frequency) in the optical spectrum that is more (for an emission line) or less (for an absorption line) intense than neighboring wavelengths or frequencies. Similarly, a radio spectral line corresponds to a frequency (or wavelength) at which the radio radiation is slightly more, or slightly less, intense. A radio station is an emission line on a home radio.

The most likely candidate for a radio spectral line that might be discovered was a line from the lowest energy levels of interstellar hydrogen atoms. This line was predicted to be at a wavelength of 21 cm. Since hydrogen is by far the most abundant element in the universe, it seems reasonable that it should produce a strong spectral line.

A hydrogen atom is basically an electron orbiting a proton. Both the electron and the proton have the property of spin, as if each were spinning on its axis. The spin of the electron can be either in the same direction as the spin of the proton or in the opposite direction. The rules of quantum mechanics prohibit intermediate orientations. The energies of the two allowed conditions are slightly different. The energy difference is equal to a photon of 21-cm radiation.

If an atom is sitting alone in space in the upper of these two energy states, with its electron and proton spins aligned in the same direction, it has a certain small probability that the spinning electron would spontaneously flip over to the lower energy state and emit a photon. We thus call this a

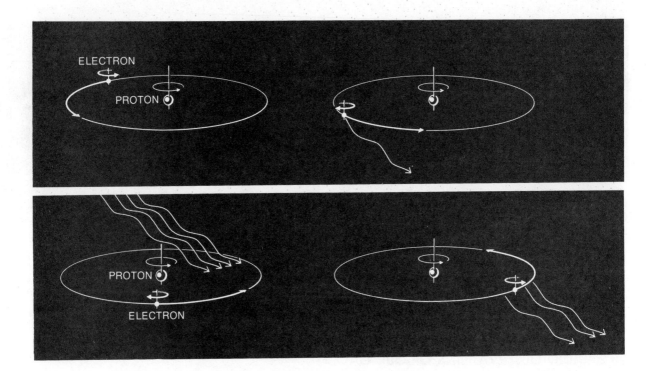

FIGURE 25–3   When the electron in a hydrogen atom flips over so that it is spinning in the opposite direction from the spin of the proton *(top)*, an emission line at a wavelength of 21 cm results. When an electron takes energy from a passing beam of radiation, causing it to flip from spinning in the opposite direction from the proton to spinning in the same direction *(bottom)*, then a 21-cm line in absorption results.

FIGURE 25–4   Harold Ewen with the horn radio telescope he and Edward Purcell used to discover the 21-cm line at Harvard in 1951. Passing undergraduates would lob snowballs into their antenna, which was stuck out the window of the physics lab.

spin-flip transition (Fig. 25–3). The photon of hydrogen's spin-flip corresponds to radiation at a wavelength of 21 cm—the *21-cm line*. If the electron flips from the higher to the lower energy state, we have an emission line. If it absorbs energy from passing continuous radiation, it can flip to the higher energy state and we have an absorption line.

If we were to watch any particular group of hydrogen atoms in the higher-energy state, we would find that it would take 11 million years before half of the electrons had undergone spin-flips; we say that the "half-life" is 11 million years for this transition. But there are so many hydrogen atoms in space that enough 21-cm radiation is given off to be detected. The existence of the line was predicted in 1944 and discovered in 1951 (Fig. 25–4). Spectral-line radio astronomy had been born.

25.2a Mapping Our Galaxy

21-cm hydrogen radiation has proved to be a very important tool for studying our galaxy because this radiation passes unimpeded through the dust that prevents optical observations very far into the plane of the galaxy. Using 21-cm observations, astronomers can study the distribution of gas in the spiral arms. We can detect this radiation from gas located anywhere in our galaxy, even on the far side, whereas light waves penetrate the dust clouds in the galactic plane only about 10 per cent of the way to the galactic center.

Astronomers have ingeniously been able to find out how far it is to the clouds of gas that emit the 21-cm radiation. They use the fact that gas nearer the center of our galaxy rotates faster than the gas farther away from the center. Though there are substantial uncertainties in interpreting the Doppler shifts in terms of distance from the galaxy's center, astronomers have succeeded in making some maps. The earliest 21-cm maps showed many narrow arms (Fig. 25–5) but no clear pattern of a few broad spiral arms like those we see in other galaxies. The question emerged: Is our galaxy really a spiral at all? Only in recent years, with the additional information from studies of molecules in space that we describe in the next section, have we made further progress.

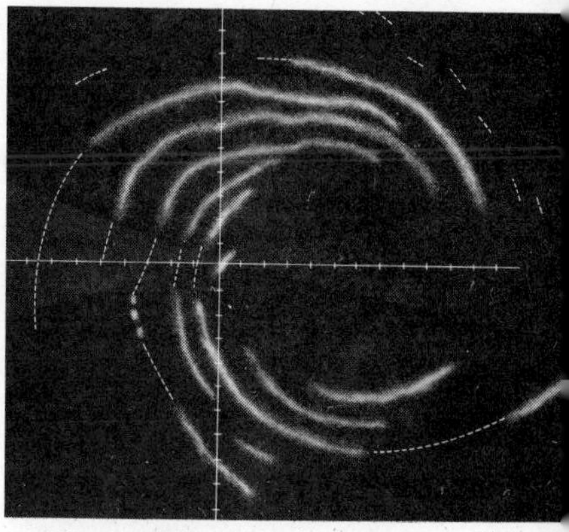

25.3 RADIO SPECTRAL LINES FROM MOLECULES

Radio astronomers had only the hydrogen spectral line to study for a dozen years, and then only the addition of one other group of lines for another five years. In 1968, however, radio spectral lines of water (H_2O) and ammonia (NH_3) were found. The spectral lines of these molecules proved surprisingly strong, and were easily detected once they were looked for. Dozens of additional molecules have since been found.

The earlier notion that it would be difficult to form molecules in space was wrong. In some cases, atoms apparently stick to interstellar dust grains, perhaps for thousands of years, and molecules build up (Fig. 25–6). Though hydrogen molecules form on dust grains, most of the other molecules may be formed in the interstellar gas without need for grains.

Studying the spectral lines provides information about physical conditions—temperature, densities, and motion, for example—in the gas clouds that emit the lines. Studies of molecular spectral lines have been used together with 21-cm observations to improve the maps of the spiral structure of our galaxy. Observations of carbon monoxide (CO) in particular have provided better information about the parts of our galaxy farther out from the galaxy's center than our sun. The result is a four-armed spiral (Fig. 25–7).

By studying the velocity of rotation of gas clouds, we can measure the mass of our galaxy. If we know how an object at the edge of the galaxy is moving, we can deduce how much mass must be closer to the center in order to keep the object in orbit. A similar method is used to find the amount of mass in the sun by studying the orbits of the planets, or the amount of a planet's mass by observing the orbits of its moons. The method was worked

FIGURE 25–5 An artist's impression of the structure of our galaxy based on 21-cm data. Because hydrogen clouds located in the directions either toward or away from the galactic center have no radial velocity with respect to us, we cannot find their distances.

FIGURE 25–6 Hydrogen molecules are formed in space with the aid of dust grains at an intermediate stage.

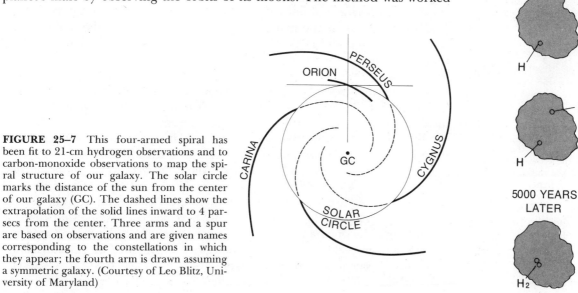

DUST GRAIN

H

H

H

5000 YEARS LATER

H_2

FIGURE 25–7 This four-armed spiral has been fit to 21-cm hydrogen observations and to carbon-monoxide observations to map the spiral structure of our galaxy. The solar circle marks the distance of the sun from the center of our galaxy (GC). The dashed lines show the extrapolation of the solid lines inward to 4 parsecs from the center. Three arms and a spur are based on observations and are given names corresponding to the constellations in which they appear; the fourth arm is drawn assuming a symmetric galaxy. (Courtesy of Leo Blitz, University of Maryland)

ORION   PERSEUS

CARINA   GC   CYGNUS

SOLAR CIRCLE

FIGURE 25–8   The globule Barnard 335. The small cloud of gas and dust appears as a "hole" in the distribution of stars. We assume that the stars are roughly uniformly distributed and that such holes result because of the extinction caused by dust in the line of sight. Emission from carbon monoxide has been observed from this region, signifying that something is really present.

out by Newton in the 17th century in his derivation and elaboration of one of Kepler's laws of orbital motion.

It now seems that our galaxy is twice as massive as the Andromeda Galaxy, and contains perhaps a trillion solar masses; this is at least twice as massive as had previously been calculated. If we make the standard assumption for this purpose that an average star has 1 solar mass, then our galaxy contains about a trillion stars.

25.4 THE FORMATION OF STARS

Giant molecular clouds are 150 to 300 light years across. There are a few thousand of them in our galaxy. The largest giant molecular clouds contain 100,000 to 1,000,000 times the mass of the sun. Their internal densities are about 100 times that of the interstellar medium around them. Since giant molecular clouds break up to form stars, they only last 10 million to 100 million years.

Most radio spectral lines seem to come only from the molecular clouds. (Carbon monoxide is the major exception, for it is widely distributed across the sky.) Infrared and radio observations together have provided us with an understanding of how stars are formed from these dense regions of gas and dust.

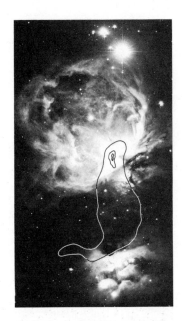

FIGURE 25–9 The contours show the molecular cloud associated with the Orion Nebula. The molecular cloud is actually on the far side of the glowing gas of the nebula, but the radio waves from the molecules penetrate the nebula. The contours of radio emission correspond roughly to regions of different density. The shape of the outermost contour indicates that the lower H II region may be expanding against the molecular cloud.

Carbon monoxide observations reveal the giant molecular clouds, but it is molecular hydrogen (H_2) rather than carbon monoxide that is significant in terms of mass. It is difficult to detect the molecular hydrogen directly, though there is over 100 times more molecular hydrogen than dust. So we must satisfy ourselves with observing the tracer, carbon monoxide.

Sometimes we see smaller objects, called *globules*. Some of the smaller globules are visible in silhouette against H II regions, as in Color Plate 61 (top right of picture, for example). Others are larger, and appear isolated against the stellar background (Fig. 25–8). It has been suggested that the globules may be on their way to becoming stars. IRAS discovered infrared radiation from at least some that indicate the presence of star formation inside (as we saw in the Color Essay on IRAS).

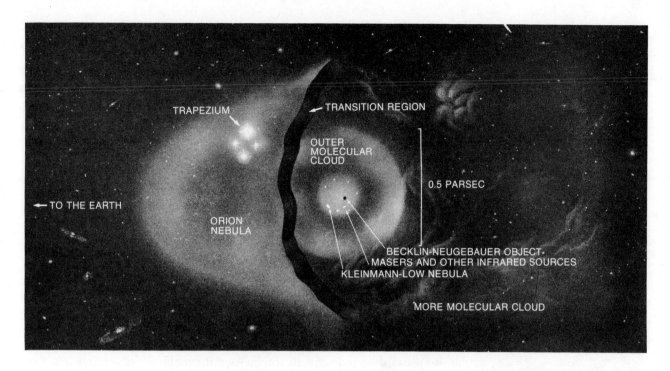

FIGURE 25–10 The structure of the Orion Nebula and the Orion Molecular Cloud, proposed by Ben Zuckerman, now of UCLA.

25.4a A Case Study: The Orion Molecular Cloud

Many radio spectral lines have been detected only in a particular cloud of gas located in the constellation Orion, not very far from the main Orion Nebula—the Orion Molecular Cloud (Fig. 25–9). It contains about 500 times the mass of the sun. It is relatively accessible to our study because it is only about 1500 light years from us. Even though less than 1 percent of the Cloud's mass is dust, that is still a sufficient amount of dust to prevent ultra-violet light from nearby stars from entering and breaking the molecules apart. Thus molecules can accumulate.

We know that young stars are found in this region—the Trapezium, a group of four hot stars readily visible in a small telescope, is the source of ionization and of energy for the Orion Nebula. The Trapezium stars are relatively young, about 100,000 years old. The Orion Nebula (Color Plate 56), prominent as it is in the visible, is an H II region located along the near side of the molecular cloud (Fig. 25–10).

The properties of the molecular cloud can be deduced by comparing the radiation from its various molecules and by studying the radiation from each molecule individually. The density increases toward the center. The cloud may actually be as dense as a million particles/cm^3 at its center. This is still billions of times less dense than our earth's atmosphere, though it is substantially denser than the average interstellar density of about 1 particle per cm^3.

Another example of star formation in a molecular cloud is M17 (Color Plate 62). The relatively sharp edge at right marks the left edge of the molecular cloud, which contains both infrared objects and tiny radio sources. M8 (Color Plate 64) is also an example.

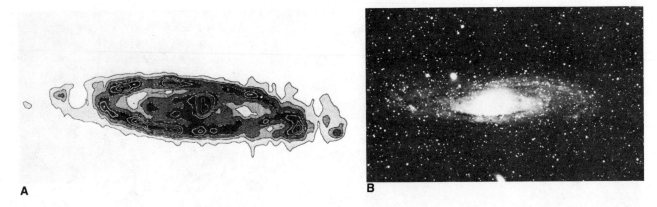

A   B

FIGURE 25–11   (A) The Andromeda Galaxy (Color Plate 68) observed at the long infrared wavelength of 100 microns from IRAS. (By comparison, red light is 0.65 microns.) The darkest regions in the image, corresponding to the strongest radiation, are regions where there are young stars. (B) An optical view, observed with an amateur (20-cm) telescope, for comparison.

25.4b IRAS and Star Formation

Not only has IRAS discovered stars forming in globules in our own galaxy but also it has detected signs of similar star formation in other galaxies. For example, its scan of the Andromeda galaxy (Fig. 25–11) showed a ring of bright infrared emission as well as a bright infrared region near the galaxy's nucleus that indicate the recent formation of stars in those locations. We don't yet know why stars should form in such a ring.

IRAS observations of the Tarantula Nebula, which glows brightly in reddish hydrogen light in the Large Magellanic Cloud (Color Plate 67), showed many stars in formation. Each infrared wavelength observed tended to show a different stage of evolution (Fig. 25–12).

25.5   AT A RADIO OBSERVATORY

What is it like to go observing at a radio telescope? First, you decide just what you want to observe, and why. You have probably worked in the field before, and your reasons might tie in with other investigations under way. Then you decide at which telescope you want to observe; let us say it is one of the telescopes of the National Radio Astronomy Observatory. You first send in a proposal to the NRAO, describing what you want to observe and why. You also list recent articles you have published, providing your track record. After all, even the best observations don't help anyone if you don't study your data and report your results.

Your proposal will be read by a panel of scientists. If the proposal is approved, it is placed in a queue waiting for observing time. You might be scheduled to observe for a five-day period to begin six months after you have submitted your proposal.

At the same time you might apply (usually to the National Science Foundation) for support to carry out the research. Your proposal possibly contains requests for some support for yourself, perhaps for a summer, and support for a student or students to work on the project with you. It might also contain requests for funds for computer time at your home institution, some travel support, and funds to support the eventual study of the data

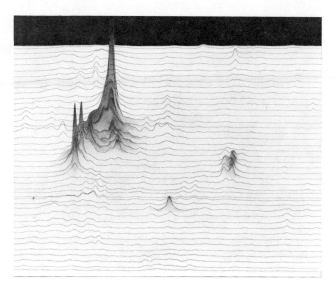

FIGURE 25–12 An IRAS scan of part of the Large Magellanic Cloud, including the Tarantula Nebula. The longest infrared wavelengths observed show the molecular clouds about to collapse, the medium wavelengths show the matter collapsing and becoming denser, and the shortest infrared wavelengths show the newly formed stars.

and the publication of the research. The NRAO does not charge you directly for the use of the telescope itself—that cost is covered in the observatory's overall budget.

The NRAO has telescopes at Green Bank in West Virginia, for study of wavelengths longer than about 1 cm, at Kitt Peak in Arizona, for study of millimeter wavelengths (Fig. 25–13), and an array of telescopes (the Very

FIGURE 25–13 The 12-m telescope of the National Radio Astronomy Observatory on Kitt Peak in Arizona, used for observation at millimeter wavelengths. Much of its time has been spent studying interstellar molecules.

FIGURE 25–14 Several of the 27 telescopes that together make up the VLA, located west of Socorro, New Mexico.

Unlike the optical sky, which is bright (and blue) in the daytime, the radio-sky background at long wavelengths remains dark even when the sun is up. As long as they don't point their telescopes to within a few degrees of the sun, radio astronomers working at long enough wavelengths can observe anywhere in the sky at any hour of the day or night.

FIGURE 25–15 The taking of spectral-line data is controlled by a scientist or observing assistant at a keyboard linked to a computer. A screen on which the spectra can be continually seen is at top left. The view here is inside the control room of the 64-m telescope at Parkes, Australia.

Large Array, Fig. 25–14) west of Socorro, New Mexico, for making high-resolution images. All the observing locations have dormitories and dining halls so that astronomers can stay near the telescopes.

The astronomers sit at a computer console monitoring the data as they come in from the telescope or array of telescopes (Fig. 25–15). A trained telescope operator actually runs the mechanical aspects of the telescope. You give the telescope operator the coordinates of the point in the sky that you want to observe, and the telescope is pointed for you. When you have finished observing one source, perhaps because it has set (gone below the horizon), you ask the operator to point the telescope to another source, and off you go again. The astronomers arrange their own schedules so that they can observe around the clock—one doesn't want to waste any observing time. This is unlike optical astronomers who can work all night and sleep for part of the day.

The electronics that are used to treat the incoming signal collected by the radio dish or dishes are particularly advanced. A computer can display the incoming data on a video screen. You can even use the computer to manipulate the data a bit, perhaps adding together the results from different five-minute exposures that you have made. At the Very Large Array, huge computers combine the output from two dozen telescopes and show you a color-coded image, with each color corresponding to a different brightness level (Color Plate 52). The data are stored on magnetic computer tape. You take home with you both copies made on paper or film and the computer tapes themselves for further study.

At the end of your observing run, you take the data back to your home institution to complete the analysis. You are expected to publish the results as soon as possible in one of the scientific journals, probably after you have given a paper about the results at a professional meeting, such as one of those held by the American Astronomical Society.

KEY WORDS

interstellar medium, H I region, H II region, spin-flip, 21-cm line, giant molecular cloud, globule

QUESTIONS

1. List two relative advantages and disadvantages of radio astronomy and optical astronomy.

2. Describe the relation of hot stars to H I and H II regions.

3. What determines whether the 21-cm lines will be observed in emission or absorption?

4. Describe how a spin-flip transition can lead to a spectral line, using hydrogen as an example. Could deuterium also have a spin-flip line?

5. Why are dust grains important for the formation of interstellar molecules?

6. Which molecule is found in the most locations in interstellar space?

7. Explain what property of young stars and/or the region of space nearby was observed with the IRAS spacecraft.

8. Describe the relation of the Orion Nebula and the Orion Molecular Cloud.

9. How do we detect star formation in another galaxy? Describe an example.

10. Optical astronomers can observe only at night. In what time period can radio astronomers observe? Why?

TOPICS FOR DISCUSSION

1. What does the discovery of fairly complex molecules in space imply to you about the existence of extraterrestrial life?

2. What does the formation of stars like the sun so commonly in our galaxy imply to you about the existence of planets in our galaxy?

PART VI Galaxies and Beyond

The individual stars that we see with the naked eye are all part of the Milky Way Galaxy, discussed in the preceding two chapters. But we cannot be so categorical about the conglomerations of gas and stars that can be seen through telescopes. Once they were all called "nebulae," but we now restrict the meaning of this word to gas and dust in our own galaxy. Some of the objects that were originally classed as nebulae turned out to be huge collections of gas, dust, and stars located far from our Milky Way Galaxy and of a scale comparable to that of our galaxy. These objects are galaxies in their own right, and are both fundamental units of the universe and the stepping stones that we use to extend our knowledge to tremendous distances.

In the 1770's, a French astronomer named Charles Messier was interested in discovering comets. To do so, he had to be able to recognize whenever a new fuzzy object appeared in the sky. He thus compiled a list of about 100 diffuse objects that could always be seen. To this day, these objects are commonly known by their *Messier numbers.* Messier's list contains the majority of the most beautiful objects in the sky, including nebulae, star clusters, and galaxies.

Soon after, William Herschel, in England, compiled a list of 1000 nebulae and clusters, which he expanded in subsequent years to include 2500 objects. Herschel's son John continued the work, incorporating observations made in the southern hemisphere. In 1864, he published the *General Cata-*

Color Plate 66 (top): The Whirlpool Galaxy, M51, in Canes Venatici, a type Sc spiral galaxy. At the end of one of its arms, a companion galaxy, NGC 5195, appears. (Canada-France-Hawaii Telescope)

Color Plate 67 (bottom): The Large Magellanic Cloud, a satellite galaxy of our own Milky Way Galaxy. It is best visible from the southern hemisphere. (Photo by Hans Vehrenberg)

Color Plate 68: The Andromeda Galaxy, also known as M31 and NGC 224, the nearest spiral galaxy to the Milky Way. It is accompanied by two elliptical galaxies, NGC 205 (below) and M32 (above). M31 is Hubble type Sb. (Palomar Observatory, California Institute of Technology photo with the 1.2-m Schmidt camera.)

Color Plate 69: NGC 2997, a spiral galaxy (Sc) in Antlia. (© 1980 Anglo-Australian Telescope Board)

Color Plate 70 (bottom): Centaurus A, a powerful radio source whose optical image, NGC5128, shows an elliptical galaxy around which a heavy zone of dust appears to be wrapped. (Cerro Tololo Inter-American Observatory with the 4-m telescope)

Color Plage 71: The spiral galaxy NGC 253 (Sc) with its light-absorbing dust lanes. (© 1980 Anglo-Australian Telescope Board)

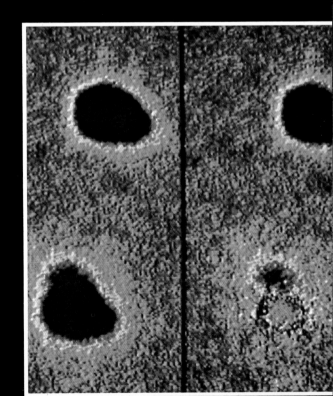

logue of Nebulae. In 1888, J. L. E. Dreyer published a still more extensive catalogue, *A New General Catalogue of Nebulae and Clusters of Stars,* the *NGC,* and later published two supplementary *Index Catalogues, IC's.*

When larger telescopes were turned to the Messier objects, especially by Lord Rosse in Ireland in about 1850, some of the objects showed traces of spiral structure, like pinwheels. They were called "spiral nebulae." But where were they located? Were they close by or relatively far away?

When such telescopes as the 0.9-m reflector at Lick in 1898, and later the 1.5-m and 2.5-m reflectors on Mount Wilson, began to photograph the "spiral nebulae," they revealed many more of them. The shapes and motions of these "nebulae" were carefully studied. Some scientists thought that they were merely in our own galaxy, while others thought that they were very far away, "island universes" in their own right, so far away that the individual stars appeared blurred together. (The name "island universes" had originated with the philosopher Immanuel Kant in 1755.)

The debate raged, and an actual debate on the scale of our galaxy and the nature of the "spiral nebulae" was held on April 16, 1920, as an after-dinner event of the National Academy of Sciences. Harlow Shapley (pronounced to rhyme with "map lee") argued that the Milky Way Galaxy was larger than had been thought, and thus implied that it could contain the spiral nebulae. Heber Curtis argued for the independence of the "spiral nebulae" from our galaxy. This famous Shapley-Curtis debate is an interesting example of the scientific process at work. (It has recently been pointed out that most reports of the Shapley-Curtis debate are based on the published transcript, while the actual words spoken on that evening were less thorough. Shapley, for one, had prepared a low-level introductory talk for this audience.)

Shapley's research on globular clusters had led him to correctly assess our own galaxy's large size. But he also argued that the "spiral nebulae" were close by because proper motion had been detected in some of them by another astronomer. These observations were subsequently shown to be incorrect. He also reasoned that an apparent nova in the Andromeda Galaxy, S Andromedae, had to be close by or it couldn't have been as bright; nobody knew about supernovae then. Curtis's conclusion that the "spiral nebulae" were external to our galaxy was based in large part on an incorrect notion of our galaxy's size. He treated S Andromedae as an anomaly and considered only "normal" novae.

So Curtis's conclusion that the "spiral nebulae" were comparable to our own galaxy was correct, but for the wrong reasons. Shapley, on the other hand, came to the wrong conclusion but followed a proper line of argument that was unfortunately based on incorrect and inadequate data.

The matter was settled in 1924, when observations made at the Mount Wilson Observatory by Edwin Hubble proved that there were indeed other galaxies in the universe besides our own. In fact, we think of galaxies and clusters of galaxies as fundamental units in the universe. The galaxies are among the most distant objects we can study. Many quasars are even farther away, and turn out to be certain types of galaxies seen at special times in the history of the universe.

Galaxies and quasars can be studied in most parts of the spectrum. Radio astronomy, in particular, has long proved a fruitful method of study. The study across the spectrum of galaxies and quasars provides tests of physical laws at the extremes of their applications and links us to cosmological consideration of the universe on the largest scale.

Chapter 26

An edge-on view of a galaxy of type Sb, NGC 4565, in the constellation Coma Berenices. Its nuclear bulge and dust lane show clearly. Several other galaxies also appear on this picture, though they are much farther away and thus much smaller. The tiny irregular objects about 2 cm upward from the center of NGC 4565, for example, are distant galaxies.

GALAXIES: BUILDING BLOCKS OF THE UNIVERSE

Since Hubble showed us sixty years ago that our Milky Way Galaxy was only one of many, our view of the universe has changed. Our place in the universe may be humble, but we must still do the best that we can with what we have.

26.1 TYPES OF GALAXIES

Spiral galaxies are in the minority; many galaxies have elliptical shapes and others are irregular or abnormal in appearance. In 1925, Hubble set up a system of classification of galaxies; we still use a modified form of this system today.

26.1a Elliptical Galaxies

Most galaxies are elliptical in shape (Fig. 26–1). Some are giant ellipticals, over 10 times more massive than our own galaxy; they are rare. Much more common are dwarf ellipticals, which contain "only" a few million solar masses, a few per cent of the mass of our own galaxy.

Elliptical galaxies range from nearly circular in shape, which Hubble called *type E0,* to very elongated, which Hubble called *type E7.* The spiral Andromeda Galaxy, M31 (Color Plate 68), is accompanied by two elliptical companions of types E2 (for the galaxy closer to M31) and E5, respectively. It is obvious on the photograph that the companions are much smaller than M31 itself. We assign types based on the optical appearance of a galaxy rather than how elliptical it actually is. After all, we can't change our point of view for such a far-off object. But even a very elliptical galaxy will appear round when seen end on.

FIGURE 26–1 M87 (NGC 4486), a galaxy of Hubble type E0(pec)—that is, E0 with a peculiarity—in the constellation Virgo. The jet that makes M87 a peculiar elliptical galaxy doesn't show in this view. Globular clusters can be seen in the outer regions.

26.1b Spiral Galaxies

Although spiral galaxies are a minority of all galaxies, they form a majority in certain groups of galaxies. Also, since they are brighter than the more abundant small ellipticals, we tend to see the spirals even though the ellipticals may make up the majority.

Sometimes the arms are tightly wound around the nucleus; Hubble called this type *Sa,* the S standing for "spiral" (Fig. 26–2). Spirals with their arms less and less tightly wound (that is, looser and looser) are called *type Sb* (Color Plate 72) and *type Sc* (Color Plate 66). The nuclear bulge as seen from edge-on is less and less prominent as we go from Sa to Sc. On the other hand, the dust lane—obscuring dust in the disk of the galaxy—becomes more prominent. Doppler shifts indicate that galaxies rotate in the sense that the arms trail.

NGC 1201   Type S0   NGC 2811   Type Sa   NGC 2841   Type Sb   NGC 628 M74   Type Sc

FIGURE 26–2 Normal spiral galaxies.

Spiral galaxies contain a billion to over a trillion solar masses. Since most stars are of less than 1 solar mass, this means that spirals contain a billion to over a trillion stars.

In about one-third of the spirals, the arms unwind not from the nucleus but rather from a straight *bar* of stars, gas, and dust that extends to both sides of the nucleus. These are similarly classified in the Hubble scheme from *a* to *c* in order of increasing openness of the arms, but with a *B* for "barred" inserted: *SBa, SBb,* and *SBc*. The more open the arms, the more interstellar gas and dust in the spiral galaxy.

Subsequently to Hubble's earliest work, it became clear that there was a transitional type between spirals and ellipticals. Type *S0* has an elongated disk but no arms.

Gerard de Vaucouleurs of the University of Texas has expanded Hubble's classification scheme. De Vaucouleurs has defined a class Sd with extremely open arms, and has further subdivided barred spirals. Those whose arms come off a ring are given an "(r)" while those whose arms come off the ends of the central bar are given an "(s)." There is actually a continuous range of intermediate types from ordinary spiral galaxies (called SA by de Vaucouleurs, parallel notation to SB) to barred spirals (Fig. 26–3), so "normal" spirals and barred spirals may not really be distinct types.

26.1c Irregular and Peculiar Galaxies

A few per cent of galaxies show no regularity. The Magellanic Clouds, for example, are basically irregular galaxies (Color Plate 65). Sometimes traces

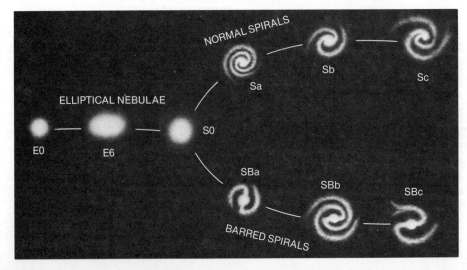

FIGURE 26–3A The Hubble tuning-fork classification of galaxies.

FIGURE 26–3*B* This set of photographs by Gerard de Vaucouleurs of regular spirals (SA) and barred-spiral galaxies (SB) shows that there is no clear demarcation between those classes. We see here arms typical of subclass "b," as in SAb (formerly Sb) and SBb.

of regularity—perhaps a bar—can be seen. De Vaucouleurs' classification scheme defines class Sm = Magellanic spiral class beyond Sd, with the Large Magellanic Cloud (see also Color Plate 67) as the prototype. Since the LMC has a bar, it is class SBm.

Some cases (Fig. 26–4) appear regular but with some peculiarity. We know of many such *peculiar galaxies*.

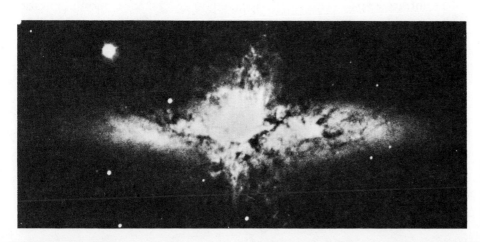

FIGURE 26–4 M82, a most unusual galaxy that is a powerful source of radio radiation. It was once thought to be exploding, but then gentler processes were thought to cause its form.

FIGURE 26–5 Many galaxies of all types are visible in this photo of part of a cluster of galaxies, Abell 1060. A bright star is at upper left.

The Andromeda Galaxy at 2 million light-years and perhaps M33 at 4 million light-years are the farthest objects you can see with your unaided eye; they appear as fuzzy blobs in the sky if you know where to look.

26.1d IRAS Observations of Galaxies

1983's infrared spacecraft IRAS observed many galaxies, mostly spirals. Since the infrared radiation observed presumably comes from dust clouds heated by young stars, the infrared brightness should tell us how fast stars are being created. About half the energy emitted by our own Milky Way Galaxy is in the infrared, which indicates that a lot of stars are being formed. But we don't know why the Andromeda Galaxy, which in optical radiation resembles the Milky Way, emits only 3 per cent of its energy in the infrared.

Some galaxies give out 10 or more times as much infrared as optical energy, which indicates that they are creating stars especially rapidly.

Some bright sources discovered by IRAS don't correspond to anything visible. Some of them could be distant galaxies that emit many times more infrared than optical radiation, but are too far away for us to see them optically. The Hubble Space Telescope and the next generation of infrared satellites may answer these questions.

26.2 CLUSTERS OF GALAXIES

Most galaxies are part of groups or clusters. Groups have just a handful of members, while *clusters of galaxies* (Fig. 26–5) may have hundreds or thousands.

The two dozen or so galaxies nearest us form the *Local Group*. It contains three spiral galaxies, four irregular galaxies, at least a dozen dwarf irregulars, and the remainder ellipticals (four regular ellipticals, the others dwarf ellipticals).

The nearest cluster of many galaxies can be observed in the constellation Virgo and surrounding regions of the sky; it is called the Virgo Cluster. It covers a region in the sky over 6° in radius, 12 times greater than the angular diameter of the moon. The Virgo Cluster contains hundreds of galaxies of all types. It is about 6 million light years across, and is about 60 million light years away from us.

Rich clusters of galaxies (clusters containing many galaxies) are generally x-ray sources. Studies with the Einstein Observatory have revealed a hot intergalactic gas—10 to 100 million K—containing as much mass as is in the galaxies themselves. The gas is clumped in some clusters while in others it is spread out more smoothly with a concentration near the center. This may be an evolutionary effect, with gas being ejected from individual galaxies in younger clusters and spreading out as the clusters age.

Every nearby very rich cluster is apparently located in a cluster of clusters, a *supercluster*. The Local Group, the several similar groupings nearby, and the Virgo Cluster form the *Local Supercluster*. This cluster of clusters contains 100 member clusters roughly in a pancake shape, on the order of 100 million light years across and 10 million light years thick. Superclusters are apparently separated by giant voids (Fig. 26–6).

Does the clustering continue in scope? Are there clusters of clusters of clusters, and clusters of clusters of clusters of clusters, and so on? The evidence at present is that this is not so.

26.3 THE EXPANSION OF THE UNIVERSE

In 1929, Hubble announced that galaxies in all directions are moving away from us. Further, he found that the distance of a galaxy from us is directly proportional to its redshift (that is, when the redshift we observe is greater by a certain factor, the distance is greater by the same factor; Fig. 26–7). The proportionality between redshift and distance is known as *Hubble's law*. The redshift is presumably caused by the Doppler effect (Section 16.4). The law is usually stated in terms of the velocity that corresponds (by the Doppler

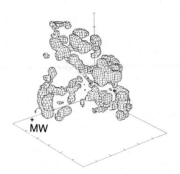

FIGURE 26–6 A computer plot of the distribution of galaxies in space, plotted on the basis of many radial velocity measurements made with new electronic techniques. Long chains of galaxies, showing their grouping into clusters, are clearly visible in this three-dimensional display. Tick marks show intervals of 50 million light-years. MW marks the position of our Milky Way Galaxy. (Courtesy of Frenk and White, and the CfA survey)

Following the style of the late George Gamow, I can thus write my address as:
Jay M. Pasachoff
Williamstown
Berkshire County
Massachusetts
United States of America
North America
Earth
Solar System
Milky Way Galaxy
Local Group
Local Supercluster
Universe

FIGURE 26–7 (A) Hubble's original diagram from 1929. Dots are individual galaxies; open circles are from groups of galaxies. The scatter to one side of the line or the other is substantial. (B) By 1931, Hubble and Milton Humason had extended the measurements to greater distances, and Hubble's law was well established. All the points shown in the 1929 work appear bunched near the origin of this graph. $v = H_0 d$ represents a straight line of slope H_0. These graphs use older distance measurements than we now use, and so give different values for H_0 than we now derive.

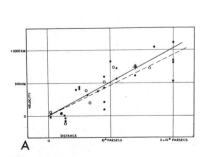

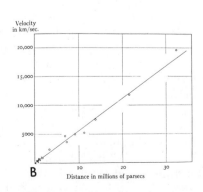

FIGURE 26–8 Spectra are shown at right for the galaxies at left, all reproduced to the same scale. Distances are based on Hubble's constant = 50 km/sec/mega parsec = 15 km/sec/million light years (one par sec = 3.26 light years). Notice how the farther away a galaxy is, the smaller it looks. The arrow below each horizontal streak of spectrum shows how far the strongly absorbing lines of ionized calcium, known as H and K, are redshifted. The spectrum of an emission-line source located inside the telescope building appears as vertical lines above and below each galactic spectrum to provide a comparison with a redshift known to be zero.

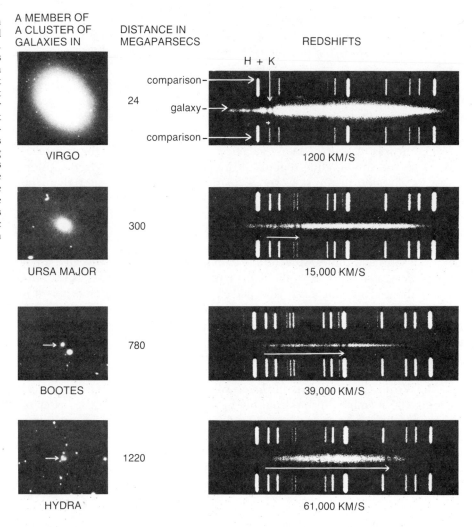

A MEMBER OF A CLUSTER OF GALAXIES IN — DISTANCE IN MEGAPARSECS — REDSHIFTS

VIRGO — 24 — 1200 KM/S

URSA MAJOR — 300 — 15,000 KM/S

BOOTES — 780 — 39,000 KM/S

HYDRA — 1220 — 61,000 KM/S

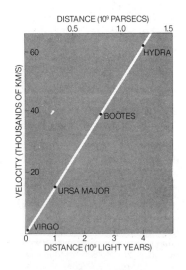

FIGURE 26–9 The Hubble diagram for the galaxies shown in Fig. 26–8.

effect) to the measured wavelength, rather than in terms of the redshift itself.

Hubble's law states that the velocity of recession of a galaxy is proportional to its distance. It is written

$$v = H_0 d,$$

where v is the velocity, d is the distance, and H_0 is the present-day value of the constant of proportionality (simply, the constant factor by which you multiply d to get v), which is known as *Hubble's constant* (Figs. 26–8 and 26–9).

One of the greatest debates going on in astronomy now is the value of Hubble's constant. One group of scientists measures 50 km/sec/megaparsec (where a parsec is about 3.3 light years), while some other groups of scientists measure 100 km/sec/megaparsec. Each group is careful about its work, and the debate is hard and sharp. But all this work involves finding the distances to some of the farthest objects in the universe, and the uncertainties are very great.

We cannot find distances by triangulation even to the edges of our galaxy. The best method we have for finding the distances to nearby galaxies is to study Cepheid variable stars in them. For more distant galaxies, a new method makes use of a link between a galaxy's brightness in the infrared and the speed at which it rotates, which we can measure with a radio telescope. For the most distant galaxies, we often make the plausible but inexact assumption that the brightest galaxy in a cluster always has the same intrinsic brightness or the same size. All methods except triangulation depend on a galaxy's intrinsic brightness. By comparing how bright it looks with how bright it really is, we can tell how far away it is. Much effort and observing time on the largest telescopes is now being devoted to applying these and other methods to assessing the "distance scale." Often, astronomers determine the distance to a faraway galaxy by measuring its redshift and applying Hubble's law; of course, they can't test Hubble's law itself in this way.

Hubble's constant is given in units that may appear strange, but they merely state that for each megaparsec (3.3×10^6 light years; mega, whose symbol is M, means million) of distance from the sun, the velocity increases by 50 or 100 km/sec. Thus the units are 50 or 100 km/sec per Mpc. (The values are equivalent to 15 km/sec/mega-light year or 30 km/sec/mega-light year, but astronomers never express them in this way.)

From Hubble's law for $H_0 = 50$, we see that a galaxy at 10 Mpc would have a redshift corresponding to 500 km/sec; at 20 Mpc the redshift of a galaxy would correspond to 1000 km/sec; and so on. The redshift for a given galaxy is the same no matter in which part of the spectrum we observe.

Note that if Hubble's constant is 100 instead of 50, then a galaxy whose redshift is measured to be 500 km/s would be (500 km/sec)/(100 km/sec/Mpc) = 5 Mpc, half the distance that was derived for the smaller Hubble's constant. So the debate over the size of Hubble's constant has broad effect on our notion of the size of the universe. The debate is often heated, and sessions of scientific meetings at which the subject is discussed are well attended. In the next few years, data from the aptly named Hubble Space Telescope may resolve some of the uncertainties.

26.3a The Expanding Universe

The major import of Hubble's law is that all but the closest galaxies in all directions are moving away from us; the universe is expanding. Since the time when Copernicus moved the earth out of the center of the universe (and the time when Shapley moved the earth and sun out of even the center of the Milky Way Galaxy), we have not liked to think that we could be at the center of the universe. Fortunately, Hubble's law can be accounted for without our having to be at any such favored location, as we see below.

Imagine a raisin cake (Fig. 26–10) about to go into the oven. The raisins are spaced a certain distance away from each other. Then, as the cake rises, the raisins spread apart from each other. If we were able to sit on any one

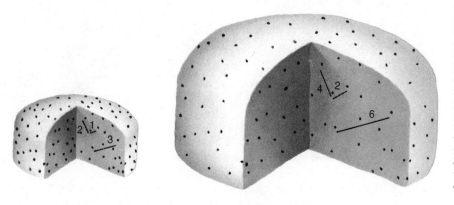

FIGURE 26–10 From every raisin in a raisin cake, every other raisin seems to be moving away from you at a speed that depends on its distance from you. This leads to a relation like the Hubble law between the velocity and the distance. Note also that each raisin would be at the center of the expansion measured from its own position, yet the cake is expanding uniformly. For a better analogy with the universe, consider an infinite cake; unlike the finite analogy pictured, even in a flat universe there is then no center to its expansion.

of those raisins, we would see our neighboring raisins move away from us at a certain speed. It is important to realize that raisins farther from us would be moving away faster, because there is more cake between them and us to expand. No matter in what direction we looked, the raisins would be receding from us, with the velocity of recession proportional to the distance.

The next important point to realize is that it doesn't matter which raisin we sit on; all the other raisins would always seem to be receding. Of course, any real raisin cake is finite in size, while the universe may have no limit so that we would never see an edge. The fact that all the galaxies appear to be receding from us does not put us in a unique spot in the universe; there is no center to the universe. Each observer at each location would observe the same effect.

26.4 ACTIVE GALAXIES

Most of the objects that we detect in the radio sky turn out not to be located in our galaxy. The study of these *extragalactic radio sources* is a major subject of this section.

The core of our galaxy, the radio source we call Sagittarius A, is one of the strongest radio sources that we can observe in our galaxy. But if the Milky Way Galaxy were at the distance of other galaxies, its radio emission would be very weak.

Some galaxies emit quite a lot of radio radiation, many orders of magnitude (that is, many powers of ten) more than "normal" galaxies. We shall use the term *radio galaxy* to mean these relatively powerful radio sources. They often appear optically as giant elliptical galaxies that show some peculiarity. Radio galaxies, and galaxies that similarly radiate much more strongly in x-rays than normal galaxies, are called *active galaxies*.

The first radio galaxy to be detected, Cygnus A (Fig. 26–11), radiates about a million times more energy in the radio region of the spectrum than does the Milky Way Galaxy. Cygnus A, and dozens of other radio galaxies,

FIGURE 26–11 *(A)* A radio map of Cygnus A, with shading indicating the intensity of the radio emission. An electronic image of the faint optical object or objects observable is superimposed at the proper scale. *(B)* The especially high resolution on this new image of Cygnus A reveals a third, less prominent feature between the two main structures. This central region contains the central radio source. The resolution was improved over past images by using a computer-controlled monitor of the seeing and telescope guiding to form an image with 2/3-arc second resolution on a CCD at Mauna Kea. The overall structure again makes the source look like galaxies in collision. (Radio image from the VLA; optical image by Laird Thompson, U. Hawaii)

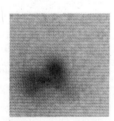

B

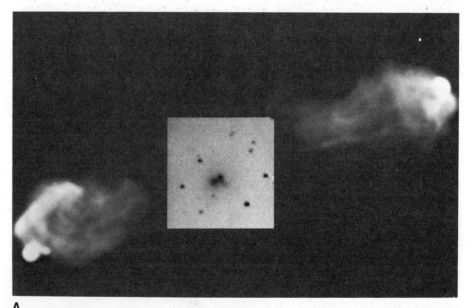

A

emit radio radiation mostly from two zones, called *lobes,* located far to either side of the optical object. Such *double-lobed structure* is typical of many radio galaxies.

The optical object that corresponds to Cygnus A—a fuzzy, divided blob or perhaps two fuzzy blobs—has been the subject of much analysis, but its makeup is not yet understood. Perhaps we see a single object partly obscured by dust.

Often the optical images that correspond to radio sources show peculiarities. For example, on short exposures of M87 (Fig. 26–12), which corresponds to the powerful radio galaxy Virgo A, we see (optically) a jet of gas. Light from the jet is polarized — the waves from a given point vibrate in the same plane instead of oriented at all angles. The high percentage of polarized light tells us that a strong magnetic field is present.

As our observational abilities in radio astronomy have increased, especially with the techniques described in the next section, lobes and jets aligned with them have become commonly known. The best current model is that a giant rotating black hole in the center of a radio galaxy is accreting matter. Twin jets carrying matter at a high velocity are given off almost continuously along the poles of rotation. These jets carry energy into the lobes. (We may see only one, depending on the alignment and Doppler shifts.) Calculations show that a not-very-hungry giant black hole would provide the right amount of energy to keep the lobes shining.

26.5 RADIO INTERFEROMETRY

The resolution of single radio telescopes is very low, because of the long wavelength of radio radiation. Single radio telescopes may be able to resolve structure only a few minutes of arc or even a degree or so (twice the diameter of the moon) across. But arrays of radio telescopes can now map the sky with resolutions far higher than the 1 arc sec or so (1/60 of an arc minute

At first, the optical object in Cygnus A was thought to be two galaxies in collision. This idea was discarded when it was realized that collisions between galaxies would not be frequent enough to account for the many similar radio sources that had by then been discovered. One distinguished astronomer lost a bet of a bottle of whiskey to a similarly distinguished astronomer on the subject. But now there is evidence that the material that fuels the giant black holes at the centers of radio galaxies may come from collisions of galaxies. The suggestion has thus been made that a bottle of whiskey might be returned (posthumously). And computer processing of a photograph to improve resolution has made it look even more as though Cygnus A is galaxies in collision after all.

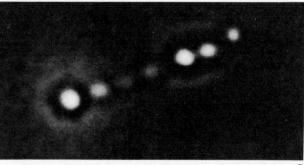

FIGURE 26–12 *(A)* The galaxy M87, which corresponds to the radio source Virgo A, as photographed with the Kitt Peak 4-m telescope. Two small background galaxies appear at the "5 o'clock" position, falsely mimicking a jet of gas. The inset shows a shorter exposure reproduced to the same scale, on which a real jet of gas is visible. *(B)* Computer enhancement of a photograph taken with the Palomar 5-m telescope brings out details of the jet. North is at top and East is at left in all photographs.

BOX 26.1
The Peculiar Galaxy
M87

The radio galaxy M87 is a giant elliptical galaxy, one of the brighter members of the Virgo Cluster. The galaxy turns out to have an odd optical appearance on short exposures, for a jet of gas can be seen. The galaxy corresponds to the powerful radio galaxy Virgo A. A radio jet 6000 light-years long has been discovered.

High-resolution radio observations have shown that the nucleus of M87 is tiny. It is a very small source in which to generate so much energy, which is emitted across the spectrum from x-rays to radio waves. Indeed, this galaxy gives off much more energy than do other galaxies of its type; the central region is as bright as a hundred million suns.

Optical studies of the motion of matter circling M87's nucleus have allowed astronomers to estimate the mass of the nucleus. The stars are moving so fast that a huge mass must be present to hold them in. It turns out that 5 billion solar masses of matter must be there. Other optical studies disclosed the presence of an extremely bright point of light in the center of the galaxy. Astronomers continue to gather spectral and spatial information to determine if the source is a giant black hole or is matter under more conventional conditions. Higher-resolution observations from Space Telescope should tell us more about this exotic source.

and 1/1800 the diameter of the moon) that we can get with optical telescopes (Fig. 26–13).

The resolution of a single-dish radio telescope at a given frequency depends on the diameter of the telescope. (A single reflecting surface of a radio telescope is known as a "dish.") If we could somehow retain only the outer zone of the dish (Fig. 26–14), the resolution would remain the same. (The area collecting incoming radiation would be decreased, though, so we would have to collect the signal for a longer time to get the same intensity.)

Let us picture radiation from a distant source as coming in wavefronts, with the peaks of the waves in step (Fig. 26–15). If we can maintain our knowledge of the relative arrival times of the wavefront at each of two dishes, we can retain the same resolution as though we had one large dish whose diameter is equal to the spacing of the two small dishes shown. For a single dish, the maximum spacing of the two most distant points from which we can detect radiation is the "diameter." For an array, we call it the *baseline*.

We study the signals by adding together the signals from the two dishes; we say the signals "interfere," hence the device is an "interferometer." Since the delay in arrival time of a wavefront at the two dishes depends on the

FIGURE 26–13 The Andromeda Galaxy as it would appear at different resolutions.

1 ARC SEC

1 ARC MIN

10 ARC MIN

30 ARC MIN

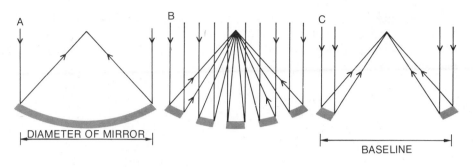

FIGURE 26–14 A large single mirror *(A)* can be thought of as a set of smaller mirrors *(B)*. Since the resolution for radiation of a certain wavelength depends only on the telescope's aperture, retaining only the outermost segments *(C)* matches the resolution of a full-aperture mirror. We can use a property of radiation called *interference* to analyze the incoming radiation. The device is then called an *interferometer*.

angular position of an object in the sky with respect to the baseline, by studying the time delay we can figure out angular information about the object.

If the source were made of two points close together, the wavefronts from the two sources would be at slight angles to each other, and the time interval between the source reaching the two dishes would be slightly different. Thus an interferometer can tell if an object is double, even if it is unresolved by each of the dishes used alone.

The first radio interferometers were separated by hundreds or thousands of meters. The signals were sent over wires to a central collecting location, where the signals were combined. The breakthrough in timekeeping came with the invention of atomic clocks, which drift only three hundred-billionths of a second in a year. A time signal from an atomic clock can be recorded by a tape recorder on one channel of the tape, while the celestial radio signal is recorded on an adjacent tape channel. The radio signal recorded can be compared at any later time with the signal from the other dish, synchronized accurately through comparison of the clock signals.

26.5a Very-Long-Baseline Interferometry

The ability to record the time so accurately freed radio astronomers of the need to have the dishes in direct contact with each other during the period of observation. Now all that is necessary is that the two telescopes observe the same object at the same period of time; the comparison of the signals

FIGURE 26–15 In the left half of the figure, a given wave peak reaches both dishes simultaneously, so the amplitudes (heights) of the waves add. In the right half, the wave peak reaches one dish while a minimum of the wave reaches the other; the amplitudes subtract and zero total intensity results. Thus "interference" results.

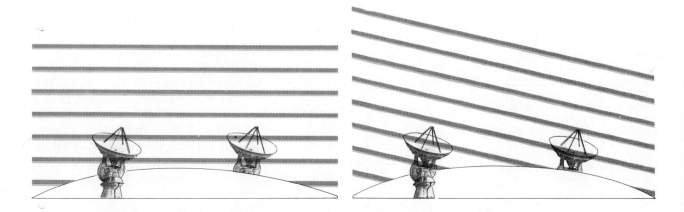

FIGURE 26–16 VLBI techniques with a baseline that is the diameter of the earth allow radio sources to be studied with extremely high resolution.

can take place in a computer weeks later. With this ability, astronomers can make up an interferometer of two or more dishes very far apart, even thousands of kilometers (Fig. 26–16). This technique is called *very-long-baseline interferometry (VLBI)*.

VLBI techniques can be applied only to very small areas of sky. But for those few areas, chosen for their special interest, our knowledge of the structure of radio sources has been fantastically improved. VLBI gives the best possible resolution.

VLBI techniques have provided an accurate measurement for the deflection of electromagnetic waves by the mass of the sun (Section 18.6), providing a confirmation of Einstein's general theory of relativity. The discovery of the extremely small sources at the nucleus of our galaxy and at the center of the galaxy M87 are also VLBI results.

The next stage in VLBI work is to set up permanent networks of radio telescopes devoted to this purpose. Britain's MERLIN (**M**ulti-**E**lement **Ra**dio-**L**inked **I**nterferometer **N**etwork) with a 64-km-maximum baseline is already in operation. Other plans are under way in the U.S., in Canada, in Australia, and elsewhere.

26.5b Aperture-Synthesis Techniques

By suitably arranging a set of radio telescopes across a landscape, one can simultaneously make measurements over a variety of baselines, because each pair of telescopes in the set has a different baseline from each other pair. With such an arrangement one can more rapidly map a radio source than one can with two-dish interferometers. Also, several dishes instead of just two gives that much more collecting area.

Since the resolution of an interferometer depends on the baseline, at any one time the resolution is quite good along the line in which the telescopes lie but is only the resolution of a single dish in the perpendicular direction. Fortunately, we can take advantage of the fact that the earth's rotation changes the orientation of the telescopes with respect to the stars (Fig. 26–17). Thus by observing over a 12-hour period, one can improve the resolution in all directions.

In effect, we have synthesized a large telescope that covers an elliptical area whose longest diameter is the same as the maximum separation of the outermost telescopes. Alternatively, one can synthesize a large telescope by distributing smaller telescopes so that the baselines between them are in different directions. This interferometric technique is known as *aperture synthesis;* it was used to make the radio image of Cygnus A we saw earlier in this chapter.

The most fantastic aperture-synthesis radio telescope has been constructed in New Mexico by the National Radio Astronomy Observatory. It is composed of 27 dishes, each 26 m in diameter, arranged in the shape of a "Y" over a flat area 27 km in diameter (Fig. 26–18 and Color Plate 7). Because the array is in the shape of a "Y" rather than a straight line, it does not need to wait for the earth to rotate in order to synthesize the aperture. The "Y" is marked by railroad tracks, on which the telescopes can be transported to 72 possible observing sites. The control room at the center of the

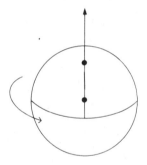

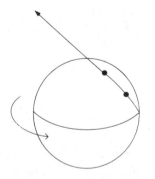

FIGURE 26–17 The earth's rotation carries around a straight line of dishes so that it maps out an ellipse in the sky. The figure shows the view of a line of dishes on the earth that we would see from the direction of the source being observed.

FIGURE 26–18 The VLA, west of Socorro, New Mexico, has 27 dishes operating together.

"Y" contains powerful computers to analyze the signals. The system is prosaically called the *Very Large Array (VLA)*.

The VLA can make pictures of a field of view a few minutes of arc across, with resolutions comparable to the 1 arc sec of optical observations from large telescopes. Though this resolution is lower than that obtainable for a limited number of sources with VLBI techniques, the ability to make pictures in such a short time is invaluable (Color Plate 52). Plans are now under way for a dedicated set of telescopes to make an array spanning the United States. This system, much larger than MERLIN, would be a *Very-Long-Baseline Array (VLBA)*.

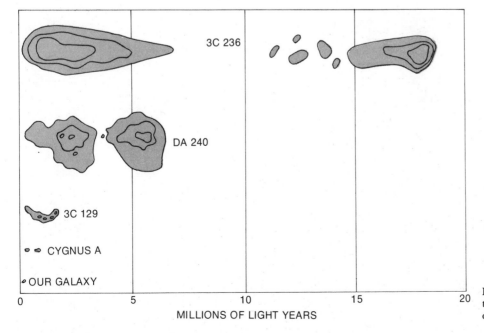

3C 236

DA 240

3C 129

CYGNUS A

OUR GALAXY

0   5   10   15   20

MILLIONS OF LIGHT YEARS

FIGURE 26–19 A comparison of the sizes of giant double-lobed radio galaxies with our own galaxy.

FIGURE 26–20 A Westerbork radio map of the head-tail radio source 3C 129, converted to a radio-photograph in which the brightness of the image corresponds to the intensity of radio emission at a wavelength of 21 cm (near but not at the hydrogen wavelength). The front end of the head of the radio galaxy corresponds to the position of a giant optical galaxy, a member of a rich cluster of galaxies.

26.5c Aperture-Synthesis Observations

At Westerbork in the Netherlands, twelve telescopes, each 25 m in diameter, are spaced over a 1.6-km baseline. Westerbork, an older aperture-synthesis system than the VLA, has discovered several giant double radio sources (Fig. 26–19), much larger than any of the double-lobed sources previously known. These are the largest single objects currently known in the universe.

Interferometer observations have revealed the existence of a class of galaxies that resemble tadpoles. They are called *head-tail galaxies* (Fig. 26–20). These galaxies expel the clouds of gas that we see as tails. They are double-lobed radio sources with the lobes bent back as the objects move through intergalactic space. High-resolution observations of one such galaxy with the VLA (Fig. 26–21) show that the source at the nucleus is less than 0.1 arc sec across, corresponding at the distance of this galaxy to a diameter of only 0.03 light years. Such observations are being used to understand both the galaxy itself and the intergalactic medium, and tie in with the x-ray observations of clusters of galaxies. Depending on the velocity of the galaxy and the density of the intergalactic medium, the lobes can be bent back by different amounts, so there is a range from lobes opposite each other to lobes slightly bent back to lobes bent back enough to make head-tail galaxies. Thus it makes sense that most head-tail galaxies are found in rich clusters of galaxies.

The discrete blobs that we can see in the tails indicate that the galaxies give off puffs of ionized gas every few million years as they chug through intergalactic space. Perhaps by studying these puffs, we can learn about the main galaxies themselves as they were at earlier stages in their lives. Head-tail galaxies seem to be a common although hitherto unknown type.

FIGURE 26–21 The head-tail galaxy NGC 1265 examined with the high resolution of the VLA at increasing resolution from left to right. The resolution of the close-up at the right is about 1 arc sec. The effective size of the telescope beam in each is shown by the ellipse in a lower corner.

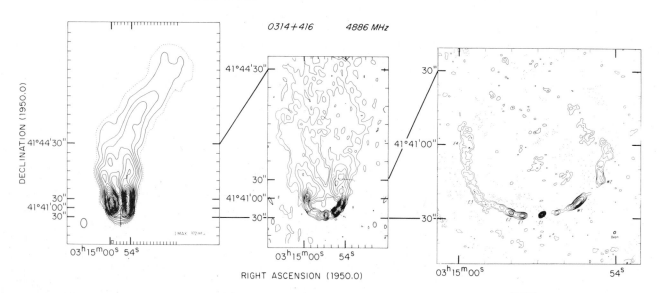

KEY WORDS

elliptical galaxies, bar, types E0, E7, Sa, Sb, Sc, SBa, SBb, SBc, S0, peculiar galaxy, cluster of galaxies, Local Group, supercluster, Local Supercluster, Hubble's law, Hubble's constant, extragalactic radio sources, radio galaxy, active galaxy, lobes, double-lobed structure, baseline, very-long-baseline interferometry, VLBI, aperture synthesis, Very Large Array, VLA, Very-Long-Baseline Array, VLBA, head-tail galaxies

QUESTIONS

1. What shape do most galaxies have?

2. Since we see only a 2-dimensional outline of an elliptical galaxy's shape, what relation does this outline have to the galaxy's actual 3-dimensional shape?

3. Sketch and compare the shapes of types Sb and SBb.

4. Draw side views of types E7, S0, Sa, Sb, and Sc, showing the extent of the nuclear bulge.

5. Which of Hubble's types of galaxies are the most likely to have new stars forming? What evidence supports this?

6. To measure the Hubble constant, you must have a means (other than the redshift) to determine the distances to galaxies. What are three methods that are used?

7. Discuss IRAS observations of galaxies.

8. Discuss the distribution of x-ray emission from rich clusters of galaxies, and why different distributions may exist.

9. At what velocity is a galaxy 3 million light years from us receding? Specify the value of Hubble's constant you use.

10. A galaxy is receding from us at a velocity of 1000 km/sec. (a) If you could travel at this rate, how long would it take you to travel from New York to California? (b) How far away is the galaxy from us? Specify the value of Hubble's constant you use.

11. (a) At what velocity in km/sec is a galaxy 100,000 parsecs away from us receding? (b) Express this velocity in km/hr, mi/sec, and mi/hr. Specify the value of Hubble's constant you use.

12. At what velocity in km/sec is a galaxy 1 million light-years away receding from us? Specify the value of Hubble's constant you use.

13. What comment can generally be made about the optical appearance of active galaxies?

14. Contrast VLA and VLBI.

15. Why does interferometry allow you to get finer detail than does a single-dish telescope?

Chapter 27

The quasi-stellar radio source 3C 273, taken with the 5-m Hale telescope of the Palomar Observatory. The object, the central white disk, was thought to look like any 13th-magnitude star except for the faint jet. Most recently, faint "fuzz" has been detected around 3C273.

QUASARS:
GIANT BLACK HOLES?

Quasars, and how they have been understood, have been one of the most exciting stories of the last twenty-five years of astronomy. First noticed as seemingly inconsequential stars, they turn out to be some of the most powerful objects in the universe, and represent violent forces at work. Our interest in them is further piqued because some of the quasars are the most distant objects we can detect in the universe. Since, as we look out, we are seeing light that was emitted farther and farther back in time, observing quasars is like using a time machine that enables us to see the universe's early years.

27.1 NOTICING QUASARS

When maps of the radio sky were first made, they turned out to look very different from maps of the optical sky. What were these radio objects? Most didn't seem to correspond to any optical object.

One problem in making identifications was that the radio positions were not precise. Only single-dish antennas were available, and very-long-baseline interferometry or even aperture synthesis was not yet invented. Some of the first interferometric measurements, using one image going directly into a single dish and another bouncing off the ocean into the same dish, started to locate a few radio sources precisely. But the most precise positions were available for the objects in the path traversed by the moon. When a radio source winked out, we knew that the moon had just covered it, and so that it was somewhere on a curved line marking the front edge of the moon. When it reappeared, we knew that the moon had just uncovered it, and so that it was somewhere on a curved line marking the moon's trailing edge. These two curves intersected at two points, so the radio source must be at one of those two points.

Many radio sources became known by their numbers in a catalogue compiled at the University of Cambridge in England. The position of the 273rd source in the third version of the catalogue—the 3C catalogue—was crossed by the moon a few times, and was thus pinpointed. But only a bluish optical object of 13th magnitude—600 times fainter than the naked eye limit and barely bright enough to be observed with medium-sized telescopes—appeared at that location.

On close inspection, the faint object appeared not completely starlike; a jet of gas seemed to be connected to the nucleus (as shown in the photograph opening this chapter). The source was not completely stellar (point-like) in appearance; it was only "quasi-stellar." The fact that the radio emission had two components, one corresponding to the jet and the other to the nucleus removed any doubt that the identification was correct.

Maarten Schmidt of Caltech photographed the spectrum of this "quasi-stellar radio source" with the 5-m telescope of the Palomar Observatory. The

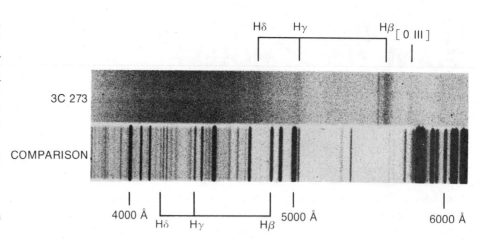

FIGURE 27–1 Spectrum of the quasar 3C 273. The lower spectrum, taken of a source in the telescope dome on earth, consists of spectral lines from hydrogen and helium; it establishes the scale of wavelength for comparison. The upper part is the spectrum of the quasar, an object of 13th magnitude. The hydrogen lines Hβ (H-beta), Hγ (H-gamma), and Hδ (H-delta) in the quasar spectrum are at longer wavelengths than in the comparison spectrum. The redshift of 16 percent corresponds, according to Hubble's law ($H_0 = 50$), to a distance of three billion light years.

spectra of both 3C 273 and 3C 48, another quasi-stellar radio source, showed emission lines (unlike normal stars), but the lines did not agree in wavelength with those of any known element.

The breakthrough came in 1963. Schmidt noticed that the emission lines barely visible in the spectrum of 3C 273 had the pattern that hydrogen lines had—a series of lines with spacing getting closer together toward shorter wavelengths (Fig. 27–1). He asked himself if he could simply be observing hydrogen that was Doppler shifted. The Doppler shift required would be huge: each wavelength would have to be redshifted by 16 per cent. This would mean that 3C 273 was receding from us at 16 per cent of the speed of light. It would take too much energy to accelerate an object in our galaxy to such a tremendous velocity. The object would have to be receding as part of the expansion of the universe, and Hubble's law showed that it was far beyond most of the galaxies that we study. But the idea looked plausible anyway. And one of Schmidt's Caltech colleagues realized that the spectrum of 3C 48 (Fig. 27–2) looked like hydrogen redshifted by a still more astounding amount: 37 per cent.

These objects, first known as "quasi-stellar radio sources," were called by some by the abbreviation QSR, which turned into *quasar*. Soon many others were found—quasi-stellar objects with huge redshifts. A class of quasars turned up from which no radio signals are detected; we call them "radio quiet" as opposed to "radio loud." We now know of over 1500 quasars. Their redshifts range up to over 350 per cent. But the rule we have used above—an object is travelling at the same percentage of the speed of light that its spectral lines are redshifted—is true only for objects travelling much slower than light. At higher velocities, we have to use Einstein's special theory of relativity to translate motion into Doppler shifts. No object can travel faster than the speed of light, but the farthest quasar (Fig. 27–3) is receding from us at more than 90 per cent of the speed of light. The light we see now left this distant quasar more than 10 billion years ago, so we are seeing back to how the universe was 10 billion years in the past.

Many of the most distant quasars we now observe have several sets of hydrogen spectral lines detectable, each set at a slightly different wavelength. We are detecting a *hydrogen forest* representing many different hydrogen clouds at that early stage of the universe. We are not certain what to make of these clouds' they seem to have relatively small percentages of elements heavier than hydrogen, so perhaps we are looking back to a time

FIGURE 27–2 An early picture of the quasar 3C 48.

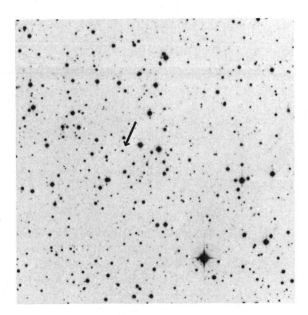

FIGURE 27–3 The farthest known quasar and thus the farthest known object in the universe, PKS 2000-330. Its redshift is 3.78. Its name comes from its position in the catalogue of radio sources compiled at the Australian National Radio Observatory at Parkes.

when little of the heavier elements had been formed. These clouds absorb radiation from the quasar, so we know that the clouds are between the quasar and us and that we are looking not quite as far back as we are to the quasar itself. Probably they are haloes of distant galaxies.

How do we detect quasars? Many of them are found by looking for faint bluish objects. Comparing pictures of the sky taken through different filters often turns up quasars in this way. Some of the newer quasars to be noticed were picked up on maps of the sky made with x-ray satellites. But in both cases, only taking spectra that show large redshifts proves that the objects are quasars.

27.2 THE ENERGY PROBLEM

Even though quasars aren't bright optically, they must be prodigiously bright intrinsically to appear even as they do, given they are so far away. Further, even though they are so far away, many quasars are among the brightest radio objects in the sky.

Further, some quasars fluctuate drastically in brightness within hours or days. So the region of the quasars that gives off most of the radiation must be small enough that light can travel across it in hours or days. How does such a small region give off so much energy? After all, we don't expect huge explosions from tiny firecrackers, but the quasars are much smaller than galaxies yet much brighter.

One reason for the widespread interest in quasars was thus "the energy problem," the question of accounting for how so much energy was given off by such a small volume. Over the years, many mechanisms were tried, such as radiation from otherwise-unknown supermassive stars or chains of supernovae going off almost all the time. But the details of none of these theories worked out. Since that time, though, we realized that black holes are probably widespread in the universe. Matter swirling around and then falling into giant black holes containing millions of times the mass of the sun could

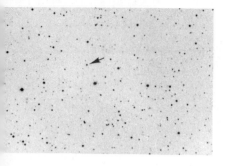

FIGURE 27–4 The object QSO 0241 + 622 appears stellar and has a luminosity far exceeding that of normal galaxies, just as quasars do. Its spectrum resembles those of many quasars and Seyferts (a type of galaxy with a bright core). Its redshift is only 4 per cent.

account for the energy from a quasar. If we had thought as much about black holes as we do now, quasars wouldn't have seemed as mysterious as they did then.

New information from observations across the spectrum carried out from the ground and from space have endorsed the model of giant black holes in the cores of quasars. The strength of ultraviolet radiation measured from the IUE spacecraft, for example, matches the prediction that the inner part of the accretion disk—where the friction of the gas is especially high—would reach 35,000 K. The matter in the accretion disk eventually falls into the black hole, so the process is called "feeding the monster." It is widely agreed, though by no means proved, that a giant black hole is the "engine" of a quasar.

As we look out in space, we seem to see few quasars close up and many more farther out. So the distribution of quasars shows us that the universe has evolved in time. In the next chapter, we shall see how this was an important observational datum for cosmology.

A few quasars have now been found with relatively small redshifts on a cosmic scale—"only" 4 to 10 per cent (Fig. 27–4). If quasars were formed early in the universe, how can these quasars still be shining? Some of the quasars near enough for us to see them relatively well seem to have small objects near them (Fig. 27–5); these small objects may be the cores of galaxies that have had their outer parts stripped away by an interaction with the quasars. The interaction may also have sent material—from either the galaxy or the region of the quasar outside its central black hole—so that it falls into the central black hole, fueling the quasar and allowing it to continue radiating so strongly. There is speculation that some quasars may have faded and then been rejuvenated. We may be seeing these quasars in a second childhood.

Certainly a few quasars will be among the first objects observed with the Hubble Space Telescope soon after its launch. Imagine being able to see them seven times more clearly!

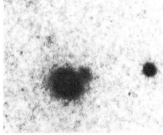

A

FIGURE 27–5A The larger circle is the image of the quasar 3C 323.1, and the smaller circle is a nearby galaxy. Stripping off the outer layers of this galaxy may have provided the fuel for the quasar. (Mauna Kea observations by Alan Stockton)

FIGURE 27–5B An interacting quasar and galaxy show in this computer scan. (Canada-France-Hawaii Telescope observations at Mauna Kea by John Hutchings and colleagues)

1747 + 684   0957 + 227

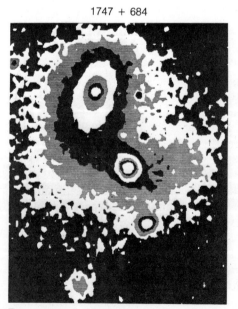

B

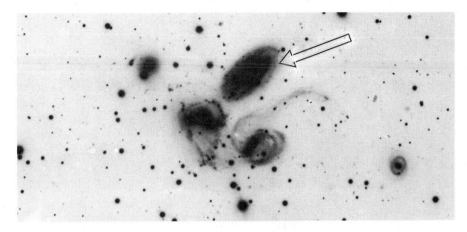

FIGURE 27–6 Stephan's Quintet, in a negative print of a photograph taken by Halton Arp with the 5-m telescope at Palomar Observatory. The redshift of the object marked with an arrow is different from the redshifts of the other four major objects present. (Less exposed photographs show that the dark area to the lower left of the galaxy with the discordant redshift has two nuclei and so is two interacting galaxies. Thus there are indeed five galaxies in this "quintet.")

27.3 ARE THE QUASARS REALLY FAR AWAY?

Using Hubble's law is the only way we have to find the distances to quasars. We have no independent way of checking. So what if Hubble's law doesn't apply to quasars?

Most astronomers would find this an unsatisfactory situation. If Hubble's law applied only in some situations and not in others, we would never know when we could rely on it. But not liking a situation doesn't mean that it isn't true. We must find out whether quasars really are at their "cosmological distances," that is, the distances that Hubble's law gives as a result of the expansion of the universe.

The conservative view, held by Schmidt and by most other astronomers, is that quasars are indeed at the distances given by Hubble's law. But Halton Arp, using the Palomar and other telescopes, has found many cases where a quasar seems associated with an object of a different redshift (Fig. 27–6). If the quasar and the other object were really linked, then they must be at the same distance and the Hubble's law distance wouldn't be valid for one of the objects. But most of the associations stem from the two objects being very close together in the sky. The argument is statistical that the odds are low that two objects would be so close together unless they are physically linked. Arp has also found some cases in which a bridge of material may be faintly visible linking objects of different redshifts.

The bridges of material are very faint, and hard to prove conclusively. The problem comes down to one of statistics: Does the propinquity of two objects prove that they are really close together? Many astronomers and statisticians have argued the point from both sides. Now you consider the following case: I have flipped a coin, and I show you that it has come up heads. What is your answer to the following: What is the probability that the coin shows heads? Do you think it is 50 per cent? Are the odds 50-50? No. I have shown you that the coin came up heads, so the probability was 100 per cent that it was heads. We can only use statistical arguments before we make an observation, not afterward.

A statistical test carried out under the best seeing conditions with telescopes at Mauna Kea in Hawaii showed that quasars indeed seemed to agree in redshift with galaxies near them (Fig. 27–7). Earlier, observational techniques had often not allowed such galaxies to be observed. Occasionally, one galaxy with a different redshift appeared close to the quasar. But one single galaxy can be just a chance superposition. It is interesting that a group of

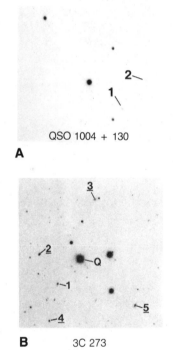

A

QSO 1004 + 130

B   3C 273

FIGURE 27–7 In these photographs taken by Alan Stockton on Mauna Kea, the central objects are quasars, whose catalogue numbers are given below the photos, and the numbered objects are galaxies. The redshifts of the galaxies and the quasars agree. Note how the quasars are brighter than galaxies at the same distances. *(A)* From his original survey. *(B)* The subsequent discovery that the first known and brightest quasar, 3C 273, is associated with a group of galaxies of the same redshift (#1–#5); one galaxy with a disparate redshift (#1, not underlined) is also present.

FIGURE 27–8 *(A)* Increasing exposure time reveals more and more galactic structure around the Seyfert galaxy NGC 4151 (redshift z = 0.001), which appears almost stellar on the least exposed photograph. *(B)* The nucleus of the Seyfert galaxy photographed from an altitude of 25,000 m with the 90-cm telescope carried aloft in 1970 by the Stratoscope II balloon. *(C)* A Seyfert galaxy, ESO 113-IG 45, with an exceptionally luminous nucleus. This is a case of a quasar (or almost quasar) in the center of a spiral galaxy.

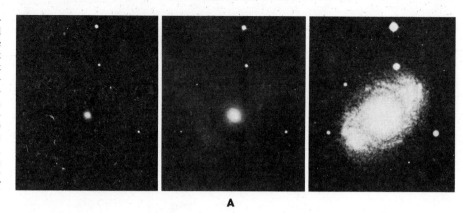

A

B

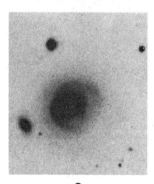

C

objects with the same redshift never has two or more exceptions. If the associations with objects of different redshift were real rather than chance occurrences, we would expect to see two or more exceptions occasionally. (The objects would then really be mixed together and there would be no restriction on how many of different redshift could be mixed in.)

In sum, the consensus is that quasars indeed follow Hubble's law. Arp hasn't given up, but has few followers on this point anymore.

27.4 WHAT ARE QUASARS?

The idea that giant black holes are the engines of quasars, linked with new observational evidence, has given us a pretty good picture of a quasar. We know of a few types of galaxies that have especially bright cores compared with their outer regions (Fig. 27–8). If these galaxies are sufficiently far away, we see only the core as a point-like object, and can't see the outer layers. We think that quasars are extreme examples of such situations (Table 27–1).

A statistical test was carried out with quasars, which, you will recall, were first defined in part by their quasi-stellar appearance. A selection of quasars, graded by redshift, were carefully examined. Faint galaxies were discovered around most of the ones with the smallest redshifts, a few of the ones with intermediate redshifts, and none of the ones with the largest redshifts. These discoveries confirmed the idea that quasars could be extreme examples of galaxies with bright cores.

In the last few years, new observational techniques—especially involving electronic detectors like CCD's—have enabled faint matter to be detected around essentially all of the lower redshift quasars (Fig. 27–9). Even 3C 273 has material around it (Fig. 27–10), so it doesn't fit the original definition of a quasar anymore.

There are still many details to be learned. Do all galaxies go through a quasar phase? How is this phase connected with interactions with other gal-

TABLE 27–1   Energies of Galaxies and Quasars

	RELATIVE LUMINOSITY		
	X-Ray	Optical	Radio
Milky Way	1	1	1
Radio galaxy	100-5,000	2	2,000–2,000,000
Bright-core galaxy	300–70,000	2	20–2,000,000
Quasar: 3C 273	2,500,000	250	6,000,000

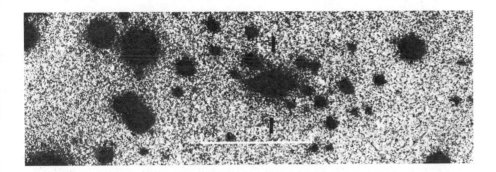

FIGURE 27–9   Special printing techniques have been applied to this photographic plate of the quasar 3C 206, receding from us at 20 per cent of the speed of light. 3C 206 is a quasar in an elliptical galaxy that is a member of a compact cluster of galaxies. All objects in the central 40 arc seconds of the photograph are galaxies in the cluster, and have the same redshift as the quasar; images of other galaxies in the cluster merge with the sides of the quasar's image. A faint nebulosity around the quasar, between the two vertical lines, is barely visible among the noise from photographic grains on this negative print. Note that the quasar image is fuzzier than the images of the many field stars visible. The fuzziness has been verified by digital study of the plates. (Observations obtained by Susan Wyckoff, Peter Wehinger, and Thomas Gehren with the 3.6-m telescope of the European Southern Observatory's station in Chile.)

axies? What more can we say about the giant black hole at the quasar's center, and about the matter orbiting it close in? How long does a quasar shine brightly? But these are relatively minor questions compared with the determination of the outline we have given. If black holes had been in people's minds back in 1963, the quasars might not have seemed so mysterious. As it is, quasars are less mysterious than they were, but galaxies are more so.

27.5 SUPERLUMINAL VELOCITIES

While quasars were, at first, studied mostly as curious objects, now they are used as tools to explore other phenomena. Very-long-baseline interferometry observations with extremely high resolution have shown that some quasar jets have a few small components. And observations over a few years have shown (Fig. 27–11) that the components are apparently separating very fast, given the conversion from the angular change in position we measure across the sky to the actual speed in km/sec at the distance of the quasar.

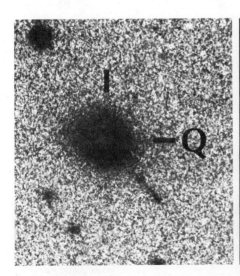
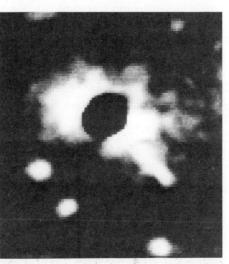

FIGURE 27–10   Fuzz (faint nebulosity), in addition to the jet can be seen around the brightest quasar, 3C 273. The redshift of the nebulosity is the same as that of the quasar. The left image is photographic. In the right image, based on data taken with a CCD camera, the image of the star has been digitally subtracted from the quasar's image; a central disk had masked the brightest middle part of the quasar and of the star on their respective plates, but some spillover light had remained to be considered. The contrast of the remainder was then enhanced to bring out the surrounding galaxy. It is 400,000 times fainter than the quasar, comparable with a giant elliptical galaxy. (*Left:* Susan Wyckoff, Peter Wehinger, and Thomas Gehren; *right:* Anthony Tyson, Bell Labs)

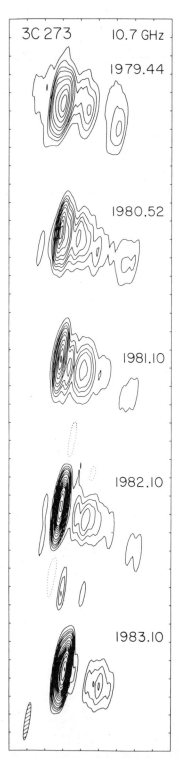

3C 273   10.7 GHz

1979.44

1980.52

1981.10

1982.10

1983.10

FIGURE 27–11 A series of photographs of the extreme center of 3C 273, made with radio telescopes spread over intercontinental distances (between California and Germany). In the earliest map, the two components are separated by less than 6/1000 arc sec. The series shows first one and then another knot of emission separating from the core at an apparently superluminal velocity of 10 times the speed of light. They go off in the direction of 3C 273's optical jet, and may become part of the jet when they move farther out.

Indeed, some of the components appear to be separating at *superluminal velocities,* that is, at velocities greater than the speed of light. But the special theory of relativity holds that velocities greater than the speed of light are not possible.

Astronomers can explain how the components only appear to be separating at greater than the speed of light even though they are actually moving at allowable speeds (speeds slower than light). As for other examples involving the theory of relativity, we must think clearly not only of where things are but of when they are observed and when we think we are observing them. If one of the components is a jet approaching us at almost the speed of light, the jet is almost keeping up with the radiation it emits. If the quasar moves a certain distance in our direction in 1 year, the radiation it emits at the end of the year reaches us sooner than it would if the quasar were not moving toward us. So in less than one year, we see the quasar's apparent motion over 1 full year. In the interval between our observations, the quasar jet had several times longer to move than we would naively think it had. So it could, without exceeding the speed of light, appear to move several times as far.

One interesting question is how often quasar jets are beamed so closely toward us. We see a few quasars with the apparent superluminal velocity, and a galaxy as well, so the phenomenon may be fairly common. Does the beaming also strengthen the intensity of quasar radiation? If it does, we may be miscalculating the amount of energy quasars give off. More research is necessary.

27.6 MULTIPLE QUASARS

It is odd enough to notice a quasar in the sky, but it is odder yet to find two almost adjacent on a photographic plate. Yet the Palomar-National Geographic Sky Survey turned up a pair of quasars extremely close together. Moreover, the properties of the quasars turned out to be almost identical—they were the same brightness and had the same redshift. So they would be at the same distance from us, physically close together.

The idea that two such quasars would be so close together seemed implausible. Perhaps it was the warping of space (Fig. 18–17) according to the general theory of relativity that causes the situation. We have seen that large masses can bend light, and that the sun is barely massive enough to test the theory. A whole galaxy can bend light a lot more, and can act as a *gravitational lens,* making two images of the same object (Fig. 27–12).

QUASAR

INTERVENING GALAXY

EARTH

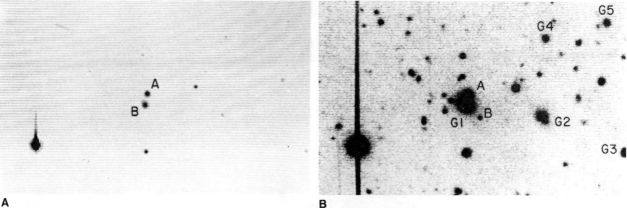

A   **B**

FIGURE 27–13 Two views of a photograph of a double quasar taken with a CCD camera at the 5-m Palomar telescope. *(A)* The frame is reproduced so that the twin images, marked A and B, stand out. *(B)* The frame is reproduced so that the fainter parts of the image show. We then see a cluster of galaxies, some of which are marked (G1–G5). 90 percent of the objects on the frame are galaxies. A and B are now overexposed and blur out. G1 provides most of the gravity for the lens effect. Though its image does not show clearly, its redshift can be measured spectroscopically.

If the two objects were really the same one seen along different paths, they should fluctuate similarly in brightness. But the paths may be different lengths, so one should mimic the pattern of brightness fluctuation the other showed earlier. Fortunately, we can calculate that a likely time delay is only a few years, so we have hope of detecting this effect. Continual monitoring is under way.

Images made with one of the new solid-state detectors called CCD's have revealed that a cluster of galaxies lies between us and the double quasar. Quasar image B (Fig. 27–13) is seen through the brightest member of the cluster. It is a giant elliptical galaxy with a redshift about one-fourth that of the quasar, making it 4 times closer to us. So we may be seeing the gravitational lens directly (Fig. 27–14 and Color Plate 73).

A handful of other "multiple quasars" have been found, but "the double quasar" remains the best case. It is an exciting verification of a prediction of Einstein's general theory of relativity.

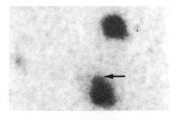

FIGURE 27–14 This optical photograph of the double quasar, 0957+61A and B, shows the intervening galaxy. The 17th magnitude quasar images are bluish and are separated by only 5.7 seconds of arc. Their redshifts are identical to the third decimal place: $z = 1.4136 \pm 0.0015$. We see the intervening galaxy only 1 second of arc from the "B" quasar image, and not resolved from it. Its redshift is about 0.37. The objects are in the constellation Ursa Major. (Photograph by Alan Stockton of the University of Hawaii)

KEY WORDS
quasar, hydrogen forest, superluminal velocities, gravitational lens

QUESTIONS

1. Why is it useful to find the optical objects that correspond in position with radio sources?

2. We observe a quasar with a spectral line at 4000 Å that we know is normally emitted at 3000 Å. (a) By what percentage is the line redshifted? (b) By approximately what percentage of the speed of light is the quasar receding? (c) At what speed is the quasar receding in km/sec? (d) Using Hubble's law, to what distance does this correspond? Specify the value of Hubble's constant you use.

3. What do radio-loud quasars and radio-quiet quasars have in common?

4. Why does the rapid time variation in some quasars help define the "energy problem"?

5. What are three differences between quasars and pulsars?

6. Explain how parts of a quasar could appear to be moving at a speed greater than that of light, without violating the special theory of relativity.

7. What features of some quasars suggest that quasars may be closely related to galaxies?

8. Describe why we think the double quasar is showing us a gravitational lens.

9. Why do we think that some nearby quasars are rejuvenated?

10. Describe some evidence that quasars are at the distances given by Hubble's law.

Chapter 28

COSMOLOGY: HOW WE BEGAN; WHERE WE ARE GOING

Stars twinkling in an inky black sky make a beautiful sight. But why is the sky dark at night? After all, if the universe is infinite, and the stars have shined forever, we should see a star eventually in whatever direction we look. So we should see the bright surface of a star in whatever direction we look. But we obviously don't, making a paradox—a conflict of a reasonable deduction with our common experience (Fig. 28–1).

This dark-sky paradox has been debated for hundreds of years. It is known as *Olbers's paradox,* though Wilhelm Olbers, whose phrasing dates from 1823, wasn't the first to realize the problem. (Note that Olbers had an "s" at the end of his name, so it is "Olbers's paradox," or "Olbers' paradox," but not "Olber's paradox"; his name wasn't Olber.)

It is not that the surfaces of stars farther away are fainter. After all, the farther they are, the smaller their surfaces appear. The surface area gets smaller at exactly the same rate that the overall brightness of the star gets fainter, so the brightness of each bit of the star's surface remains constant.

Only in this century do we realize how Olbers's paradox is resolved. First, Hubble showed us that the universe is expanding. (We must consider galaxies instead of stars, since individual galaxies are not expanding.) As each galaxy moves away from us, its energy is redshifted, and redshifted light has less energy than it previously had. So Hubble's law solves Olbers's paradox. Further, for the paradox to exist we had to assume that the stars were infinitely old. We now know that the universe is less than 20 billion years old, far younger than necessary for us to see a star in every direction. So modern ideas have solved an old problem. And sophisticated reasoning has been necessary to provide the answer to a question that is very simply stated.

In this chapter we shall study *cosmology,* the study of the universe as a whole, including its past and its future.

FIGURE 28–1 *(A)* If we look far enough in any direction in an infinite universe, our line of sight will hit the surface of a star. This leads to Olbers's paradox. *(B)* This painting by the Belgian surrealist René Magritte shows a situation that is the equivalent to the opposite of Olbers's paradox.

A

B

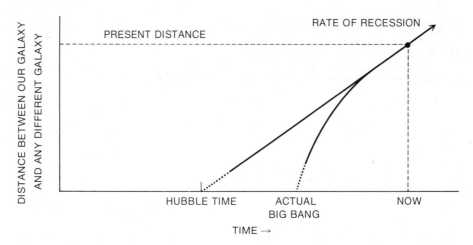

FIGURE 28–2   If we could ignore the effect of gravity, then we could trace back in time very simply; the Hubble time corresponds to the inverse of the Hubble constant ($1/H_0$). Actually, gravity has been slowing down the expansion. The vertical axis represents a scale factor.

28.1 TRACING THE UNIVERSE BACK IN TIME

Since we have likened our universe to an expanding raisin cake, what happens if we look back in time? The raisin cake, and the universe, would have been more dense, with individual objects packed closer and closer together. If we assume that the universe's expansion is not slowed by gravity, we can trace it back until all the galaxies in it would have been jammed together. If we take a value for Hubble's constant of 50 km/sec/megaparsec, as many astronomers do, we find the universe would have been compressed some 18 billion years ago. (A stack of 18 billion pennies would be 30,000 km high!) If we allow for the gravity of all the mass of the universe slowing down the expansion, then the universe could be somewhat younger, but probably younger by only a few billion years (Fig. 28–2).

What happened 18 billion years or so ago? Current theory indicates that the universe was exceedingly dense and hot. The detailed theories are worked out based on Einstein's general theory of relativity (Fig. 28–3), since it is a theory of gravity and of how gravity affects space. Something—known as the *big bang*—started the universe expanding, and it has been expanding ever since. Big bang is the technical name as well as the popular name for the class of theories that are now current.

On the whole, the universe is *homogeneous* (that is, it is about the same everywhere) and *isotropic* (that is, it looks the same in all directions from our vantage point). These two conditions, known as the *cosmological principle,* are basic to most big-bang theories. They are true only on large-scale averages—large enough even to average out recently discovered giant voids where no galaxies are found.

Big-bang theories show us a universe that is evolving. It may look about the same in all places and in all directions, but it is thinning out. Almost forty years ago, a group of scientists proposed that the universe didn't evolve. They had to suppose for their *steady-state theory* (Fig. 28–4) that matter was created as the universe expanded, so as to keep the density constant. Their idea led to much valuable analysis, but has now been discarded on the

FIGURE 28–3   Einstein and Willem de Sitter, who worked out some of the early cosmological solutions to Einstein's equations, at the Mount Wilson Observatory in 1931.

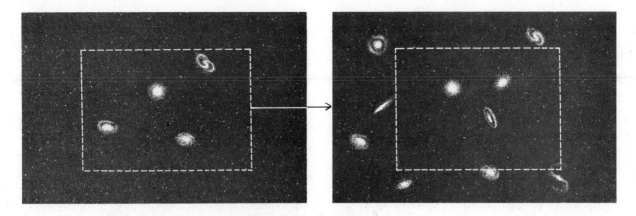

FIGURE 28–4   In the steady state theory, as the dotted box at left expands to fill the full box at right, new matter is created to keep the density constant. In the picture, the four galaxies shown at left can all still be seen at right, but new galaxies have been added so that the number of galaxies inside the dotted box is about the same as it was before.

basis of overwhelming observational evidence. For example, the quasars are clearly more numerous far from us, which means they were more numerous far back in time, which is not allowable by the steady-state theory.

Does the idea that the universe expanded from a hot big bang mean that the universe had a center? No. Imagine that our raisin cake is infinite, extending forever in every direction. Going back in time, it would be compressed (and any part would thus feel dense), but it would still extend forever in every direction. After all, half of infinity is still infinity. Similarly, the universe would have been more compressed, but would not have edges so would not have a center. Wherever we are, we can go infinitely far in any direction.

Alternatively, the general theory of relativity shows that the universe can have a "curvature," in that if we travelled far enough, we could come back on ourselves. To understand this, picture the analogy of the surface of a balloon (note: the surface, not the center). The surface is a two-dimensional space that is curved onto itself. If an ant crawls far enough on the surface, it comes around completely, and we would do the same on the earth. The curvature can take place because we are visualizing a two-dimensional surface curved into three-dimensional space. The ant on the balloon itself probably doesn't realize that it is curved. For the universe, we would have to visualize a three-dimensional surface curved into four-dimensional space. I can't do it. Can you? But the equations can be followed.

A surface that curves back onto itself, like the two-dimensional analogy of the surface of a balloon, can have a finite volume but still be unbounded. Such a three-dimensional surface need not have a center, just as the two-dimensional surface of a balloon has no center. Another type of curvature is more like the surface of a saddle than the surface of a sphere. A universe with this kind of curvature is infinite.

28.2 WHAT IS BEYOND OUR FUTURE?

It is relatively direct to consider two different cases for the future of our big-bang universe: the universe will expand forever, or it will expand for a

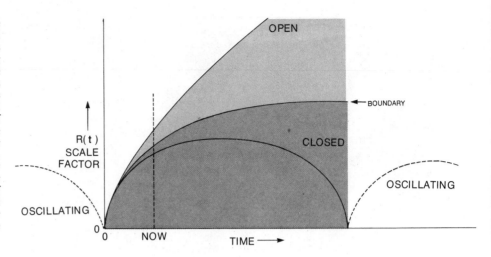

FIGURE 28–5 To trace back the growth of our universe, we would like to know the rate at which its rate of expansion is changing. Big bang models of the universe are shown; the vertical axis represents a scale factor, R, that represents some measure of distances and how they change as a function of time, t. The universe could be open and expand forever or closed and begin to contract again. If it is closed, we do not know whether it would oscillate or whether we are in the only cycle of expansion or contraction it will ever undergo. We will see later that current evidence favors the open model.

while more and then begin to contract (Fig. 28–5). Until the last few years, these were the major two cases being debated; at the end of this chapter, we will see that we may indeed be on the fine line separating them.

If the universe doesn't contain enough matter to generate enough gravity, the universe will keep expanding forever. This case is called an *open universe*. In an open universe, you can keep going on forever in any direction without curving around on yourself.

If the universe contains enough matter to generate enough gravity, the universe will eventually stop its expansion and begin contracting. This case is called a *closed universe*. Our expansion is not slowing down very much, if at all. So we know that the universe won't stop its expansion for at least another 50 billion years, so there is nothing to worry about. But we would like to know if it will eventually collapse into a *big crunch*. Cosmologists probably should be optimists to spend their time thinking about times 50 or 100 billion years into the future. Nobody knows, in any case, whether a big crunch would be followed by a new big bang, or whether the universe would end there.

How can we predict the future? The rate at which the expansion of the universe is slowing down is so slight that we can't measure it. The observations are very difficult, and our measurements of distances to faraway galaxies independently of Hubble's law are too imprecise. Perhaps the Hubble Space Telescope will help on this point. But major questions like these have a way of getting more complicated rather than simplified by the addition of new data. In any case, at present we cannot tell about the universe's future from looking at the slowdown rate.

Perhaps we can find out how much mass there is in the universe, and use that information to find out if there is enough gravity to cause the universe's expansion to end. Actually, if the universe is infinite, it would have infinite mass, so we actually want to measure the density of an average region, the amount of mass in a given volume. If we add up the amount of mass in planets, stars, interstellar matter, and other types of matter we can detect, we find only about 1 per cent of the amount necessary for the universe to be closed. But what have we missed? How many kinds of invisible and undetectable matter are there? Until the observations of the 21-cm line a few decades ago, interstellar hydrogen was invisible. And until the last decade, much of the hot matter whose x-rays we have observed from spacecraft was undetectable.

When we observe clusters of galaxies, we can measure how fast the galaxies are moving around within the cluster (aside from the overall velocity of the whole cluster, which fits Hubble's law). The faster the individual galaxies within a cluster are moving about, the more gravity must be present to keep them from escaping. In that way, we can find out how much matter is present whether it is visible or not. It turns out that there seems to be 10 to 100 times more matter in clusters of galaxies than we can see. This discrepancy, known as the *missing-mass problem* (though it is really a missing-light problem since the mass is there), could mean that the universe is closed after all. The debate rages. Maybe the matter is even in strange forms that have yet to be discovered on earth, the kinds of subnuclear particle that are occasionally discovered by giant atom smashers.

One of the best ways of determining the density of matter in the universe is to study the relative amounts of the elements present. Theoretical calculations show that whatever the conditions of temperature and density were in the first minutes after the big bang, about 10 per cent helium is formed. But the calculations show that one form of hydrogen is especially sensitive to these early conditions. This "heavy-hydrogen" form—deuterium—contains a proton and a neutron in its nucleus, and so differs from ordinary hydrogen, which contains a proton alone. Since a deuterium nucleus consists of only two particles, it is the first form heavier than ordinary hydrogen to form. But if the universe is relatively dense, the deuterium immediately combines with protons to form helium. Only if the universe is not dense during the first 15 minutes after the big bang (!) can enough deuterium build up to be measurable now (Fig. 28–6). And deuterium has another useful property—we know of no processes that form it in the inter-

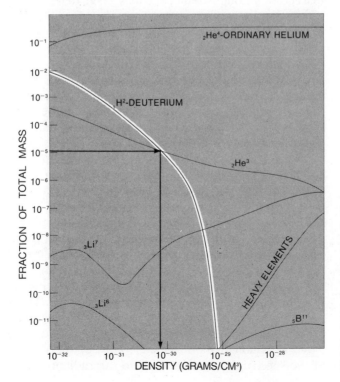

FIGURE 28–6 The horizontal axis shows the current day cosmic density of matter. From our knowledge of the approximate rate of expansion of the universe, we can deduce what the density was long ago. The abundance of deuterium, outlined in white, is particularly sensitive to the time when the deuterium was formed. Thus present-day observations of the deuterium abundance tell us what the cosmic density is, by following the arrow on the graph.

A

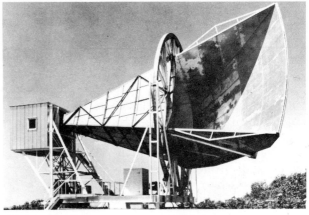

B

FIGURE 28–7 *(A)* Arno Penzias *(left)* and Robert W. Wilson *(right)* with their antenna in the background. Penzias and Wilson won the 1978 Nobel Prize in Physics for their discovery. *(B)* The large horn-shaped antenna they used at the Bell Laboratories' space communication center in Holmdel, New Jersey. Penzias and Wilson found more radio noise than they expected at the wavelength of 7 cm at which they were observing. After removing all possible sources of noise (faulty connections, loose antenna joints, contributions from nesting pigeons), a certain amount of radiation remained. It was the 3° background radiation.

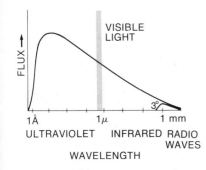

FIGURE 28–8 Curves for black bodies at different temperatures showing the distribution of radiation given off. Radiation from a 3° black body *(the inner curve at lower right)* peaks at very long wavelengths. On the other hand, radiation from a very hot black body *(the upper curve)* peaks at very short wavelengths.

stellar medium other than the big bang. Stars eat up deuterium, rather than producing more.

Deuterium measurements are being pursued in various ways, including radio measurements from the ground and ultraviolet measurements from spacecraft. Enough deuterium has been detected to indicate that the universe's density is relatively low, and therefore that it will expand forever.

28.3 THE BACKGROUND RADIATION

In 1965, two young Bell Laboratories scientists were testing a sensitive antenna for possible sources of interference. After eliminating all possible contributions to the radio signals they were gathering, they were left with a small residual. Try as they might, the faint signal persisted, and did not vary with time of day or with direction in the sky.

Soon, they learned that Princeton scientists had predicted that there might be a faint signal detectable from the era soon after the big bang. Such a signal would not vary with time of day or with direction. The Bell Labs pair—Arno Penzias and Robert Wilson (Fig. 28–7)— had found their signal, and eventually received the Nobel Prize in Physics for the discovery.

For they had heard the echo of the big bang itself. When the elements were formed, the universe was so hot and dense that no electrons could combine with nuclei to form atoms. Light couldn't travel through the bright sea of radiation and particles. But when the universe cooled off enough a few thousand years later, it became transparent. The radiation that had been present was freed to travel great distances. As the universe expanded, the distribution of energies of the photons shifted redward in the spectrum. Now, 18 billion years or so later, most of the radiation is in the radio part of the spectrum. It corresponds to the radiation that would be given off by gas at a temperature of only 3° (3 K), 3° above absolute zero and −270°C. Since it started off as radiation filling the universe, it is easy to understand why it now comes equally from every direction. The radiation follows a smooth distribution that is typical of hypothetically perfect radiating objects known as "black bodies," so we speak of *3° black-body radiation* (Fig. 28–8).

The existence of this radiation, also often called simply the *background radiation,* seems to prove that the big bang took place, or at least that the universe had a hot dense period. The steady-state theory (Section 28.1) has thus been completely ruled out. In recent years, a stream of measurements of the background radiation over a very wide range of wavelengths (Fig. 28–9) has given us detailed information about it. We now even can measure a slight deviation from perfect isotropy, which we can interpret as the result of a motion of our galaxy in space, similar to a Doppler shift. So we are using the background radiation to tell us fundamental things about our relation with the universe.

COBE (Fig. 28–10), the **CO**smic **B**ackground **E**xplorer spacecraft, is set for launch by NASA in 1987 to make especially detailed observations of the background radiation. Because it will be above the earth's atmosphere, it will be able to make precise measurements of parts of the spectrum that do not pass through our atmosphere.

28.4 THE EARLY UNIVERSE

We have traced the universe back to the time a few thousand years after the big bang when the background radiation was set free, and we have even seen how forms of the light elements were formed between one and fifteen minutes after the big bang. Using recent discoveries from the realm of nuclear physics, astronomers and physicists working together have now pushed our understanding of the universe even further back in time: earlier than the first second of time after the big bang. This period is known as *the early universe.*

The study of the early universe involves a fundamental understanding of the forces that act in the universe. We now consider that there are three fundamental types of forces: (1) the "strong force"; (2) the "electroweak force"; and (3) the "gravitational force." Let us briefly discuss them.

1. The "strong force," also known as the "strong nuclear force," holds protons, neutrons, and certain other types of particles together to form nuclei. This force got its name because it is so powerful, at least at the extremely close ranges inside nuclei. In Section 19.3 we saw that the common nuclear particles like protons and neutrons are made up of more fundamental particles called quarks. Quarks are held together to make particles by a force known as the "color force"; the strong force is really a version of the color force.

2. A much weaker force that also works inside nuclei is called, not surprisingly, the "weak force." We detect its effect only in certain radioactivity.

A force with which we are very familiar is the "electromagnetic force." It causes the electricity and magnetism that are so familiar to us in our daily lives. Since electromagnetic forces hold atoms to each other, this force even holds our bodies together. And the electromagnetic force causes the electromagnetic radiation—light, x-rays, radio waves, etc.—that provides just about our only contact with the universe around us.

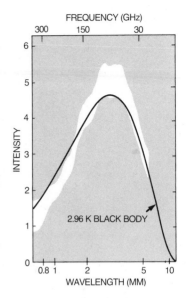

FIGURE 28–9 Balloon observations showing the agreement of the data *(white, unshaded region)* with the theoretical curve for a black body *(solid line)*. The radio observations we have discussed previously are all located in the extreme right-hand lower portion of the graph.

FIGURE 28–10 An artist's conception of the COBE (**CO**smic **B**ackground **E**xplorer spacecraft (in NASA's Explorer series), scheduled for launch in 1987.

FIGURE 28–11   Carlo Rubbia with the apparatus with which he and his colleagues discovered the W⁺, W⁻, and Z⁰ particles. These discoveries verified predictions of the electroweak theory. Rubbia and Simon van der Meer, whose instrumental work was vital to the project, received the 1984 Nobel Prize in Physics as a result.

The theoretical physicists Sheldon Glashow, Steven Weinberg, and Abdus Salam have worked out a theory that "unifies" the electromagnetic and the weak forces. In the theory, these two forces are but different aspects of a single "electroweak force." The three physicists won the 1979 Nobel Prize for their work. The unification of forces is not new: the electric force and the magnetic force were thought to be separate until James Clerk Maxwell showed theoretically a century ago that they were a single "electromagnetic force." This discovery of Maxwell's led fairly directly to the discovery of radio waves; who knows to what the electroweak unification will lead in years to come.

The electroweak theory predicted that new particles, then undiscovered, would exist: a pair of W particles and a Z particle. An international team of scientists at the European Center for Nuclear Research (CERN) discovered the particles in 1983 (Fig. 28–11); the Nobel Prize followed only one year later. The find is strong confirmation of the electroweak theory.

3. The weakest of the fundamental forces of nature is the "gravitational force." But the gravitational force always acts to attract other objects, never to repel them. (There is no "antigravity.") So the gravitational force adds up, and the gravitational force from massive objects is large.

The intriguing aspect for cosmology is that the theories indicate that the three forces are not fundamentally different from each other. They are basically different aspects of the same "unified" fundamental force. Theoretically, at very high temperatures—10^{28} K, far higher than anything we can conceive of on earth—the forces would be the same. We say that they are "symmetric." As the temperature drops, the forces separate from each other and the symmetry is "broken." Similarly, ice forms in a solid asymmetric shape when shapeless water cools enough. The early universe is the only place and time of which we can conceive where the forces were unified. So studying the early universe tells us about the most basic aspects of nature.

28.5 THE INFLATIONARY UNIVERSE

One of the most major revisions of cosmological thinking in the last sixty years seems to be under way. The new liaison between astrophysicists studying the universe and physicists studying the constituents of atoms and the forces of nature has led to new ideas of what went on in the early universe. The result is that the universe expanded extremely rapidly for a short period—"inflated"; the theory is called the *inflationary universe.*

The inflationary universe theory takes us back to the first 10^{-32} second—that is, a decimal point followed by 31 zeroes and a 1:0.000 000 000 000 000 000 000 000 000 01. During the fraction of a second up to that time, the universe apparently expanded 10^{100} (a 1 followed by a hundred zeroes) times faster than it would have by the standard big-bang theories (Fig. 28–12).

One outstanding problem in cosmology had been to explain why the universe is apparently so uniform. Distant parts of the universe in opposite directions from us seem to be uniform even though they are too far apart for light ever to have been able to travel between them. And if information can't travel from one to the other at the speed of light or less (Fig. 28–13), we don't know how they could ever have had the same set of conditions. But with the inflationary universe, the apparently distant parts would have once been much closer together than we had thought. At that time, they were close enough to reach uniformity; the inflationary universe's rapid expansion brought them to their current locations. Before the inflation, the entire part of the universe that we can now observe would have been smaller than a single proton.

Another interesting prediction of the inflationary universe solves a problem that had recently arisen because of work in basic physics that predicts the formation of *magnetic monopoles,* particles that have only north or south magnetic polarities. We now know only of pairs of north and south magnetic polarities, and never find either alone. But the inflationary universe predicts that the universe has expanded so much that the magnetic monopoles were extremely spread out, so spread out that we don't expect to find more than a handful in our part of the universe.

Further, the inflationary universe has a new answer to the question astronomers have been asking for years: Is the universe open or closed? Indeed, astronomers had come to wonder why the observations indicated that we were so close to the dividing line. Even a factor of 10 or 100 more mass than necessary or too little mass to close the universe is only a small variation, given the wide range possible. The inflationary universe theory indicates that the universe is extremely close to the dividing line between open and closed. The universe, indeed, would be so close to the dividing line that we cannot tell on which side it is. The universe would expand for a very long time but the rate of its expansion would gradually slow down.

So the inflationary universe has potentially solved some important questions in cosmology. And as a bonus, it explains where almost all the matter in the universe came from! The details of the theory involve an expansion

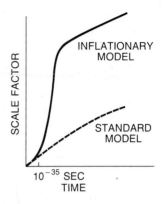

FIGURE 28–12   The rate at which the universe expands, for both inflationary and non-inflationary models.

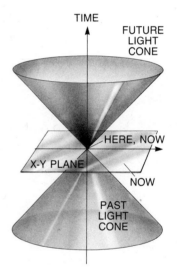

FIGURE 28–13   If we plot a three-dimensional graph with two spatial dimensions (omitting the third dimension) and a time axis, our location can receive signals only from regions close enough for signals to reach us at the speed of light or less, our "past light cone." We can never be in touch with regions outside our light cone; we say that we are "not causally connected" with those regions. In the inflationary universe, the volume of space with which we are causally connected is much greater than the observable universe. It is easy to explain how a volume that is causally connected can be homogeneous—collisions, for example, can smooth things out.

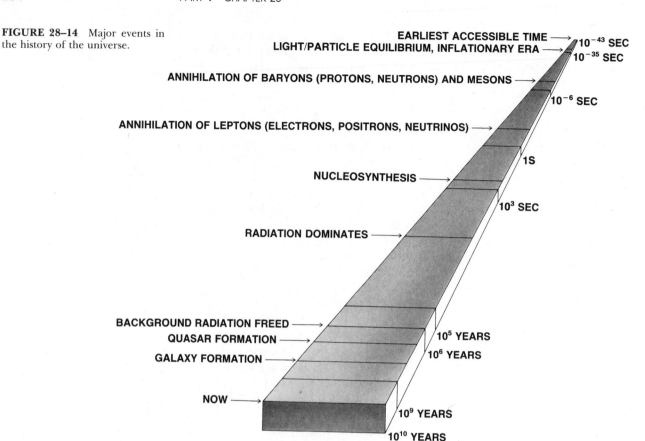

FIGURE 28-14 Major events in the history of the universe.

EARLIEST ACCESSIBLE TIME ——→ 10^{-43} SEC
LIGHT/PARTICLE EQUILIBRIUM, INFLATIONARY ERA ——→ 10^{-35} SEC

ANNIHILATION OF BARYONS (PROTONS, NEUTRONS) AND MESONS ——→

10^{-6} SEC

ANNIHILATION OF LEPTONS (ELECTRONS, POSITRONS, NEUTRINOS) ——→

1S

NUCLEOSYNTHESIS ——→

10^3 SEC

RADIATION DOMINATES ——→

BACKGROUND RADIATION FREED ——→
QUASAR FORMATION ——→ 10^5 YEARS
GALAXY FORMATION ——→ 10^6 YEARS

NOW ——→

10^9 YEARS
10^{10} YEARS

of space with the universe "frozen" in one of its earliest states. The process created the energy from which almost all the matter in the universe was transformed (Fig. 28–14). One of the inventors of the theory has called this "the ultimate free lunch."

So cosmology has made some progress in understanding the universe we live in, and is a field alive with new ideas. We look for the next decade—especially with the elaboration of theoretical ideas, with new results from the physics of subnuclear particles, and with the launch of the Hubble Space Telescope and the building of new ground-based telescopes—to bring new results as exciting as those in the decade past.

KEY WORDS

Olbers's Paradox, cosmology, big bang, homogeneous, isotropic, cosmological principle, steady-state theory, open universe, closed universe, big crunch, missing-mass problem, 3° black-body radiation, background radiation, the early universe, inflationary universe, magnetic monopoles

QUESTIONS

1. We say in the chapter that the universe is uniform on a large scale. Referring to the discussions in Chapter 26, what is the largest scale on which the universe does not seem to appear uniform?

2. List observational evidence in favor of and against

each of the following: (a) the big-bang theory, and (b) the steady-state theory.

3. What is the relation of Einstein's general theory of relativity to the big bang?

4. For a Hubble constant of 50 km/sec/Mpc, calculate

the Hubble time, the age of the universe ignoring the effect of gravity. To do so, take $1/H_0$, and simplify units so that only units of time are left.

5. What is the Hubble time if the Hubble constant is 100 km/sec/Mpc? Compare with the answer from Question 4. Comment on the additional effect that gravity would have.

6. In actuality, if current interpretations are correct, the 3° background radiation is only indirectly the remnant of the big bang, but is directly the remnant of an "event" in the early universe. What event was that?

7. What do we learn from the deviation of the background radiation from isotropy?

8. What are two pieces of evidence against the steady-state theory? How would the explanation of Olbers's paradox differ if the steady-state theory instead of the big-bang theory were true?

9. In your own words, explain why measurements of the abundance of interstellar deuterium give us important information about the future of the universe.

10. What are the advantages of the inflationary-universe theory?

EPILOGUE

We have had our tour of the universe. We have seen the stars and planets, the matter between the stars, and the distant objects in our universe like quasars and clusters of galaxies. We have learned how our universe is expanding and that it is bathed in a glow of radio waves.

Further, we have seen the vitality of contemporary science in general and astronomy in particular. The individual scientists who call themselves astronomers are engaged in fascinating studies, often pushing new technologies to their limits. New telescopes on the ground and in space, new computer capabilities for studying data and carrying out calculations, and new theoretical ideas are linked in research about the universe.

Our knowledge of the universe changes so rapidly that within a few years much of what you read here will be revised. So your study of astronomy shouldn't end here; I hope that, over the years, you will keep up by following astronomical articles and stories in newspapers, magazines, and books (some are listed in the bibliography), and on television. I hope that you will consider the role of scientific research as you vote. And I hope you will remember the methods of science—the mixture of logic and standards of proof by which scientists operate—that you have seen illustrated in this book.

A space-shuttle astronaut, set free from the orbiter, becomes a human satellite in space and orbits the Earth at 27,000 km/hr. Our increasing capabilities in space join with ground-based and theoretical studies to improve our knowledge of the universe.

Appendix 1
Measurement Systems

Metric units			SI abbrev.	Other abbrev.
SI units				
	length	meter	m	
	volume	liter	L	ℓ
	mass	kilogram	kg	kgm
	time	second	s	sec
	temperature	kelvin	K	°K
Other metric units				
1 micron (μ) = 1 micrometer (μm) $= 10^{-6}$ meter			μm	μ
1 Ångstrom (Å or A); $= 10^{-10}$ meter $= 10^{-8}$ cm			0.1 nm	Å

Prefixes for use with basic units of metric system				
Prefix	Symbol	Power		Equivalent
exa	E	10^{18} =	1,000,000,000,000,000,000	
peta	P	10^{15} =	1,000,000,000,000,000	
tera	T	10^{12} =	1,000,000,000,000	Trillion
giga	G	10^{9} =	1,000,000,000	Billion
mega	M	10^{6} =	1,000,000	Million
kilo	k	10^{3} =	1,000	Thousand
hecto	h	10^{2} =	100	Hundred
deca	da	10^{1} =	10	Ten
———	——	10^{0} =	1	One
deci	d	10^{-1} =	.1	Tenth
centi	c	10^{-2} =	.01	Hundredth
milli	m	10^{-3} =	.001	Thousandth
micro	μ	10^{-6} =	.000001	Millionth
nano	n	10^{-9} =	.000000001	Billionth
pico	p	10^{-12} =	.000000000001	Trillionth
femto	f	10^{-15} =	.000000000000001	
atto	a	10^{-18} =	.000000000000000001	

Appendix 2
Basic Constants

Physical constants		
Speed of light	c	$= 299\ 792\ 458$ m/s
Constant of gravitation	G	$= 6.6726 \pm 0.0005 \times 10^{-11}$ m^3/kg/s^2
Mass of hydrogen atom	m_H	$= 1.6735 \times 10^{-24}$ gm
Mass of neutron	m_n	$= 1.6749 \times 10^{-24}$ gm
Mass of proton	m_p	$= 1.6726 \times 10^{-24}$ gm
Mass of electron	m_e	$= 9.1096 \times 10^{-28}$ gm
Astronomical constants		
Astronomical unit	1 A.U.	$= 1.495\ 978\ 70 \times 10^{11}$ m
Parsec	pc	$= 206\ 264.806$ A.U.
		$= 3.261633$ light years
		$= 3.085678 \times 10^{18}$ cm
Light year	ly	$= 9.460530 \times 10^{17}$ cm
		$= 6.324 \times 10^4$ A.U.
Mass of sun	$M_\odot$	$= 1.9891 \times 10^{33}$ gm
Radius of sun	$R_\odot$	$= 696000$ km
Luminosity of sun	$L_\odot$	$= 3.827 \times 10^{33}$ erg/s
Mass of earth	M_E	$= 5.9742 \times 10^{27}$ gm
Equatorial radius of earth	R_E	$= 6\ 378.140$ km
Mean distance center of earth to center of moon		$= 384\ 403$ km
Radius of moon	R_M	$= 1\ 738$ km
Mass of moon	M_M	$= 7.35 \times 10^{25}$ gm

Appendix 3
The Planets

Appendix 3a. Intrinsic and Rotational Properties

Name	Equatorial Radius km	Equatorial Radius ÷ Earth's	Mass ÷ Earth's	Mean Density (g/cm³)	Surface Gravity (Earth = 1)	Sidereal Rotation Period	Inclination of Equator to Orbit	Brightest Apparent Magnitude in 1984
Mercury	2,439	0.3824	0.0553	5.43	0.378	58.646^d	0°	−1.8
Venus	6,052	0.9489	0.8150	5.24	0.894	243.01^dR	117.3	−4.3
Earth	6,378.140	1	1	5.515	1	$23^h56^h04.1^s$	23.45	—
Mars	3,397.2	0.5326	0.1074	3.93	0.379	$24^h37^m22.662^s$	25.19	−1.9
Jupiter	71,398	11.194	317.89	1.36	2.54	9^h50^m to $> 9^h55^m$	3.12	−2.7
Saturn	60,000	9.41	95.17	0.71	1.07	$10^h39.9^m$	26.73	+0.1
Uranus	26,145	4.1	14.56	1.30	0.8	16^h to 17^h	97.86	+5.5
Neptune	24,300	3.8	17.24	1.8	1.2	$18^h \pm 10^m$	29.56	+7.9
Pluto	1,500–1,800	0.4	0.02	0.5–0.8	~0.03	$6^d9^h17^m$	118?	+13.7

R signifies retrograde rotation.

Appendix 3b. Orbital Properties

Name	Semimajor Axis A.U.	Semimajor Axis 10^6 km	Sidereal Period Years	Sidereal Period Days	Eccentricity	Inclination to Ecliptic
Mercury	0.3871	57.9	0.24084	87.96	0.2056	7°00′26″
Venus	0.7233	108.2	0.61515	224.68	0.0068	3°23′40″
Earth	1	149.6	1.00004	365.26	0.0167	0°00′14″
Mars	1.5237	227.9	1.8808	686.95	0.0934	1°51′09″
Jupiter	5.2028	778.3	11.862	4,337	0.0483	1°18′29″
Saturn	9.5388	1427.0	29.456	10,760	0.0560	2°29′17″
Uranus	19.1914	2871.0	84.07	30,700	0.0461	0°48′26″
Neptune	30.0611	4497.1	164.81	60,200	0.0100	1°46′27″
Pluto	39.5294	5913.5	248.53	90,780	0.2484	17°09′03″

Appendix 4
Planetary Satellites

Planet		Satellite	Semimajor Axis of Orbit (km)	Sidereal Revolution Period (d)	(h)	(m)	Diameter (km)	Mean Density (g/cm³)	Discoverer	Visible Magnitude at Mean Opposition Distance
Earth		The Moon	384,500	27	07	43	3476	3.34	—	−12.7
Mars		Phobos	9,378	0	07	39	27 × 21 × 18	2	Hall (1877)	11.3
		Deimos	23,459	1	06	18	15 × 12 × 10	2	Hall (1877)	12.4
Jupiter	XIV	1979 J1 Adrastea	127,000	0	07	04	(40)		Synnott/Voyager 1 (1980)	
	XVI	1979 J3 Metis	129,000	0	07	06	(20)		Jewett/Voyager 2 (1979)	
	V	Amalthea	180,000	0	11	57	270 × 165 × 153		Barnard (1892)	14.1
	XV	Thebe	222,000	0	16	11	(80)		Synnott/Voyager 1 (1980)	
	I	Io	422,000	1	18	28	3632 ± 60	3.5	Galileo (1610)	5.0
	II	Europa	671,000	3	13	14	3126 ± 60	3.0	Galileo (1610)	5.3
	III	Ganymede	1,070,000	7	03	43	5276 ± 60	1.9	Galileo (1610)	4.6
	IV	Callisto	1,885,000	16	16	32	4820 ± 60	1.8	Galileo (1610)	5.6
	XIII	Leda	11,110,000	240			(10)		Kowal (1974)	20
	VI	Himalia	11,470,000	251			170		Perrine (1904)	14.7
	X	Lysithea	11,710,000	260			(20)		Nicholson (1938)	18.4
	VII	Elara	11,740,000	260			80		Perrine (1905)	16.4
	XII	Ananke	20,700,000	617 R			(20)		Nicholson (1951)	18.9
	XI	Carme	22,350,000	692 R			(30)		Nicholson (1938)	18.0
	VIII	Pasiphae	23,330,000	735 R			(40)		Melotte (1908)	17.7
	IX	Sinope	23,700,000	758 R			(30)		Nicholson (1914)	18.3
	—								Kowal (1975)	20
Saturn		Atlas	137,670		14	27	80 × 60 × 40		Voyager 1	17
		Inner F ring shepherd	139,353		14	43	140 × 100 × 80		Voyager 1	16
		Outer F ring shepherd	141,700		15	05	110 × 90 × 70		Voyager 1	16
		Epimetheus	151,422		16	40	220 × 200 × 160		Smith, Reitsema, Larson and Fountain	15
		Janus	151,472		16	40	14 × 120 × 100		Cruikshank/Pioneer 11	
	1	Mimas	185,600		22	37	390	1.2	W. Herschel (1789)	12.9
	2	Enceladus	238,100	1	08	32	510	1.1	W. Herschel (1789)	11.7
	3	Tethys	294,700	1	22	15	1,050	1.0	Cassini (1684)	10.3
		Telesto	294,700	1	22	15	34 × 28 × 26			19
		Calypso	294,700	1	22	15	34 × 22 × 22			19
	4	Dione	377,500	2	17	36	1,120	1.4	Cassini (1684)	10.4
		Dione B	378,060	2	17	45	36 × 22 × 30		Laques and Lecacheux (1980)	19
	5	Rhea	527,200	4	12	16	1,530	1.3	Cassini (1672)	9.7
	6	Titan	1,221,600	15	21	51	5,150	1.9	Huygens (1665)	8.3
	7	Hyperion	1,483,000	21	06	45	400 × 250 × 220	1.9	Bond (1848)	14.2
	8	Iapetus	3,560,100	79	03	43	1,440	1.2	Cassini (1671)	11.2
	9	Phoebe	12,950,000	549	03	33	200		W. Pickering (1898)	16.3
Uranus	5	Miranda	130,000	1	09	56	484		Kuiper (1948)	16.5
	1	Ariel	191,000	2	12	29R	1178		Lassell (1851)	14.4
	2	Umbriel	266,000	4	03	27R	1192		Lassell (1851)	15.3
	3	Titania	436,000	8	16	56R	1610		W. Herschel (1787)	14.0
	4	Oberon	583,000	13	11	07R	1546		W. Herschel (1787)	14.2
	10	Small satellites							Voyager 2 (1985/86)	
Neptune	3		75,000						Reitsema et al. (1981)	
		Triton	354,000	5	21	03R	3200 ± 400		Lassell (1846)	13.6
		Nereid	5,570,000	365	5		(600)		Kuiper (1949)	18.7
Pluto		Charon	20,000?	6	9	17	(1000)		Christy (1978)	16–17

R signifies retrograde rotation; parentheses signify uncertainty.

A.4

Appendix 5
Brightest Stars

Star	Name	Position 2000.0		Apparent Magnitude (V)	Spectral Type*		Absolute Magnitude	Approx. Distance (ly)
		R.A.	Dec.					
1. α CMa A	Sirius	06 45.1	− 16 43	− 1.46	A1	V	+ 1.42	9
2. α Car	Canopus	06 24.0	− 52 42	− 0.72	F0	Ib-II	− 3.1	98
3. α Boo	Arcturus	14 15.7	+ 19 11	− 0.06	K2	IIIp	− 0.3	36
4. α Cen A	Rigil Kentaurus	14 39.6	− 60 50	0.01	G2	V	+ 4.37	4.2
5. α Lyr	Vega	18 36.9	+ 38 47	0.04	A0	V	+ 0.5	27
6. α Aur	Capella	05 16.7	+ 46 00	0.05	G8III? + F		− 0.6	45
7. β Ori A	Rigel	05 14.5	− 08 12	0.14v	B8	Ia	− 7.1	900
8. α CMi A	Procyon	07 39.3	+ 05 14	0.37	F5	IV-V	+ 2.6	11
9. α Ori	Betelgeuse	05 55.2	+ 07 24	0.41v	M2	Iab	− 5.6	520
10. α Eri	Achernar	01 37.7	− 57 14	0.51	B3	Vp	− 2.3	118
11. β Cen AB	Hadar	14 03.8	− 60 22	0.63v	B1	III	− 5.2	490
12. α Aql	Altair	19 50.8	+ 08 52	0.76	A7	IV-V	+ 2.2	17
13. α Tau A	Aldebaran	04 35.9	+ 16 31	0.86v	K5	III	− 0.7	68
14. α Vir	Spica	13 25.2	− 11 10	0.91v	B1	V	− 3.3	220
15. α Sco A	Antares	16 29.4	− 26 26	0.92v	M1 Ib + B		− 5.1	520
16. α PsA	Fomalhaut	22 57.6	− 29 37	1.15	A3	V	+ 2.0	23
17. β Gem	Pollux	07 45.3	+ 28 02	1.16	K0	III	+ 1.0	35
18. α Cyg	Deneb	20 41.4	+ 45 17	1.26	A2	Ia	− 7.1	1600
19. β Cru	Beta Crucis	12 47.7	− 59 41	1.28v	B0.5	III	− 4.6	490
20. α Leo A	Regulus	10 08.4	+ 11 58	1.36	B7	V	− 0.7	84
21. α Cru A	Acrux	12 26.6	− 63 06	1.39	B0.5	IV	− 3.9	370
22. ε CMa A	Adhara	06 58.6	− 28 58	1.48	B2	II	− 5.1	680
23. λ Sco	Shaula	17 33.6	− 37 06	1.60v	B1	V	− 3.3	310
24. γ Ori	Bellatrix	05 25.1	+ 06 21	1.64	B2	III	− 4.2	470
25. β Tau	Elnath	05 26.3	+ 28 36	1.65	B7	III	− 3.2	300

*In Spectral Type column: I = supergiants, II = bright giants, III = giants, IV = subgiants, V = dwarfs.

Appendix 6
Greek Alphabet

Upper Case	Lower Case		Upper Case	Lower case	
A	α	alpha	N	ν	nu
B	β	beta	Ξ	ξ	xi
Γ	γ	gamma	O	o	omicron
Δ	δ	delta	Π	π	pi
E	ε	epsilon	P	ρ	rho
Z	ζ	zeta	Σ	σ	sigma
H	η	eta	T	τ	tau
Θ	θ	theta	Υ	υ	upsilon
I	ι	iota	Φ	φ	phi
K	κ	kappa	X	χ	chi
Λ	λ	lambda	Ψ	ψ	psi
M	μ	mu	Ω	ω	omega

Appendix 7
The Nearest Stars

Name	R.A. h	(2000.0) m	Dec. °	'	Parallax π "	Distance ly	Spectral Type	Apparent Magnitude V	Absolute Magnitude M_V	Luminosity ($L_\odot = 1$)
1 Sun							G2 V	−26.72	4.85	1.0
2 Proxima Cen	14	30.0	−62	40	0.763	4.3	M5.5 V	11.05	15.49	0.00006
α Cen A	14	39.6	−60	50	.750	4.3	G2 V	−0.01	4.37	1.6
α Cen B							K1 V	1.33	5.71	.45
3 Barnard's star	17	57.9	+04	41	.549	5.9	M3.8 V	9.54	13.22	0.00045
4 Wolf 359 (CN Leo)	10	56.7	+07	00	.421	7.8	M5.8 V	13.53	16.65	0.00002
5 BD + 36°2147 = HD95735 (Lalande 21185)	11	03.4	+35	58	.400	8.2	M2.1 V	7.50	10.50	0.0055
6 Sirius A	6	45.1	−16	43	.376	8.7	A1 V	−1.46	1.42	23.5
Sirius B							DA	8.3	11.2	0.003
7 L 726-8 = A	1	38.8	−17	57	.372	8.8	M5.6 V	12.52	15.46	0.00006
UV Cet = B								13.02	15.96	0.00004
8 Ross 154 (V1216 Sgr)	18	49.7	−23	49	.346	9.4	M3.6 V	10.45	13.14	0.00048
9 Ross 248 (HH And)	23	41.9	+44	10	.313	10.4	M4.9 V	12.29	14.78	0.00011
10 ε Eri	3	32.9	−09	28	.303	10.8	K2 V	3.73	6.14	0.30
11 Ross 128 (FI Vir)	11	47.6	+00	48	.301	10.8	M4.1 V	11.10	13.47	0.00036
12 61 Cyg A	21	06.9	+38	45	.295	11.1	K3.5 V	5.22	7.56	0.082
61 Cyg B							K4.7 V	6.03	8.37	0.039
13 Procyon A	7	39.3	+05	14	.292	11.2	F5 IV-V	0.37	2.64	7.65
Procyon B							DF	10.7	13.0	0.00055
14 L 789-6	22	38.4	−15	18	.291	11.2	M5.5 V	12.18	14.49	0.00014
15 ε Ind	22	03.4	−56	47	.290	11.3	K3 V	4.68	7.00	0.14
16 BD + 43°44 A (GX And)	00	18.1	+44	00	.290	11.3	M1.3 V	8.08	10.39	0.0061
+ 43°44 B (GQ And)		(Groombridge 34 AB)					M3.8 V	11.06	13.37	0.00039
17 τ Ceti	1	44.1	−15	56	.288	11.3	G8 V	3.50	5.72	0.45
18 BD + 59°1915 A	18	42.9	+59	37	.287	11.4	M3.0 V	8.90	11.15	0.0030
+ 59°1915 B		(HD173739/40 = Struve 2398AB = ADS 11632AB)					M3.5 V	9.69	11.94	0.0015
19 CD −36°15693 = HD217987	23	05.9	−35	51	.283	11.5	M1.3 V	7.35	9.58	0.013
20 G 51−15	8	29.9	+26	46	.276	11.8	M6.6 V	14.81	17.03	0.00001
21 L 725−32 (YZ Cet)	1	12.4	−16	59	.270	12.1	M4.5 V	12.04	14.12	0.00020
22 BD + 5°1668	7	27.4	+05	13	.267	12.2	M3.7 V	9.82	11.94	0.0015
23 CD −30°14192 = HD202560	21	17.3	−38	52	.261	12.5	K5.5 V	6.66	8.74	0.028
24 Kapteyn's star	5	11.2	−45	01	.259	12.6	M0.0 V	8.84	10.88	0.0039
25 Krüger 60 A	22	28.1	+57	42	.253	12.9	M3.3 V	9.85	11.87	0.0016
Krüger 60 B (DO Cep)							M5 V	11.3	13.3	0.0004

Appendix 8
Messier Catalogue

M	NGC	α h m	(1980.0)	δ °	m_v	Description
1	1952	5 33.3		+22 01	11.3	Crab Nebula in Taurus
2	7089	21 32.4		−00 54	6.3	Globular cluster in Aquarius
3	5272	13 41.3		+28 29	6.2	Globular cluster in Canes Venatici
4	6121	16 22.4		−26 27	6.1	Globular cluster in Scorpius
5	5904	15 17.5		+02 11	6	Globular cluster in Serpens
6	6402	17 38.9		−32 11	6	Open cluster in Scorpius
7	6475	17 52.6		−34 48	5	Open cluster in Scorpius
8	6523	18 02.4		−24 23		Lagoon Nebula in Sagittarius
9	6333	17 18.1		−18 30	7.6	Globular cluster in Ophiuchus
10	6254	16 56.0		−04 05	6.4	Globular cluster in Ophiuchus
11	6705	18 50.0		−06 18	7	Open cluster in Scutum
12	6218	16 46.1		−01 55	6.7	Globular cluster in Ophiuchus
13	6205	16 41.0		+36 30	5.8	Globular cluster in Hercules
14	6402	17 36.5		−03 14	7.8	Globular cluster in Ophiuchus
15	7078	21 29.1		+12 05	6.3	Globular cluster in Pegasus
16	6611	18 17.8		−13 48	7	Open cluster and nebula in Serpens
17	6618	18 19.7		−16 12	7	Omega Nebula in Sagittarius
18	6613	18 18.8		−17 09	7	Open cluster in Sagittarius
19	6273	17 01.3		−26 14	6.9	Globular cluster in Ophiuchus
20	6514	18 01.2		−23 02		Trifid Nebula in Sagittarius
21	6531	18 03.4		−22 30	7	Open cluster in Sagittarius
22	6656	18 35.2		−23 55	5.2	Globular cluster in Sagittarius
23	6494	17 55.7		−19 00	6	Open cluster in Sagittarius
24	6603	18 17.3		−18 27	6	Open cluster in Sagittarius
25	IC4725	18 30.5		−19 16	6	Open cluster in Sagittarius
26	6694	18 44.1		−09 25	9	Open cluster in Scutum
27	6853	19 58.8		+22 40	8.2	Dumbbell Nebula; planetary nebula in Vulpecula
28	6626	18 23.2		−24 52	7.1	Globular cluster in Sagittarius
29	6913	20 23.3		+38 27	8	Open cluster in Cygnus
30	7099	21 39.2		−23 15	7.6	Globular cluster in Capricornus
31	224	0 41.6		+41 09	3.7	Andromeda Galaxy (Sb)
32	221	0 41.6		+40 45	8.5	Elliptical galaxy in Andromeda; companion to M31
33	598	1 32.8		+30 33	5.9	Spiral galaxy (Sc) in Triangulum
34	1039	2 40.7		+42 43	6	Open cluster in Perseus
35	2168	6 07.6		+24 21	6	Open cluster in Gemini
36	1960	5 35.0		+34 05	6	Open cluster in Auriga
37	2099	5 51.5		+32 33	6	Open cluster in Auriga
38	1912	5 27.3		+35 48	6	Open cluster in Auriga
39	7092	21 31.5		+48 21	6	Open cluster in Cygnus
40	—	—		—		Double star in Ursa Major (Messier mistake)
41	2287	6 46.2		−20 43	6	Open cluster in Canis Major
42	1976	5 34.4		−05 24		Orion Nebula
43	1982	5 34.6		−05 18		Orion Nebula; smaller part
44	2632	8 38.8		+20 04	4	Praesepe; open cluster in Cancer
45	—	3 46.3		+24 03	2	The Pleiades; open cluster in Taurus

Appendix 8 continues on following page

M	NGC	α h m (1980.0)	δ ° ′	m_v	Description
46	2437	7 40.9	− 14 46	7	Open cluster in Puppis
47	2422	7 35.6	− 14 27	5	Open cluster in Puppis
48	2548	8 12.5	− 05 43	6	Open cluster in Hydra
49	4472	12 28.8	+ 08 07	8.9	Elliptical galaxy in Virgo
50	2323	7 02.0	− 08 19	7	Open cluster in Monoceros
51	5194	13 29.0	+ 47 18	8.4	Whirlpool Galaxy; spiral galaxy (Sc) in Canes Venatici
52	7654	23 23.3	+ 61 29	7	Open cluster in Cassiopeia
53	5024	13 12.0	+ 18 17	7.7	Globular cluster in Coma Berenices
54	6715	18 35.8	− 30 30	7.7	Globular cluster in Sagittarius
55	6809	19 38.7	− 31 00	6.1	Globular cluster in Sagittarius
56	6779	19 15.8	+ 30 08	8.3	Globular cluster in Lyra
57	6720	18 52.9	+ 33 01	9.0	Ring Nebula; planetary nebula in Lyra
58	4579	12 36.7	+ 11 56	9.9	Spiral galaxy (SBb) in Virgo
59	4621	12 41.0	+ 11 47	10.3	Elliptical galaxy in Virgo
60	4649	12 42.6	+ 11 41	9.3	Elliptical galaxy in Virgo
61	4303	12 20.8	+ 04 36	9.7	Spiral galaxy (Sc) in Virgo
62	6266	16 59.9	− 30 05	7.2	Globular cluster in Scorpius
63	5055	13 14.8	+ 42 08	8.8	Spiral galaxy (Sb) in Canes Venatici
64	4826	12 55.7	+ 21 48	8.7	Spiral galaxy (Sb) in Coma Berenices
65	3623	11 17.8	+ 13 13	9.6	Spiral galaxy (Sa) in Leo
66	3627	11 19.1	+ 13 07	9.2	Spiral galaxy (Sb) in Leo; companion to M65
67	2682	8 50.0	+ 11 54	7	Open cluster in Cancer
68	4590	12 38.3	− 26 38	8	Globular cluster in Hydra
69	6637	18 30.1	− 32 23	7.7	Globular cluster in Sagittarius
70	6681	18 42.0	− 32 18	8.2	Globular cluster in Sagittarius
71	6838	19 52.8	+ 18 44	6.9	Globular cluster in Sagittarius
72	6981	20 52.3	− 12 39	9.2	Globular cluster in Aquarius
73	6994	20 57.8	− 12 44		Open cluster in Aquarius
74	628	1 35.6	+ 15 41	9.5	Spiral galaxy (Sc) in Pisces
75	6864	20 04.9	− 21 59	8.3	Globular cluster in Sagittarius
76	650	1 40.9	+ 51 28	11.4	Planetary nebula in Perseus
77	1068	2 41.6	− 00 04	9.1	Spiral galaxy (Sb) in Cetus
78	2068	5 45.8	+ 00 02		Small emission nebula in Orion
79	1904	5 23.3	− 24 32	7.3	Globular cluster in Lepus
80	6093	16 15.8	− 22 56	7.2	Globular cluster in Scorpius
81	3031	9 54.2	+ 69 09	6.9	Spiral galaxy (Sb) in Ursa Major
82	3034	9 54.4	+ 69 47	8.7	Irregular galaxy (Irr) in Ursa Major
83	5236	13 35.9	− 29 46	7.5	Spiral galaxy (Sc) in Hydra
84	4374	12 24.1	+ 13 00	9.8	Elliptical galaxy in Virgo
85	4382	12 24.3	+ 18 18	9.5	Elliptical galaxy (S0) in Coma Berenices
86	4406	12 25.1	+ 13 03	9.8	Elliptical galaxy in Virgo
87	4486	12 29.7	+ 12 30	9.3	Elliptical galaxy (Ep) in Virgo
88	4501	12 30.9	+ 14 32	9.7	Spiral galaxy (Sb) in Coma Berenices
89	4552	12 34.6	+ 12 40	10.3	Elliptical galaxy in Virgo
90	4569	12 35.8	+ 13 16	9.7	Spiral galaxy (Sb) in Virgo
91	—	—	—		M58?
92	6341	17 16.5	+ 43 10	6.3	Globular cluster in Hercules
93	2447	7 43.6	− 23 49	6	Open cluster in Puppis
94	4736	12 50.1	+ 41 14	8.1	Spiral galaxy (Sb) in Canes Venatici
95	3351	10 42.8	+ 11 49	9.9	Barred spiral galaxy (SBb) in Leo

Appendix 8 continues on facing page

M	NGC	α h m (1980.0)	δ ° '	m_v	Description
96	3368	10 45.6	+11 56	9.4	Spiral galaxy (Sa) in Leo
97	3587	11 13.7	+55 08	11.1	Owl Nebula; planetary nebula in Ursa Major
98	4192	12 12.7	+15 01	10.4	Spiral galaxy (Sb) in Coma Berenices
99	4254	12 17.8	+14 32	9.9	Spiral galaxy (Sc) in Coma Berenices
100	4321	12 21.9	+15 56	9.6	Spiral galaxy (Sc) in Coma Berenices
101	5457	14 02.5	+54 27	8.1	Spiral galaxy (Sc) in Ursa Major
102	—	—	—		M101; duplication (Messier mistake)
103	581	1 31.9	+60 35	7	Open cluster in Cassiopeia
104	4594	12 39.0	−11 35	8	Sombrero Nebula; spiral galaxy (Sa) in Virgo
105	3379	10 46.8	+12 51	9.5	Elliptical galaxy in Leo
106	4258	12 18.0	+47 25	9	Spiral galaxy (Sb) in Canes Venatici
107	6171	16 31.8	−13 01	9	Globular cluster in Ophiuchus
108	3556	11 10.5	+55 47	10.5	Spiral galaxy (Sb) in Ursa Major
109	3992	11 56.6	+53 29	10.6	Barred spiral galaxy (SBc) in Ursa Major

Appendix 9
The Constellations

Latin Name	Genitive	Abbreviation	Translation
Andromeda	Andromedae	And	Andromeda*
Antlia	Antliae	Ant	Pump
Apus	Apodis	Aps	Bird of Paradise
Aquarius	Aquarii	Aqr	Water Bearer
Aquila	Aquilae	Aql	Eagle
Ara	Arae	Ara	Altar
Aries	Arietis	Ari	Ram
Auriga	Aurigae	Aur	Charioteer
Boötes	Boötis	Boo	Herdsman
Caelum	Caeli	Cae	Chisel
Camelopardalis	Camelopardalis	Cam	Giraffe
Cancer	Cancri	Cnc	Crab
Canes Venatici	Canum Venaticorum	CVn	Hunting Dogs
Canis Major	Canis Majoris	CMa	Big Dog
Canis Minor	Canis Minoris	CMi	Little Dog
Capricornus	Capricorni	Cap	Goat
Carina	Carinae	Car	Ship's Keel**
Cassiopeia	Cassiopeiae	Cas	Cassiopeia*
Centaurus	Centauri	Cen	Centaur*
Cepheus	Cephei	Cep	Cepheus*
Cetus	Ceti	Cet	Whale
Chamaeleon	Chamaeleonis	Cha	Chameleon
Circinus	Circini	Cir	Compass
Columba	Columbae	Col	Dove
Coma Berenices	Comae Berenices	Com	Berenice's Hair*
Corona Australis	Coronae Australis	CrA	Southern Crown
Corona Borealis	Coronae Borealis	CrB	Northern Crown
Corvus	Corvi	Crv	Crow
Crater	Crateris	Crt	Cup
Crux	Crucis	Cru	Southern Cross
Cygnus	Cygni	Cyg	Swan
Delphinus	Delphini	Del	Dolphin
Dorado	Doradus	Dor	Swordfish

Appendix 9 continues on following page

Latin Name	Genitive	Abbreviation	Translation
Draco	Draconis	Dra	Dragon
Equuleus	Equulei	Equ	Little Horse
Eridanus	Eridani	Eri	River Eridanus*
Fornax	Fornacis	For	Furnace
Gemini	Geminorum	Gem	Twins
Grus	Gruis	Gru	Crane
Hercules	Herculis	Her	Hercules*
Horologium	Horologii	Hor	Clock
Hydra	Hydrae	Hya	Hydra* (water monster)
Hydrus	Hydri	Hyi	Sea serpent
Indus	Indi	Ind	Indian
Lacerta	Lacertae	Lac	Lizard
Leo	Leonis	Leo	Lion
Leo Minor	Leonis Minoris	LMi	Little Lion
Lepus	Leporis	Lep	Hare
Libra	Librae	Lib	Scales
Lupus	Lupi	Lup	Wolf
Lynx	Lyncis	Lyn	Lynx
Lyra	Lyrae	Lyr	Harp
Mensa	Mensae	Men	Table (mountain)
Microscopium	Microscopii	Mic	Microscope
Monoceros	Monocerotis	Mon	Unicorn
Musca	Muscae	Mus	Fly
Norma	Normae	Nor	Carpenter's square
Octans	Octantis	Oct	Octant
Ophiuchus	Ophiuchi	Oph	Ophiuchus* (serpent bearer)
Orion	Orionis	Ori	Orion*
Pavo	Pavonis	Pav	Peacock
Pegasus	Pegasi	Peg	Pegasus* (winged horse)
Perseus	Persei	Per	Perseus*
Phoenix	Phoenicis	Phe	Phoenix
Pictor	Pictoris	Pic	Easel
Pisces	Piscium	Psc	Fish
Piscis Austrinus	Piscis Austrini	PsA	Southern Fish
Puppis	Puppis	Pup	Ship's Stern**
Pyxis	Pyxidis	Pyx	Ship's Compass**
Reticulum	Reticuli	Ret	Net
Sagitta	Sagittae	Sge	Arrow
Sagittarius	Sagittarii	Sgr	Archer
Scorpius	Scorpii	Sco	Scorpion
Sculptor	Sculptoris	Scl	Sculptor
Scutum	Scuti	Sct	Shield
Serpens	Serpentis	Ser	Serpent
Sextans	Sextantis	Sex	Sextant
Taurus	Tauri	Tau	Bull
Telescopium	Telescopii	Tel	Telescope
Triangulum	Trianguli	Tri	Triangle
Triangulum Australe	Trianguli Australis	TrA	Southern Triangle
Tucana	Tucanae	Tuc	Toucan
Ursa Major	Ursae Majoris	UMa	Big Bear
Ursa Minor	Ursae Minoris	UMi	Little Bear
Vela	Velorum	Vel	Ship's Sails**
Virgo	Virginis	Vir	Virgin
Volans	Volantis	Vol	Flying Fish
Vulpecula	Vulpeculae	Vul	Little Fox

*Proper names
**Formerly formed the constellation Argo Navis, the Argonauts' ship.

Appendix 10
Temperature Conversions

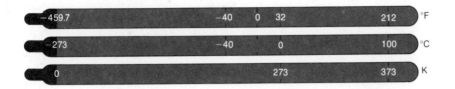

FIGURE A10–1 Temperature scales in the range useful for studies of planets and interstellar space.

Though most Americans use the Fahrenheit temperature scale, in which water freezes at 32°F and boils at 212°F, most of the rest of the world uses the Celsius scale, where water freezes at 0°C and boils at 100°C. Note that the difference between freezing and boiling points for each scale is 180°F and 100°C respectively, so a change of 180°F equals a change of 100°C and a change of 9°F equals a change of 5°C.

There is no water on the stars or on most planets, so astronomers use a more fundamental scale. Their scale, the kelvin scale (whose symbol is K) begins at absolute zero, the coldest temperature that can ever be approached. Since absolute zero is about −273.16°C, and one kelvin (1 K) is the same size as 1°C, the freezing point of water is about 273 K (Fig. A10–1).

To convert from kelvins to °C, simply subtract 273. To change from °C to °F, we must first multiply by 9/5 and then add 32. (Remember: times two, minus point two, plus thirty-two: multiply by 2, subtract a tenth of that, and then add 32, to get °F.) For stellar temperatures, the 32° is too small to notice (Fig. A10–2).

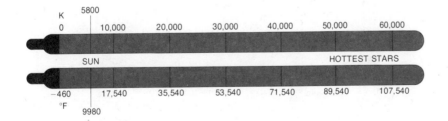

FIGURE A10–2 Temperature scales in the range useful for stellar studies.

Appendix 11
Elements and Solar-System Abundances

		Name	Atomic Weight	Abundance*			Name	Atomic Weight	Abundance*
1	H	hydrogen	1.01	2.72×10^{10}	55	Cs	cesium	123.91	0.37
2	He	helium	4.00	2.18×10^{9}	56	Ba	barium	137.34	4.36
3	Li	lithium	6.94	59.7	57	La	lanthanum	138.91	0.45
4	Be	beryllium	9.01	0.78	58	Ce	cerium	140.12	1.16
5	B	boron	10.81	24	59	Pr	praseodymium	140.91	0.17
6	C	carbon	12.01	1.21×10^{7}	60	Nd	neodymium	144.24	0.84
7	N	nitrogen	14.01	2.48×10^{6}	61	Pm	promethium	146	
8	O	oxygen	16.00	2.01×10^{7}	62	Sm	samarium	150.4	0.26
9	F	fluorine	19.00	843	63	Eu	europium	151.96	0.10
10	Ne	neon	20.18	3.76×10^{6}	64	Gd	gadolinium	157.25	0.33
11	Na	sodium	22.29	5.70×10^{4}	65	Tb	terbium	158.93	0.06
12	Mg	magnesium	24.31	1.08×10^{6}	66	Dy	dysprosium	162.50	0.40
13	Al	aluminum	26.98	8.49×10^{4}	67	Ho	holmium	164.93	0.09
14	Si	silicon	28.09	1.00×10^{6}	68	Er	erbium	167.26	0.25
15	P	phosphorus	30.97	1.04×10^{4}	69	Tm	thulium	168.93	0.04
16	S	sulphur	32.06	5.15×10^{5}	70	Yb	ytterbium	170.04	0.24
17	Cl	chlorine	35.45	5240	71	Lu	lutetium	174.97	0.04
18	Ar	argon	39.95	1.04×10^{5}	72	Hf	hafnium	178.49	0.18
19	K	potassium	39.10	3770	73	Ta	tantalum	180.95	0.02
20	Ca	calcium	40.08	6.11×10^{4}	74	W	tungsten	183.85	0.14
21	Sc	scandium	44.96	33.8	75	Re	rhenium	186.2	0.05
22	Ti	titanium	47.90	2400	76	Os	osmium	190.2	0.72
23	V	vanadium	50.94	295	77	Ir	iridium	192.2	0.66
24	Cr	chromium	52.00	1.34×10^{4}	78	Pt	platinum	195.09	1.37
25	Mn	manganese	54.94	9510	79	Au	gold	196.97	0.19
26	Fe	iron	55.85	9.00×10^{5}	80	Hg	mercury	200.59	0.52
27	Co	cobalt	58.93	2250	81	Tl	thallium	204.37	0.18
28	Ni	nickel	58.71	4.93×10^{4}	82	Pb	lead	207.19	3.15
29	Cu	copper	63.55	514	83	Bi	bismuth	208.98	0.14
30	Zn	zinc	65.37	1260	84	Po	polonium	210	
31	Ga	gallium	69.72	37.8	85	At	astatine	210	
32	Ge	germanium	72.59	118	86	Rn	radon	222	
33	As	arsenic	74.92	6.79	87	Fr	francium	223	
34	Se	selenium	78.96	62.1	88	Ra	radium	226.03	
35	Br	bromine	79.90	11.8	89	Ac	actinium	227	
36	Kr	krypton	83.80	45.3	90	Th	thorium	232.04	0.05
37	Rb	rubidium	85.47	7.09	91	Pa	protactinium	230.04	
38	Sr	strontium	87.62	23.8	92	U	uranium	238.03	0.03
39	Y	yttrium	88.91	4.64	93	Np	neptunium	237.05	
40	Zr	zirconium	91.22	10.7	94	Pu	plutonium	242	
41	Nb	niobium	92.91	0.71	95	Am	americium	242	
42	Mo	molybdenum	95.94	2.52	96	Cm	curium	245	
43	Tc	technetium	98.91		97	Bk	berkelium	248	
44	Ru	ruthenium	101.07	1.86	98	Cf	californium	252	
45	Rh	rhodium	102.91	0.34	99	Es	einsteinium	253	
46	Pd	palladium	106.4	1.39	100	Fm	fermium	257	
47	Ag	silver	107.87	0.53	101	Md	mendelevium	257	
48	Cd	cadmium	112.40	1.69	102	No	nobelium	255	
49	In	indium	114.82	0.18	103	Lr	lawrencium	256	
50	Sn	tin	118.69	3.82	104	Rf	rutherfordium	261	
51	Sb	antimony	121.75	0.35	105	Ha	hahnium	257–262	
52	Te	tellurium	127.60	4.91	106		not named; 1974	259–263	
53	I	iodine	126.90	0.90	107		not named; 1981	262	
54	Xe	xenon	131.30	4.35	108		not named; 1984	265	
					109		not named; 1982	266	

*Abundances relative to 10^{6} for silicon.

SELECTED READINGS

Monthly Non-Technical Magazines on Astronomy

Sky and Telescope, 49 Bay State Road, Cambridge, MA 02238.

Astronomy, 625 E. St. Paul Avenue, P.O. Box 92788, Milwaukee, WI 53202.

Mercury, Astronomical Society of the Pacific, 1290 24th Avenue, San Francisco, CA 94122.

The Griffith Observer, 2800 East Observatory Road, Los Angeles, CA 90027.

Magazines and Annuals Carrying Articles on Astronomy

Science 86, 87, 88, etc., P.O. Box 10790, Des Moines, IA 50340.

Science News, 1719 N Street, N.W., Washington, DC 20036. Published weekly.

Scientific American, 415 Madison Avenue, New York, NY 10017.

National Geographic, Washington, DC 20036.

Natural History, Membership Services, Box 4300, Bergenfield, N.J. 07621.

Physics Today, American Institute of Physics, 335 East 45 Street, New York, NY 10017.

Science Year (Chicago, IL: Field Enterprises Educational Corp.). The World Book Science Annual.

Smithsonian, 900 Jefferson Drive, Washington, DC 20560.

Yearbook of Science and the Future (Chicago, IL 60611: Encyclopaedia Britannica).

Science Digest, P.O. Box 10076, Des Moines, IA 50350.

Discover, 3435 Wilshire Blvd, Los Angeles, CA 90010.

Observing Reference Books

Donald H. Menzel and Jay M. Pasachoff, *A Field Guide to the Stars and Planets,* 2nd ed. (Boston: Houghton Mifflin Co., 1983). All kinds of observing information, including monthly maps and the 2000.0 sky atlas by Wil Tirion, and Graphic Timetables to locate planets and special objects like clusters.

Charles A. Whitney, *Whitney's Star Finder,* 3rd ed. (New York: Alfred A. Knopf, Inc., 1981).

Wil Tirion, *Star Atlas 2000.0* (Cambridge, MA: Sky Publishing Corp., 1981). Accurate and up-to-date. Available in black-on-white, white-on-black, and color verions.

Ben Mayer, *Starwatch* (New York: Perigee/Putnam, 1984). An inspirational introduction to observing.

Arthur P. Norton, *Norton's Star Atlas and Reference Handbook,* regularly revised by the successors of the late Mr. Norton (Cambridge, MA: Sky Publishing Corp.).

The Observer's Handbook (yearly), Royal Astronomical Society of Canada, 252 College Street, Toronto M5T 1R7, Canada.

Guy Ottewell, *Astronomical Calendar* (yearly), Department of Physics, Furman University, Greenville, SC 29613.

The Astronomical Almanac (yearly), U.S. Government Printing Office, Washington, DC 20402.

Hans Vehrenberg, *Atlas of Deep Sky Splendors* (Cambridge, MA: Sky Publishing Corp., 4th ed., 1983). Photographs, descriptions, and finding charts for hundreds of beautiful objects.

For Information about Amateur Societies

American Association of Variable Star Observers (AAVSO), 187 Concord Avenue, Cambridge, MA 02138.

Association of Lunar and Planetary Observers (ALPO), 8930 Raven Drive, Waco, TX 76710.

Careers in Astronomy

Charles R. Tolbert, Education Officer, American Astronomical Society, Leander McCormick Observatory, University of Virginia, P.O. Box 3818, University Station, Charlottesville, VA. 22903-0818. A free booklet, *Careers in Astronomy,* is available on request.

Space for Women, derived from a symposium for women on careers. For free copies, write to: Center for Astrophysics, 60 Garden Street, Cambridge, MA 02138.

General Reading

Herbert Friedman, *The Amazing Universe* (Washington, DC: National Geographic Society, 1975). National Geographic's survey of modern astronomy, written by a pioneer in space science.

David Malin and Paul Murdin, *Colours of the Stars* (Cambridge University Press, 1984). Fantastic color photographs and interesting descriptions of color in celestial objects.

Simon Mitton, ed., *The Cambridge Encyclopedia of Astronomy* (New York: Crown Publishers, 1977).

Harlow Shapley, ed., *Source Book in Astronomy 1900–1950* (Cambridge, MA: Harvard University Press, 1960). Reprints.

Otto Struve and Velta Zebergs, *Astronomy of the Twentieth Century* (New York: Macmillan, 1962). A historical view.

SOME ADDITIONAL BOOKS

Observatories and Observing

H. T. Kirby-Smith, *U.S. Observatories: A Directory and Travel Guide* (New York: Van Nostrand Reinhold, 1976).

David O. Woodbury, *The Glass Giant of Palomar* (New York: Dodd Mead, 1970). The story of the construction of the 5-m telescope.

Helen Wright, Joan N. Warnow, and Charles Weiner, eds., *The Legacy of George Ellery Hale* (Cambridge, MA: MIT Press, 1972). A beautifully illustrated historical treatment.

W. M. Smart, *Text-Book on Spherical Astronomy* (Cambridge: Cambridge University Press, 1977). The old standard, revised from the first (1931) edition and the later (1944) edition.

Martin Cohen, *In Quest of Telescopes* (Cambridge, MA: Sky Publishing Corp., 1981). What it is like to be an astronomer.

Stars

Lawrence H. Aller, *Atoms, Stars, and Nebulae*, Revised ed. (Cambridge, MA: Harvard University Press, 1971). One of the Harvard Books on Astronomy series of popular works.

Bart J. Bok and Priscilla F. Bok, *The Milky Way*, 5th ed. (Cambridge, MA: Harvard University Press, 1981). A readable and well-illustrated survey from the Harvard Books on Astronomy series; includes good discussions of star clusters and H-R diagrams.

John A. Eddy, with Rein Ise, ed., *A New Sun: The Solar Results from Skylab* (NASA SP-402, 1979, GPO 033-000-00742-6).

Donald H. Menzel, *Our Sun*, Revised ed. (Cambridge, MA: Harvard University Press, 1959). A delightful general survey in the Harvard Books on Astronomy series.

Robert W. Noyes, *The Sun* (Cambridge, MA: Harvard University Press, 1982). A new member of the Harvard Books.

Fire of Life, The Smithsonian Book of the Sun (Smithsonian Exposition Press, W. W. Norton Co., 1981).

Walter Sullivan, *Black Holes* (New York: Anchor Press/Doubleday, 1979).

Henry L. Shipman, *Black Holes, Quasars, and the Universe,* second edition (Boston: Houghton Mifflin, 1980). A careful discussion of several topics of great current interest.

Cecilia Payne-Gaposchkin, *Stars and Clusters* (Cambridge, MA: Harvard University Press, 1979).

Donald A. Cooke, *The Life and Death of Stars* (New York: Crown, 1985). A lavishly illustrated description of stellar evolution.

Ronald Giovanelli, *Secrets of the Sun* (Cambridge University Press, 1984).

Solar System

Clark R. Chapman, *Planets of Rock and Ice: from Mercury to the Moons of Saturn* (New York: Charles Scribner's Sons, 1982). A nontechnical discussion of the planets and their moons.

Fred L. Whipple, *Orbiting the Sun: Planets and Satellites of the Solar System,* enlarged edition of *Earth, Moon, and Planets* (Cambridge, MA: Harvard University Press, 1981).

J. Kelly Beatty, Brian O'Leary, and Andrew Chaikin, *The New Solar System,* 2nd ed. (Cambridge, MA: Sky Publishing Co., 1982). Each chapter written by a different expert.

David D. Morrison and Jane Samz, *Voyage to Jupiter* (NASA SP 439, 1980). The story of the Voyager missions to Jupiter and what we learned, masterfully told and profusely illustrated.

David D. Morrison, *Voyages to Saturn* (NASA SP 451, 1982). On to Saturn.

Bevan M. French, *The Moon Book* (New York: Penguin, 1977). A clear, authoritative, thorough, and interesting to read report on the lunar program and its results.

Donald Goldsmith and Tobias Owen, *The Search for Life in the Universe* (Menlo Park, CA: Benjamin/Cummings, 1980).

James Elliot and Richard Kerr, *Rings* (Cambridge, MA: MIT Press, 1984). Includes first-person and other stories of the discoveries.

Galaxies and the Outer Universe

Richard Berendzen, Richard Hart, and Daniel Seeley, *Man Discovers the Galaxies* (New York: Neale Watson Academic Publications, 1976). A historical review.

Charles A. Whitney, *The Discovery of Our Galaxy* (New York: Alfred A. Knopf, 1971). A historical discussion on a more popular level.

Gerrit L. Verschuur, *The Invisible Universe* (New York: Springer-Verlag, 1974). A non-technical treatment of radio astronomy.

David A. Allen, *Infrared, the New Astronomy* (New York: John Wiley & Sons, 1985). Includes a personal narrative.

Halton C. Arp, *Atlas of Peculiar Galaxies* (Pasadena, CA: California Institute of Technology, 1966). Worth poring over.

Allan Sandage, *The Hubble Atlas of Galaxies* (Washington, DC: Carnegie Institution of Washington, 1961), Publication No. 618. Beautiful photographs of galaxies and thorough descriptions. Everyone should examine this carefully.

Harlow Shapley, *Galaxies*, 3rd ed., revised by Paul W. Hodge (Cambridge, MA: Harvard University Press, 1972). A non-technical study of galaxies, written by the master. In the Harvard series.

Timothy Ferris, *The Red Limit* (New York: William Morrow & Co., 1977). Written for the general reader.

Nigel Calder, *Einstein's Universe* (New York: Viking, 1979). Mostly relativity.

William J. Kaufmann, *The Cosmic Frontiers of General Relativity* (Boston: Little, Brown, 1977).

George Gamow, *One, Two, Three . . . Infinity* (New York: Bantam Books, 1971). A reprinting of a wonderful description of the structure of space that has introduced at an early age many a contemporary astronomer to his or her profession.

Steven Weinberg, *The First Three Minutes* (New York: Basic Books, 1977). A readable discussion of the first minutes after the big bang, including a discussion of the background radiation.

Timothy Ferris, *Galaxies* (New York: Stewart, Tabori, and Chang, 1982). Beautifully illustrated; paperback edition.

John D. Barrow and Joseph Silk, *The Left Hand of Creation* (New York: Basic Books, 1983). Cosmology elucidated.

Wallace Tucker and Riccardo Giacconi, *The X-Ray Universe* (Cambridge, MA: Harvard University Press, 1985). In the Harvard non-technical series.

GLOSSARY

absolute magnitude   The magnitude that a star would appear to have if it were at a distance of ten parsecs (about 32.6 light years) from us.

absorption line   Wavelengths at which the intensity of radiation is less than it is at neighboring wavelengths.

absorption nebula   Gas and dust seen in silhouette.

accretion disk   Matter that an object has taken up and which has formed a disk around the object.

active galaxy   A galaxy radiating much more than average in some part of the non-optical spectrum, revealing high-energy processes.

active sun   The group of solar phenomena that vary with time, such as active regions and their phenomena.

albedo   The fraction of light reflected by a body.

alpha particle   A helium nucleus; consists of two protons and two neutrons.

alt-azimuth   A two-axis telescope mounting in which motion around one of the axes, which is vertical, provides motion in azimuth, and motion around the perpendicular axis provides up-and-down (altitude) motion.

amino acid   A type of molecule containing the group NH_2 (the amino group). Amino acids are fundamental building blocks of life.

analemma   The figure-8 on a globe representing how far the true sun is ahead of or behind the mean sun for each declination through the year.

angstrom   A unit of length equal to 10^{-8} cm.

angular momentum   An intrinsic property of a system corresponding to the amount of its revolution or spin. The amount of angular momentum of a body orbiting around a point is the mass of the orbiting body times its (linear) velocity of revolution times its distance from the point. The amount of angular momentum of a spinning sphere is the moment of inertia, an intrinsic property of the distribution of mass, times the angular velocity of spin.

annular eclipse   A type of solar eclipse in which a ring (annulus) of solar photosphere remains visible.

antimatter   A type of matter in which each particle (antiproton, antineutron, etc.) is opposite in charge and certain other properties to a corresponding particle (proton, neutron, etc.) of the same mass of the ordinary type of matter from which the solar system is made.

aperture   The diameter of the lens or mirror that defines the amount of light focused by an optical system.

aperture synthesis   The use of several smaller telescopes together to give some of the properties, such as resolution, of a single larger aperture.

Apollo asteroids   A group of asteroids, with semimajor axes greater than earth's and less than 1.017 A.U., whose orbits overlap the earth's.

apparent magnitude   The brightness of a star as seen by an observer, given in a specific system in which a difference of five magnitudes corresponds to a brightness ratio of one hundred times; the scale is fixed by correspondence with a historical background.

asterism   A special apparent grouping of stars, part of a constellation.

asteroid   A "minor planet"; a non-luminous chunk of rock smaller than planet-size but larger than a meteoroid, in orbit around a star.

asteroid belt   A region of the solar system, between the orbits of Mars and Jupiter, in which most of the asteroids orbit.

astrometric binary   A system of two stars in which the existence of one star can be deduced by study of its gravitational effect on the proper motion of the other star.

astrometry   The branch of astronomy that involves the detailed measurement of the positions and motions of stars and other celestial bodies.

Astronomical Unit   The average distance from the earth to the sun.

Aten asteroids   A group of asteroids with semimajor axes smaller than 1 A.U.

atom   The smallest possible unit of a chemical element. When an atom is subdivided, the parts no longer have properties of any chemical element.

atomic number   The number of protons in an atom.

atomic weight   The number of protons and neutrons in an atom, averaged over the abundances of the different isotopes.

aurora   Glowing lights visible in the sky, resulting from processes in the earth's upper atmosphere and linked with the earth's magnetic field.

aurora australis   The southern aurora.

aurora borealis   The northern aurora.

autumnal equinox   Of the two locations in the sky where the ecliptic crosses the celestial equator, the one that the sun passes each year when moving from northern to southern declinations.

A.U. Astronomical Unit.

azimuth The angular distance around the horizon from the northern direction, usually expressed in angular measure from 0° for an object in the northern direction, to 180° for an object in the southern direction, around to 360°.

background radiation See *primordial background radiation.*

Baily's beads Beads of light visible around the rim of the moon at the beginning and end of a total solar eclipse. They result from the solar photosphere shining through valleys at the edge of the moon.

bar The straight structure across the center of some spiral galaxies, from which the arms unwind.

basalt A type of rock resulting from the cooling of lava.

belts Dark bands around certain planets, notably Jupiter.

beta particle An electron or positron outside an atom.

big-bang theory A cosmological model, based on Einstein's general theory of relativity, in which the universe was once compressed to infinite density and has been expanding ever since.

big crunch The end of the universe, if it eventually collapses.

binary star Two stars revolving around each other.

black body A hypothetical object that, if it existed, would absorb all radiation that hit it and would emit radiation that exactly followed Planck's law.

black hole A region of space from which, according to the general theory of relativity, neither radiation nor matter can escape.

blueshifted Wavelengths shifted to the blue; when the shift is caused by motion, from a velocity of approach.

breccia A type of rock made up of fragments of several types of rocks. Breccias are common on the moon.

capture A model in which a moon was formed elsewhere and then was captured gravitationally by its planet.

carbon-nitrogen cycle The carbon cycle, acknowledging that nitrogen also plays an intermediary role.

Cassegrainian (Cassegrain) A type of reflecting telescope in which the light focused by the primary mirror is intercepted short of its focal point and refocused and reflected by a secondary mirror through a hole in the center of the primary mirror.

Cassini's division The major division in the rings of Saturn.

CCD Charge-coupled device, a solid-state imaging device.

celestial equator The intersection of the celestial sphere with the plane that passes through the earth's equator.

celestial poles The intersections of the celestial sphere with the axis of rotation of the earth.

celestial sphere The hypothetical sphere centered at the center of the earth to which it appears that the stars are affixed.

central star The hot object at the center of a planetary nebula, which is the remaining core of the original star.

Cepheid variable A type of supergiant star that oscillates in brightness in a manner similar to the star δ Cephei. The periods of Cepheid variables, which are between 1 and 100 days, are linked to the absolute magnitude of the stars by known relationships; this allows the distances to Cepheids to be found.

chromosphere The part of the atmosphere of the sun (or another star) between the photosphere and the corona. It is probably entirely composed of spicules and probably roughly corresponds to the region in which mechanical energy is deposited.

circumpolar stars For a given observing location, stars that are close enough to the celestial pole that they never set.

closed universe A big-bang universe with positive curvature; it has finite volume and will eventually contract.

cluster (a) Of stars, a physical grouping of many stars; (b) of galaxies, a physical grouping of at least a few galaxies.

color-magnitude diagram A Hertzsprung-Russell diagram in which the temperature on the horizontal axis is expressed in terms of color index and the vertical axis is in magnitudes.

coma (a) Of a comet, the region surrounding the head; (b) of an optical system, an off-axis aberration in which the images of points appear with comet-like asymmetries.

comet A type of object orbiting the sun, often in a very elongated orbit, that when relatively near to the sun shows a coma and may show a tail.

comet cloud The group of incipient comets surrounding the solar system; also known as the Oort comet cloud.

comparative planetology Studying the properties of solar-system bodies by comparing them.

condensation A region of unusually high mass or brightness.

constellation One of 88 areas into which the sky has been divided for convenience in referring to the stars or other objects therein.

continental drift The slow motion of the continents across the Earth's surface, explained in the theory of plate tectonics as a set of shifting regions called plates.

continuous spectrum A spectrum with radiation at all wavelengths but with neither absorption nor emission lines.

core   The central region of a star or planet.

corona, galactic   The outermost region of our current model of the galaxy, containing most of the mass in some unknown way.

corona, solar *or* **stellar**   The outermost region of the sun (or of other stars), characterized by temperatures of millions of kelvins.

coronal holes   Relatively dark regions of the corona having low density; they result from open field lines.

cosmic rays   Nuclear particles or nuclei travelling through space at high velocity.

cosmological principle   The principle that on the whole the universe looks the same in all directions and in all regions.

cosmology   The study of the universe as a whole.

crust   The outermost solid layer of some objects, including neutron stars and some planets.

dark nebula   Dust and gas seen in silhouette.

daughter molecules   Relatively simple molecules in comets resulting from the breakup of more complex molecules.

deceleration parameter (q_0)   A particular measure of the rate at which the expansion of the universe is slowing down. $q_0 < \frac{1}{2}$ corresponds to an open universe and $q_0 > \frac{1}{2}$ corresponds to a closed universe.

declination   Celestial latitude, measured in degrees north or south of the celestial equator.

density   Mass divided by volume.

density wave   A circulating region of relatively high density, important, for example, in models of spiral arms of galaxies.

diamond-ring effect   The last Baily's bead glowing brightly at the beginning of the total phase of a solar eclipse, or its counterpart at the end of totality.

dirty snowball   A theory explaining comets as amalgams of ices, dust, and rocks.

disk   (a) Of a galaxy, the disk-like flat portion, as opposed to the nucleus or the halo; (b) of a star or planet, the two-dimensional projection of its surface.

Doppler effect (*Doppler shift*)   A change in wavelength that results when a source of waves and the observer are moving relative to each other.

double-lobed structure   An object in which radio emission comes from a pair of regions on opposite sides.

double star   A binary star; two or more stars orbiting each other.

dust tail   The dust left behind a comet, reflecting sunlight.

dwarf ellipticals   Small, low-mass elliptical galaxies.

dwarfs   Dwarf stars.

dwarf stars   Main-sequence stars.

E = mc²   Einstein's formula (special theory of relativity) for the equivalence of mass and energy.

early universe   The universe during its first minutes.

eclipsing binary   A binary star in which one member periodically hides the other.

ecliptic   The path followed by the sun across the celestial sphere in the course of a year.

electromagnetic force   One of the four fundamental forces of nature, giving rise to electromagnetic radiation.

electromagnetic radiation   Radiation resulting from changing electric and magnetic fields.

electron   A particle of one negative charge, 1/1830 the mass of a proton, that is not affected by the strong force. It is a lepton.

electroweak force   The unified electromagnetic and weak forces, according to a recent theory.

element   A kind of atom characterized by a certain number of protons in its nucleus. All atoms of a given element have similar chemical properties.

elementary particle   One of the constituents of an atom.

elliptical galaxy   A type of galaxy characterized by elliptical appearance.

emission line   Wavelengths (or frequencies) at which the intensity of radiation is greater than it is at neighboring wavelengths (or frequencies).

emission nebula   A glowing cloud of interstellar gas.

energy level   A state corresponding to an amount of energy that an atom is allowed to have by the laws of quantum mechanics.

equator   (a) Of the earth, a great circle on the earth, midway between the poles; (b) celestial, the projection of the earth's equator onto the celestial sphere; (c) galactic, the plane of the disk as projected onto a map.

equatorial mount   A type of telescope mounting in which one axis, called the polar axis, points toward the celestial pole and the other axis is perpendicular. Motion around only the polar axis is sufficient to completely counterbalance the effect of the earth's rotation.

equinox   An intersection of the ecliptic and the celestial equator. The center of the sun is geometrically above and below the horizon for equal lengths of time on the two days of the year when the sun passes the equinoxes; if the sun were a point and atmospheric refraction were absent, then day and night would be of equal length on those days.

ergosphere   A region surrounding a rotating black hole from which work can be extracted.

escape velocity   The velocity that an object must have to escape the gravitational pull of a mass.

event horizon   The sphere around a black hole from within which nothing can escape; the place at which the exit cones close.

exit cone   The cone that, for each point within the photon sphere of a black hole, defines the directions of rays of radiation that escape.

extragalactic   Exterior to the Milky Way Galaxy.

extragalactic radio sources   Radio sources outside our galaxy.

filament   A feature of the solar surface seen in Hα as a dark wavy line; a prominence projected on the solar disk.

fireball   An exceptionally bright meteor.

fission, nuclear   The splitting of an atomic nucleus.

flare   An extremely rapid brightening of a small area of the surface of the sun, usually observed in hydrogen-alpha and other strong spectral lines and accompanied by x-ray and radio emission.

focus (*pl: foci*)   (a) A point to which radiation is made to converge; (b) of an ellipse, one of the two points the sum of the distances to which remains constant.

galactic cannibalism   The apparent incorporation of one galaxy into another; may be a projection effect.

galactic corona   The outermost part of our galaxy.

Galilean satellites   The four brightest satellites of Jupiter.

gamma rays   Electromagnetic radiation with wavelengths shorter than approximately 1 Å.

gas tail   The puffs of ionized gas trailing a comet.

general theory of relativity   Einstein's 1916 theory of gravity.

geocentric   Earth-centered.

geology   The study of the earth, or of other solid bodies.

geothermal energy   Energy from under the earth's surface.

giant molecular cloud   A basic building block of our galaxy, containing dust, which shields the molecules present.

giant planets   Jupiter, Saturn, Uranus, and Neptune.

giants   Stars that are larger and brighter than main-sequence stars of the same color.

globular cluster   A spherically symmetric type of collection of stars that shared a common origin.

globule   A dense dust cloud that appears in silhouette.

grand unified theories (*GUT's*)   Theories unifying the electroweak force and the strong force.

granulation   Convection cells on the sun about 1 arc sec across.

gravitational force   One of the four fundamental forces of nature, the force by which two masses attract each other.

gravitational interlock   One body controlling the orbit or rotation of another by gravitational attraction.

gravitational lens   In the gravitational-lens phenomenon, a massive body changes the path of electromagnetic radiation passing near it so as to make more than one image of an object. The double quasar was the first example to be discovered.

gravitational radius   The radius that, according to Schwarzschild's solutions to Einstein's equations of the general theory of relativity, corresponds to the event horizon of a black hole.

gravitational waves   Waves that many scientists consider to be a consequence, according to the general theory of relativity, of changing distributions of mass.

Great Red Spot   A giant circulating region on Jupiter.

greenhouse effect   The effect by which the atmosphere of a planet heats up above its equilibrium temperature because it is transparent to incoming visible radiation but opaque to the infrared radiation that is emitted by the surface of the planet.

H_0   The Hubble constant.

H I region   An interstellar region of neutral hydrogen.

H II region   An interstellar region of ionized hydrogen.

half-life   The length of time for half a set of particles to decay through radioactivity or instability.

halo   Of a galaxy, the region of the galaxy that extends far above and below the plane of the galaxy, containing globular clusters.

head   Of a comet, the nucleus and coma together.

head-tail galaxies   Double-lobed radio galaxies whose lobes are so bent that they look like a tadpole.

heliocentric   Sun-centered; using the sun rather than the earth as the point to which we refer. A heliocentric measurement, for example, omits the effect of the Doppler shift caused by the earth's orbital motion.

Herbig-Haro objects   Blobs of gas ejected in star formation.

hertz   The measure of frequency, with units of /sec (per second); formerly called cycles per second.

high-energy astrophysics   The study of x-rays, gamma rays, and cosmic rays, and of the processes that make them.

homogeneous   Uniform throughout.

Hubble constant (*H_0*)   The constant of proportionality in Hubble's law linking the velocity of recession of a distant object and its distance from us.

Hubble's law   The linear relation between the velocity of recession of a distant object and its distance from us, $V = H_0 d$.

Hubble type   Hubble's galaxy classification scheme: E0, E7, Sa, SBa, etc.

hydrogen forest   The many redshifted lines of hydrogen visible in the spectrum of some quasars.

IC   Index Catalogue, one of the supplements to Dreyer's *New General Catalogue*.

igneous   Rock cooled from lava.

Index Catalogue   See *IC*.

inflationary universe   A model of the expanding universe involving a brief period of extremely rapid expansion.

infrared   Radiation beyond the red, about 7000 Å to 1 mm.

interference The property of radiation, explainable by the wave theory, in which waves in phase can add (constructive interference) and waves out of phase can subtract (destructive interference); for light, this gives alternate light and dark bands.

interferometer A device that uses the property of interference to measure such properties of objects as their positions or structure.

interior The inside of an object.

International Date Line A crooked imaginary line on the earth's surface, roughly corresponding to 180° longitude, at which, when crossed from east to west, the date jumps forward by one day.

interstellar medium Gas and dust between the stars.

inverse-square law Decreasing with the square of increasing distance.

ion An atom that has lost one or more electrons. See also *negative hydrogen ion.*

ionosphere The highest region of the earth's atmosphere.

iron meteorites Meteorites with a high iron content (about 90%); most of the rest is nickel.

isotope A form of chemical element with a specific number of neutrons.

isotropic Being the same in all directions.

Julian calendar The calendar with 365-day years and leap years every fourth year without exception; the predecessor to the Gregorian calendar.

laser An acronym for "**l**ight **a**mplification by **s**timulated **e**mission of **r**adiation," a device by which certain energy levels are populated by more electrons than normal, resulting in an especially intense emission of light at a certain frequency when the electrons drop to a lower energy level.

latitude Number of degrees north or south of the equator measured from the center of a coordinate system.

leap year A year in which a 366th day is added.

light Electromagnetic radiation between about 3000 and 7000 Å.

light curve The graph of the magnitude of an object vs. time.

lighthouse model The explanation of a pulsar as a spinning neutron star whose beam we see as it comes around.

light-year The distance that light travels in a year.

limb The edge of a star or planet.

lobes Of a radio source, the regions to the sides of the center from which high-energy particles are radiating.

Local Group The two dozen or so galaxies, including the Milky Way Galaxy, that form a subcluster.

Local Supercluster The supercluster of galaxies in which the Virgo Cluster, the Local Group, and other clusters reside.

longitude The angular distance around a body measured along the equator from some particular point; for a point not on the equator, it is the angular distance along the equator to a great circle that passes through the poles and through the point.

long-period variables Mira variables.

luminosity class Different regions of the H-R diagram separating objects of the same spectral type: supergiants (I), bright giants (II), giants (III), subgiants (IV), dwarfs (V).

lunar eclipse The passage of the moon into the earth's shadow.

Magellanic Clouds Two small irregular galaxies, satellites of the Milky Way Galaxy, visible in the southern sky; the SMC may be split.

magnetic-field lines Directions mapping out the direction of the force between magnetic poles; the packing of the lines shows the strength of the force.

magnetic monopole A single magnetic charge of only one polarity; may or may not exist.

magnitude A factor of $\sqrt[5]{100} = 2.511886\ldots$ in brightness. See *absolute magnitude* and *apparent magnitude.* An *order of magnitude* is a power of ten.

main sequence A band on a Hertzsprung-Russell diagram in which stars fall during the main, hydrogen-burning phase of their lifetimes.

mantle The shell of rock separating the core of a differentiated planet from its thin surface crust.

mare *(pl: maria)* One of the smooth areas on the moon or on some of the other planets.

mascon A concentration of mass under the surface of the moon, discovered from its gravitational effect on spacecraft orbiting the moon.

maser An acronym for "**m**icrowave **a**mplification by **s**timulated **e**mission of **r**adiation," a device by which certain energy levels are more populated than normal, resulting in an especially intense emission of radio radiation at a certain frequency when the system drops to a lower energy level.

mass number The total number of protons and neutrons in a nucleus.

Maunder minimum The period 1645–1715, when there were very few sunspots, and no periodicity, visible.

mean solar day A solar day for the "mean sun," which moves at a constant rate during the year.

meridian The great circle on the celestial sphere that passes through the celestial poles and the observer's zenith.

Messier numbers Numbers of non-stellar objects in the 18th-century list of Charles Messier.

meteor A track of light in the sky from rock or dust burning up as it falls through the earth's atmosphere.

meteorite An interplanetary chunk of rock after it impacts on a planet or moon, especially on the earth.

meteoroid An interplanetary chunk of rock smaller than an asteroid.

micrometeorite A tiny meteorite. The micrometeorites that hit the earth's surface are sufficiently slowed down that they can reach the ground without being vaporized.

mid-Atlantic ridge The spreading of the sea floor in the middle of the Atlantic Ocean as upwelling material forces the plates to move apart.

middleweight stars Stars between about 4 and 8 solar masses.

midnight sun The sun seen around the clock from locations sufficiently far north or south at the suitable season.

Milky Way The band of light across the sky from the stars and gas in the plane of the Milky Way Galaxy.

minor planets Asteroids.

missing-mass problem The discrepancy between the mass visible and the mass derived from calculating the gravity acting on members of clusters of galaxies.

naked singularity A singularity that is not surrounded by an event horizon and therefore kept from our view.

nebula (pl: nebulae) Interstellar regions of dust or gas.

negative hydrogen ion A hydrogen atom with an extra electron, H^-.

neutrino A spinning, neutral elementary particle with little or no rest mass, formed in certain radioactive decays.

neutron A massive, neutral elementary particle, one of the fundamental constituents of an atom.

neutron star A star that has collapsed to the point where it is supported against gravity by neutrons resisting getting packed together more closely.

New General Catalogue "A New General Catalogue of Nebulae and Clusters of Stars" by J. L. E. Dreyer, 1888.

Newtonian (telescope) A reflecting telescope where the beam from the primary mirror is reflected by a flat secondary mirror to the side.

NGC New General Catalogue.

nova (pl: novae) A star that suddenly increases in brightness; an event in a binary system when matter from the giant component falls on the white dwarf component.

nuclear bulge The central region of spiral galaxies.

nuclear fusion The amalgamation of lighter nuclei into heavier ones.

nucleosynthesis The formation of the elements.

nucleus (a) Of an atom, the core, which has a positive charge, contains most of the mass, and takes up only a small part of the volume; (b) of a comet, the chunks of matter, no more than a few km across, at the center of the head; (c) of a galaxy, the innermost region.

open cluster An irregularly shaped type of star cluster, also known as a galactic cluster.

open universe A big-bang cosmology in which the universe has infinite volume and will expand forever.

optical double A pair of stars that appear extremely close together in the sky even though they are at different distances from us and are not physically linked.

organic Containing carbon in its molecular structure.

paraboloid A 3-dimensional surface formed by revolving a parabola around its axis.

parallax (a) Trigonometric parallax, half the angle through which a star appears to be displaced when the earth moves from one side of the sun to the other (2 A.U.); it is inversely proportional to the distance. (b) Other ways of measuring distance, as in spectroscopic parallax.

parallel light Light that is neither converging nor diverging.

parsec The distance from which 1 A.U. subtends one second of arc. (Approximately 3.26 l-y.)

penumbra (a) For an eclipse, the part of the shadow from which the sun is only partially occulted; (b) of a sunspot, the outer region, not as dark as the umbra.

photon A packet of energy that can be thought of as a particle travelling at the speed of light.

photon sphere The sphere around a black hole, 3/2 the size of the event horizon, within which exit cones open and in which light can orbit.

photosphere The region of a star from which most of its light is radiated.

planet A celestial body of substantial size (more than about 1000 km across), basically non-radiating and of insufficient mass for nuclear reactions ever to begin, ordinarily in orbit around a star.

planetary nebulae Shells of matter ejected by low-mass stars after their main-sequence lifetime, ionized by ultraviolet radiation from the star's remaining core.

planetesimal One of the small bodies into which the primeval solar nebula condensed and from which the planets formed.

plasma An electrically neutral gas composed of approximately equal numbers of ions and electrons.

plates Large flat structures making up a planet's crust.

plate tectonics The theory of the earth's crust, explaining it as plates moving because of processes beneath.

pole star A star approximately at a celestial pole; Polaris is now the pole star; there is no south pole star.

primary cosmic rays The cosmic rays arriving at the top of the earth's atmosphere.

primordial background radiation Isotropic radiation corresponding to matter at a temperature of about 3K; interpreted as a remnant of the big bang.

prominence Solar gas protruding over the limb, visible to the naked eye only at eclipses but also observed outside of eclipses by its emission-line spectrum.

proper motion   Angular motion across the sky with respect to a framework of galaxies or fixed stars.

proton   Elementary particle with positive charge 1, one of the fundamental constituents of an atom.

proton-proton chain   A set of nuclear reactions by which four hydrogen nuclei combine one after the other to form one helium nucleus, with a resulting release of energy.

protoplanets   The loose collections of particles from which the planets formed.

protosun   The sun in formation.

pulsar   A celestial object that gives off pulses of radio waves.

quantum *(pl: quanta)*   A bundle of energy.

quantum theory   The current theory explaining radiation in terms of quanta of energy emitted from atoms, especially as elaborated in quantum mechanics (1926).

quark   One of the subatomic particles of which modern theoreticians believe such elementary particles as protons and neutrons are composed. The various kinds of quarks have positive or negative charges of ⅓ or ⅔.

quasar   One of the very-large-redshift objects that are almost stellar (point-like) in appearance.

quiet sun   The collection of solar phenomena that do not vary with the solar activity cycle.

radar   The acronym for **ra**dio **d**etection **a**nd **r**anging, an active rather than passive radio technique in which radio signals are transmitted and their reflections received and studied.

radial velocity   The velocity of an object along a line (the radius) joining the object and the observer; the component of velocity toward or away from the observer.

radiant   The point in the sky from which all the meteors in a meteor shower appear to be coming.

radiation   Electromagnetic radiation. Sometimes also particles such as alpha (helium nuclei) or beta (electrons).

radiation belts   Belts of charged particles surrounding planets.

radioactive   Having the property of spontaneously changing into another isotope or element.

radio galaxy   A galaxy that emits radio radiation orders of magnitude stronger than that from normal galaxies.

radio telescope   An antenna or set of antennas, often together with a focusing reflecting dish, that is used to detect radio radiation from space.

radio waves   Electromagnetic radiation with wavelengths longer than about one millimeter.

red giant   A post-main-sequence stage of the lifetime of a star; the star becomes relatively bright and cool.

redshifted   When a spectrum is shifted to longer wavelengths.

red supergiant   Extremely bright, cool, and large stars; a post-main-sequence phase of evolution of stars of more than about 4 solar masses.

reflecting telescope   A type of telescope that uses a mirror or mirrors to form the primary image.

reflection nebula   Interstellar gas and dust that reflect light from a nearby star.

refracting telescope   A type of telescope in which the primary image is formed by a lens or lenses.

resolution   The ability of an optical system to distinguish detail.

right ascension   Celestial longitude, measured eastward along the celestial equator in hours of time from the vernal equinox.

Roche's limit *(Roche limit)*   The sphere for each mass inside of which blobs of gas cannot agglomerate by gravitational interaction without being torn apart by tidal forces; normally about 2½ times the radius of a planet.

scarps   Lines of cliffs; found on Mercury, earth, the moon, and Mars.

scattered   Light absorbed and then reemitted in all directions.

Schmidt camera   A telescope that uses a spherical mirror and a thin lens to provide photographs of a wide field.

scientific method   No easy definition is possible, but it has to do with a way of testing and verifying hypotheses.

secondary cosmic rays   High-energy particles generated in the earth's atmosphere by primary cosmic rays.

seismology   The study of waves propagating through a body and the resulting deduction of the internal properties of the body. "Seismo-" comes from the Greek for earthquake.

shooting stars   Meteors.

showers   A time of many meteors from a common cause.

sidereal   With respect to the stars.

sidereal day   A day with respect to the stars.

sidereal rotation period   A rotation with respect to the stars.

sidereal time   The hour angle of the vernal equinox; equal to the right ascension of objects on your meridian.

sidereal year   A circuit of the sun with respect to the stars.

singularity   A point in space where quantities become exactly zero or infinitely large; one is present in a black hole.

solar activity cycle   The 11-year cycle with which solar activity like sunspots, flares, and prominences varies.

solar atmosphere   The photosphere, chromosphere, and corona.

solar flare   An explosive release of energy on the sun.

solar time   A system of time-keeping with respect to the sun such that the sun is overhead of a given location at noon.

solar wind   An outflow of particles from the sun representing the expansion of the corona.

solar year (*tropical year*)   An object's complete circuit of the sun; a tropical year is between vernal equinoxes.

solstice   The point on the celestial sphere of northernmost or southernmost declination of the sun in the course of a year; colloquially, the time when the sun reaches that point.

special theory of relativity   Einstein's 1905 theory of relative motion.

spectral lines   Wavelengths at which the intensity is abruptly different from intensity at neighboring wavelengths.

spectral type   One of the categories O, B, A, F, G, K, M, C, S, into which stars can be classified from study of their spectral lines, or extensions of this system. The sequence of spectral types corresponds to a sequence of temperature.

spectroscopic binary   A type of binary star that is known to have more than one component because of the changing Doppler shifts of the spectral lines that are observed.

spectrum   A display of electromagnetic radiation spread out by wavelength or frequency.

spin-flip   A change in the relative orientation of the spins of an electron and the nucleus it is orbiting.

spiral arms   Bright regions looking like a pinwheel.

spiral galaxy   A class of galaxy characterized by arms that appear as though they are unwinding like a pinwheel.

sporadic   Not regularly.

star   A self-luminous ball of gas that shines or has shone because of nuclear reaction in its interior.

star clouds   The regions of the Milky Way where the stars are so densely packed that they cannot be seen as separate.

star trails   Tracks on a film left by stars when a long exposure by a stationary camera allows their motion to be seen.

stationary limit   In a rotating black hole, the location where space-time is flowing at the speed of light, making stationary particles that would be travelling at that speed.

steady-state theory   The cosmological theory based on the perfect cosmological principle, in which the universe is unchanging over time.

stellar evolution   The changes of a star's properties with time.

streamers   Coronal structures at low solar latitudes.

strong force   The nuclear force, the strongest of the four fundamental forces of nature.

strong nuclear force   The strong force.

subtend   The angle that an object appears to take up in your field of view; for example, the full moon subtends ½°.

sunspot   A region of the solar surface that is dark and relatively cool; it has an extremely high magnetic field.

sunspot cycle   The 11-year cycle of variation of the number of sunspots visible on the sun.

supercluster   A cluster of clusters of galaxies.

supergiant   A post-main-sequence phase of evolution of stars of more than about 4 solar masses. They fall in the upper right of the H-R diagram; luminosity class I.

supergranulation   Convection cells on the solar surface about 20,000 km across and vaguely polygonal in shape.

superluminal velocity   An apparent velocity greater than that of light.

supernova (*pl: supernovae*)   The explosion of a star with the resulting release of tremendous amounts of radiation.

supernova remnant   The gaseous remainder of the star destroyed in a supernova.

synchronous orbit   An orbit of the same period; a satellite in geosynchronous orbit has the same period as the earth's rotation and so appears to hover.

synthetic-aperture radar   A radar mounted on a moving object, used with analysis taking the changing perspective into account to give the effect of a larger dish.

syzygy   An alignment of three celestial bodies.

tail   Gas and dust left behind as a comet orbits sufficiently close to the sun, illuminated by sunlight.

terminator   The line between night and day on a moon or planet; the edge of the part that is lighted by the sun.

terrestrial planets   Mercury, Venus, Earth, and Mars.

thermal pressure   Pressure generated by the motion of particles that can be characterized by a temperature.

3° black-body radiation   The isotropic black-body radiation at 3 K, thought to be a remnant of the big bang.

tidal force   A force caused by the differential effect of the gravity from one body being greater on the near side of a second body than on the far side.

tidal theory   An explanation of solar-system formation in terms of matter being tidally drawn out of the sun by a passing star.

transit   The passage of one celestial body in front of another celestial body. When a planet is *in transit*, we understand that it is passing in front of the sun. Also, *transit* is the moment when a celestial body crosses an observer's meridian, or the special type of telescope used to study such events.

triple-alpha process A chain of fusion processes by which three helium nuclei (alpha particles) combine to form a carbon nucleus.

troposphere The lowest level of the atmosphere of the Earth and some other planets, in which all weather takes place.

T Tauri star A type of irregularly varying star, like T Tauri, whose spectrum shows broad and very intense emission lines. T Tauri stars have presumably not yet reached the main sequence and are thus very young.

tuning-fork diagram Hubble's arrangement of types of elliptical, spiral, and barred spiral galaxies.

21-cm line The 1420-MHz line from neutral hydrogen's spin-flip.

twinkle Change rapidly in brightness.

Type I supernova A supernova whose distribution in all types of galaxies, and whose lack of hydrogen in its spectrum, make us think that it is an event in low-mass stars, probably resulting from the collapse and incineration of a white dwarf in a binary system.

Type II supernova A supernova associated with spiral arms, and which has hydrogen in its spectrum, making us think that it is the explosion of a massive star.

ultraviolet The region of the spectrum 100–4000 Å, also used in the restricted sense of ultraviolet radiation that reaches the ground, namely, 3000–4000 Å.

umbra (*pl: umbrae*) (a) Of a sunspot, the dark central region; (b) of an eclipse shadow, the part from which the sun cannot be seen at all.

Van Allen belts Regions of high-energy particles trapped by the magnetic field of the earth.

variable star A star whose brightness changes over time.

vernal equinox The equinox crossed by the sun as it moves to northern declinations.

Very Large Array The National Radio Astronomy Observatory's set of radio telescopes in New Mexico, used together for interferometry.

Very-Long-Baseline Array A proposed set of telescopes spread across North America.

very-long-baseline-interferometry The technique using simultaneous measurements made with radio telescopes at widely separated locations to obtain extremely high resolution.

visible light Light to which the eye is sensitive, 3900–6600 Å.

visual binary A binary star that can be seen through a telescope to be double.

VLA See *Very Large Array*.

VLBA See *Very-Long-Baseline Array*.

VLBI See *very-long-baseline interferometry*.

weak force One of the four fundamental forces of nature, weaker than the strong force and the electromagnetic force. It is important only in the decay of certain elementary particles.

weight The force of the gravitational pull on a mass.

white dwarf The final stage of the evolution of a star of between 0.07 and 1.4 solar masses; a star supported by electron degeneracy. White dwarfs are found to the lower left of the main sequence of the H-R diagram.

white light All the light of the visible spectrum together.

x-rays Electromagnetic radiation between 1–100 Å.

year The period of revolution of a planet around its central star; more particularly, the earth's period of revolution around the sun.

zenith The point in the sky directly overhead an observer.

zodiac The band of constellations through which the sun, moon, and planets move in the course of the year.

zones Bright bands in the clouds of a planet, notably Jupiter's.

ILLUSTRATION ACKNOWLEDGMENTS

STAR CHARTS—Wil Tirion

COLOR PLATES—Cover Lowell Observatory photo from 1910, digitized and displayed by National Optical Astronomy Observatory; IRAS Color Essay JPL/NASA; Plates 1 and 64 © National Optical Astronomy Observatory/Kitt Peak; Plate 2 Courtesy of I. M. Kopylov, Special Astrophysical Observatory, U.S.S.R.; Plates 3, 4, 15, and 37 Jay M. Pasachoff; Plate 5 Deutsches Museum; Plate 6 NASA and TRW Systems Group; Plate 7 National Radio Astronomy Observatory; Plate 8 Mario Grassi; Plate 9 © 1972 Gary Ladd; Plate 10 Institute for Astronomy, University of Hawaii, photo by Jay M. Pasachoff; Plate 12 Charles Eames; Plate 13 Martin Grossmann; Plates 14 and 16–18 NASA; Plate 19 International Planetary Patrol; Plates 20–22, 27–34, and 36 Jet Propulsion Laboratory/NASA, with the assistance of Jurrie van der Woude; Plates 23 and 24 NASA; color-corrected version courtesy of Friedrich O. Huck; Plate 25 Experiment and data—Massachusetts Institute of Technology; maps—U.S. Geological Survey; NASA/Ames spacecraft; courtesy of Gordon H. Pettengill; Plate 26 Gustav Lamprecht; Plate 35 Lunar and Planetary Laboratory, U. Arizona; Plate 38 Dennis di Cicco photographs, Williams College Expedition; Plates 39 and 40 Naval Research Laboratory/NASA; Plate 41 NASA/JSC; Plate 42 High Altitude Observatory/NASA; Plates 43 and 44 Lewis House, Ernest Hildner, William Wagner, and Constance Sawyer/High Altitude Observatory, National Center for Atmospheric Research, National Science Foundation, and NASA; Plates 45 and 46 Courtesy of Bruce E. Woodgate, Einar Tandberg-Hanssen, and colleagues at NASA's Marshall Space Flight Center; Plates 47, 50, 54, and 66 Canada-France-Hawaii Telescope Corp., © Regents of the University of Hawaii, photo by Laird Thompson, Institute for Astronomy, University of Hawaii; Plate 48 Photography by D. F. Malin of the Anglo-Australian Observatory. Original negative by U.K. Schmidt Telescope Unit, © 1985 Royal Observatory, Edinburgh; Plate 52 Courtesy of Philip E. Angerhofer, Richard A. Perley, Bruce Balick, and Douglas Milne with the VLA of NRAO; Plate 53 Courtesy of S. S. Murray and colleagues at the Harvard-Smithsonian Center for Astrophysics; Plates 57 and 70 © National Optical Astronomy Observatory/Cerro Tololo; Plates 49, 51, 55, 56, 60–62, and 68 © The California Institute of Technology 1959, 1961, and 1965; Plates 59 and 71 Courtesy of the U.K. Schmidt Telescope Unit, Royal Observatory, Edinburgh, © 1980; Plate 58 Photography by D. F. Malin of the Anglo-Australian Observatory. Original negative by U.K. Schmidt Telescope Unit, © 1979 Royal Observatory, Edinburgh; Plate 63 Photography by D. F. Malin, © 1977 Anglo-Australian Telescope Board; Plates 65 and 72 Photography by D. F. Malin, © 1984 Royal Observatory, Edinburgh; Plate 67 Hans Vehrenberg; Plate 69 © 1980 Anglo-Australian Telescope Board; Plate 71 © 1980 Anglo-Australian Telescope Board; Plate 73 Alan Stockton, Institute for Astronomy, U. Hawaii.

FRONTISPIECE—David Malin, © 1979 Anglo-Australian Telescope Board.

CHAPTER 1—Opener Lick Observatory photo; Fig. 1–1 © David Scharf, 1977. All rights reserved; Figs. 1–2 and 1–3 Jay M. Pasachoff; Fig. 1–4 Skyviews Survey, Inc.; Figs. 1–5, 1–6, 1–7, and 1–9 NASA; Fig. 1–12 Harvard College Observatory; Figs. 1–13, 1–14, and 1–15 Palomar Observatory photograph.

CHAPTER 2—Opener Perkin-Elmer Corp. and Lockheed Missiles & Space Co., Inc. Fig. 2–1 Jay M. Pasachoff and Chapin Library; Fig. 2–2 U. Michigan Library, Dept. of Rare Books and Special Collections, translation by Stillman Drake, reprinted courtesy of Scientific American; Fig. 2–3 New Mexico State University Observatory; Fig. 2–10A Yerkes Observatory; Fig. 2–10B American Institute of Physics, Niels Bohr Library; Figs. 2–11 and 2–13A Palomar Observatory photograph; Fig. 2–12 Institute for Astronomy, University of Hawaii, photo by Jay M. Pasachoff; Fig. 2–13B National Optical Astronomy Observatory/ADP; Fig. 2–14 Photo courtesy of the Fred Lawrence Whipple Observatory, a joint facility of U. Arizona and the Smithsonian Institution; Fig. 2–15 From the historical collection of the Mt. Wilson and Las Campanas Observatories, Carnegie Institution of Washington; Fig. 2–17 Dennis Milon; Fig. 2–18 Rod E. McConnell photo; drawing after one by Celestron International; Fig. 2–19 Williams College—Hopkins Observatory; Fig. 2–20A NASA; Fig. 2–20B Perkin-Elmer; Fig. 2–23 NASA; TRW; Figs. 2–24 to 2–29 Jay M. Pasachoff.

CHAPTER 3—Opener © 1977 Anglo-Australian Telescope Board; Figs. 3–1 and 3–10 through 3–13 Jay M. Pasachoff and Chapin Library; Fig. 3–4 Lick Observatory photograph; Fig. 3–5 Richard E. Hill from the site of the Case Western Reserve Universities Burrell Schmidt telescope on Kitt Peak; 3-hr exposure, f/5.6, Ektachrome 400 film; Fig. 3–6 Dennis di Cicco, Sky & Telescope; Puzzle: courtesy of David Allen, Anglo-Australian Observatory.

CHAPTER 4—Opener © 1984 Jay M. Pasachoff; Fig. 4–2 Dennis di Cicco, Sky & Telescope; Fig. 4–3 Emil Schulthess, Black Star; Fig. 4–4 Lick Observatory photos; Fig. 4–6 NASA; Fig. 4–12 Bryan Brewer, Earth View, Inc.; Figs. 4–13, 4–14, and 4–17 © 1984 Jay M. Pasachoff; Fig. 4–15 Photo by Daniel Fischer, Eclipse Staff, People's Observatory Bonn; Fig. 4–16 © 1983 Jay M. Pasachoff; Fig. 4–18 Drawing by Handelsman, © 1978 The New Yorker Magazine, Inc.; Fig. 4–19 Chapin Library, Williams College.

PART II

CHAPTER 5—Opener Produced by Société Européenne de Propulsion (S.E.P) (France) VISIR Laser Beam Recorder (Courtesy of Allied International Corporation, New York);

Fig. 5–1 From Raymond Siever, "The Earth," *Scientific American*, September 1975, and *The Solar System* (San Francisco: W. H. Freeman and Co., 1975), reprinted courtesy of *Scientific American*; **Fig. 5–2** Jay M. Pasachoff; **Fig. 5–3** NASA; **Fig. 5–4** U.S. Geological Survey, photograph by R. E. Wallace; **Fig. 5–5** Courtesy of Wilbur Rinehart, National Geophysical and Solar-Terrestrial Data Section, Environmental Data Service, National Oceanic and Atmospheric Administration; **Fig. 5–6A** Courtesy of Stanley N. Williams, Dartmouth College, from *Science*, 13 June 1980, © by the American Association for the Advancement of Science; **Fig. 5–6B** Photograph by James Zollweg; **Fig. 5–6C** Dale P. Cruikshank, U. Hawaii; **Fig. 5–7** © NFB Photothèque ONF (NFB-P-ONF), photo by G. Blouin, 1949; **Fig. 5–11** L. A. Frank, U. Iowa.

CHAPTER 6—Opener NASA; **Figs. 6–1 to 6–2** Lick Observatory Photographs; **Figs. 6–3, 6–4A, 6—7, 6–8** NASA; **Fig. 6–4B** Drawing by Alan Dunn; © 1971 The New Yorker Magazine, Inc.; **Figs. 6–5 and 6–6** Lunar Receiving Laboratory, Johnson Space Center, NASA; **Fig. 6–9** Drawing by Donald E. Davis under the guidance of Don E. Wilhelms of the U.S. Geological Survey; **Fig. 6–11** NASA Lunar and Planetary Institute.

CHAPTER 7—Opener NASA; **Fig. 7–1** New Mexico State U. Observatory; **Figs. 7–4 to 7–9** NASA.

CHAPTER 8—Opener and Figs. 8–6, 8–7, and 8–9 NASA/Ames; **Fig. 8–1** Palomar Observatory Photograph; **Fig. 8–5** Courtesy of D. B. Campbell, Arecibo Observatory; **Fig. 8–8** James W. Head, III, Brown Univ.; **Figs. 8–10 and 8–11** Courtesy of Valeriy Barsukov and Yuri Surkov; **Fig. 8–12** Tass; **Fig. 8–13** NASA.

CHAPTER 9—Opener NASA/JPL; **Fig. 9–1** Lick Observatory Photograph; **Figs. 9–3 to 9–6, 9–8 to 9–13** NASA/JPL; **Fig. 9–7** U.S. Geological Survey; **Fig. 9–14** NASA, courtesy of Joseph Veverka.

CHAPTER 10—Opener NASA/JPL; **Fig. 10–2** *Sky & Telescope*; **Figs. 10–3 to 10–14** NASA/JPL; **Fig. 10–15** Hughes Aircraft Company.

CHAPTER 11—Opener NASA/JPL; **Fig. 11–2** Lowell Observatory; **Fig. 11–3** NASA/Ames; **Figs. 11–4 to 11–15, 11–21** JPL/NASA; obtained with the assistance of Jurrie van der Woude and Stephen Edberg; **Figs. 11–15 to 11–20** Raymond Batson, U.S. Geological Survey.

CHAPTER 12—Opener NASA/JPL, drawing by Don Davis; **Fig. 12–1** Lunar and Planetary Laboratory, U. Arizona; **Fig. 12–3** Data from J. L. Elliot, E. Dunham, L. H. Wasserman, R. L. Millis, and J. Churms; **Figs. 12–4 and 12–8** R. J. Terrile, JPL, and B. A. Smith, U. Arizona, at Carnegie Institution's Las Campanas Observatory; **Fig. 12–5** NASA/JPL; **Fig. 12–6** Master and Fellows of St. John's College of Cambridge; **Fig. 12–7** Courtesy of Charles T. Kowal; **Fig. 12–9** Lick Observatory Photograph; **Fig. 12–10** Lowell Observatory Photographs; **Fig. 12–11** U.S. Naval Observatory/James W. Christy, U.S. Navy Photo; **Fig. 12–12** with permission of James W. Beletic, Harvard U.

CHAPTER 13—Opener Institut d'Astrophysique de Liège, Belgium, with the Schmidt camera at the Observatoire de Haute Provence, France; **Fig. 13–1** William Liller; **Figs. 13–2, 13–3, and 13–6** Palomar Observatory Photograph; **Fig. 13–4** Naval Research Laboratory, courtesy of Neil R. Sheeley, Jr.; **Fig. 13–5** National Portrait Gallery, London, painted by R. Phillips prior to 1721; **Fig. 13–7** (first printing) *Sky & Telescope* by Roger Sinnott from orbital elements by Joseph Brady and Edna Carpenter, Lawrence Radiation Laboratory, U. California; **Fig. 13–8** (first printing) Courtesy of Donald Yeomans, JPL; **Fig. 13–9** Peter Bloomer, *Horizons West,* by permission of Meteor Crater Enterprises, Inc.; **Fig. 13–10** Arthur A. Griffin; **Fig. 13–11** Dan Haar, © 1982 Hartford Courant; **Fig. 13–12** Photo by Ursula B. Marvin; **Fig. 13–13** NASA; **Fig. 13–14** Harvard-Smithsonian Center for Astrophysics; **Fig. 13–16** NASA/JPL; **Fig. 13–17** data from Lucy McFadden; **Fig. 13–18** Palomar Observatory Photograph/E. Helin.

CHAPTER 14—Opener and Fig. 14–11 © Lucasfilm, Ltd. (LFL) 1980. All rights reserved. From the motion picture *The Empire Strikes Back,* courtesy of Lucasfilm, Ltd.; **Fig. 14–1** Drawing by Dana Fradon; reprinted with permission of The New Yorker Magazine, Inc., © 1980; **Fig. 14–2** Cyril Ponnamperuma, U. Maryland; **Fig. 14–3** Sproul Observatory; **Fig. 14–4** NASA/JPL; **Fig. 14–5** R. J. Terrile, JPL, and B. A. Smith, U. Arizona, at Carnegie Institution's Las Campanas Observatory; **Fig. 14–6** E. Imre Friedmann; **Fig. 14–7** From *The Day the Earth Stood Still,* © 1951 Twentieth Century-Fox Film Corporation. All rights reserved; **Fig. 14–8A** NASA; **Fig. 14–8B** Jay M. Pasachoff; **Fig. 14–8C** from the MGM release *2010,* © 1984 MGM/UA Entertainment Co.; **Fig. 14–9** Cornell U. Photograph; **Fig. 14–10** From *Glinda of Oz,* by L. Frank Baum, illustrated by John R. Neill, © 1920; **Fig. 14–12** © 1984 George Long/LPI; **Fig. 14–13** Robert M. Sheaffer; **Fig. 14–14** By permission of the Royal Greenwich Observatory, Sussex, England.

PART III—Opener Institute for Astronomy, U. Hawaii

CHAPTER 15—Opener Jay M. Pasachoff and Chapin Library, Williams College; **Fig. 15–1** American Institute of Physics, Niels Bohr Library, Margrethe Bohr Collection; **Fig. 15–8** Harvard College Observatory; **Fig. 15–9** NOAO/Kitt Peak.

CHAPTER 16—Opener Dennis Milon; **Fig. 16–3** After Bok and Bok, *The Milky Way,* courtesy Harvard University Press; **Fig. 16–4** Dorrit Hoffleit—Yale U. Observatory, courtesy of American Institute of Physics, Niels Bohr Library; **Fig. 16–5** American Institute of Physics, Niels Bohr Library, Margaret Russell Edmondson Collection; **Fig. 16–6** Canada-France-Hawaii Telescope Corp., photo by Laird Thompson, © Regents of the University of Hawaii.

CHAPTER 17—Opener Palomar Observatory Photograph; **Fig. 17–1** Courtesy of Sproul Observatory; **Fig. 17–2** Mt. Wilson and Las Campanas Observatories, Carnegie Institution of Washington; **Fig. 17–3** Lick Observatory Photograph; **Fig. 17–6** Peter van de Kamp; **Figs. 17–7** American Association of Variable Star Observers/Janet Mattei; **Figs. 17–8 and 17–15** After Bok and Bok, *The Milky Way,* courtesy Harvard University Press; **Fig. 17–10** Harvard College Observatory; **Fig. 17–11** Shigetsugu Fujinami; **Fig. 17–12** Jay M. Pasachoff and Chapin Library; **Fig. 17–13** Williams College—Hopkins Observatory; **Fig. 17–14** Courtesy of the U. K. Schmidt Telescope Unit, Royal Observatory, Edinburgh; **Fig. 17–16** Bernhard M. Haisch.

CHAPTER 18—**Opener and 18–11***B* William C. Livingston, NOAO/NSO; **Fig. 18–1** Anne Norcia; **Fig. 18–2** Courtesy of Instituto de Astrofisica de Canarias, Tenerife, and Kiepenheuer-Instituto für Sonnenphysik, Freiburg; **Fig. 18–3** Mt. Wilson and Las Campanas Observatories, Carnegie Institution of Washington; **Figs. 18–4 and 18–10** NOAO/NSO; **Fig. 18–5** The Aerospace Corporation/David K. Lynch; **Fig. 18–6** High Altitude Observatory/Richard R. Fisher, T. Baur, Lee Lacey, NCAR; **Fig. 18–7** Institute for Astronomy, U. Hawaii; **Fig. 18–8** Naval Research Laboratory (*interior*) and High Altitude Observatory (*exterior*)/NASA; **Fig. 18–9** American Science and Engineering, Inc./NASA; **Fig. 18–12** A plot of Zurich sunspot numbers; updated from data provided by M. Waldmeier, Swiss Federal Observatory, with values from A. Koecklenbergh of the Sunspot Index Center, Brussels, Belgium, quoted in *Sky and Telescope*; **Figs. 18–13***A* and **18–13***B* NASA; **Fig. 18–13***C* Lewis House, Ernest Hildner, William Wagner, and Constance Sawyer, HAO/NCAR/NSF and NASA; **Figs. 18–14 and 18–15** Courtesy of Harold Zirin, California Institute of Technology, Big Bear Solar Observatory photographs; **Fig. 18–16** Courtesy of the Director of the Mt. Wilson and Las Campanas Observatories, of Otto Nathan, and of the Hebrew University, Jerusalem; **Fig. 18–18** © 1919 by The New York Times Company. Reprinted by permission.

PART IV—**Opener** Palomar Observatory Photograph.

CHAPTER 19—**Opener** Bart Bok, labelling after R. D. Schwartz; **Fig. 19–1** George H. Herbig, Lick Observatory; **Fig. 19–7** Fermi National Accelerator Laboratory.

CHAPTER 20—**Opener** © 1980 Anglo-Australian Telescope Board; **Fig. 20–1** After Richard L. Sears, *J. Royal Astron. Soc. Canada*, No. 1, Feb. 1984; originally from B. E. Paczyński, *Acta Astron. 20*, 47, 1970; **Fig. 20–2** Courtesy of R. R. Howell, C. Pilcher, and R. Hlivak, Institute for Astronomy, U. Hawaii; **Figs. 20–3 and 20–8***A* Palomar Observatory Photograph; **Fig. 20–5** Irving Lindenblad, U.S. Naval Observatory; **Fig. 20–6** © Ben Mayer, Los Angeles; **Fig. 20–7** Lick Observatory Photograph; **Fig. 20–8***B* Jay M. Pasachoff, Dale P. Cruikshank, and Clark R. Chapman with 2.2-m telescope of the Institute for Astronomy, U. Hawaii; **Fig. 20–9** NASA; **Figs. 20–10 and 20–11** Jay M. Pasachoff.

CHAPTER 21—**Opener** Lick Observatory Photograph; **Fig. 21–1** Palomar Observatory Photographs; **Fig. 21–2** Palomar Observatory Photograph, courtesy of Sidney van den Bergh, photograph by R. Minkowski; **Fig. 21–3** Paul Griboval, McDonald Observatory; **Fig. 21–4** Jay M. Pasachoff and Chapin Library; **Fig. 21–5***A* Stephen S. Murray and colleagues, Harvard-Smithsonian Center for Astrophysics; **Fig. 21–5***B* A. R. Thompson and colleagues, VLA/NRAO, courtesy of Robert M. Hjellming; **Fig. 21–6** NASA; **Fig. 21–7** Jay M. Pasachoff.

CHAPTER 22—**Opener** Lick Observatory Photograph; **Fig. 22–2** Jocelyn Bell Burnell; **Fig. 22–3** Joseph H. Taylor, Jr., Marc Damashek, and Peter Backus, then U. Mass.—Amherst at the NRAO; **Fig. 22–4** Joseph H. Taylor, Jr.; **Fig. 22–6***A* H. Y. Chiu, R. Lynds, and S. P. Maran, photographed at NOAO/Kitt Peak; **Fig. 22–6***B* F. R. Harnden, Jr., and colleagues, Harvard-Smithsonian Center for Astrophysics; **Fig. 22–7** Daniel R. Stinebring, then NRAO, now Princeton U.; **Fig. 22–9***A* Joseph Weber, U. Maryland; **Fig. 22–9***B* Peter Kramer; **Fig. 22–11***A* VLA/NRAO, operated by Associated Universities, Inc., under contract with the NSF, courtesy of H.

Hjellming; **Fig. 22–11***B* Courtesy of M. Watson, R. Willingale, Jonathan E. Grindlay, and Frederick D. Seward, *Astrophys. J. 273*, 688 (1983), courtesy of University of Chicago Press.

CHAPTER 23—**Opener and Fig. 23–9** Palomar Observatory Photograph/Jerome Kristian; **Fig. 23–5** Drawing by Chas. Addams; © 1974 The New Yorker Magazine, Inc.; **Fig. 23–6** After E. H. Harrison, U.Mass—Amherst; **Fig. 23–7** Chapin Library, Williams College; **Fig. 23–8***A* Jean-Pierre Luminet, Groupe d'Astrophysique Relativiste, Observatoire de Paris; **Fig. 23–8***B* John F. Hawley and Larry Smarr, U. Illinois at Urbana-Champaign; **Fig. 23–10** Lois Cohen—Griffith Observatory.

PART V—**Opener** Reproduced by courtesy of The Trustees, The National Gallery, London.

CHAPTER 24—**Opener and Fig. 24–2** Palomar Observatory Photograph; **Fig. 24–1** Lick Observatory Photograph; **Fig. 24–4** NASA/JPL; **Fig. 24–5***A* R. D. Ekers (NRAO) and U. J. Schwarz and W. M. Goss (Kapteyn Laboratories, the Netherlands) with the VLA of NRAO; **Fig. 24–5***B* Mark Morris (U.C.L.A.) and Farhad Yusef-Zadeh and Don Chance (Columbia U.) with the VLA of NRAO; **Fig. 24–6** Lund Observatory, Sweden; **Fig. 24–7** Kent S. Wood, Naval Research Laboratory; **Fig. 24–8** Michael Watson, Paul Hertz, and colleagues, Harvard-Smithsonian Center for Astrophysics; **Fig. 24–9** H. A. Mayer-Hasselwander, K. Bennett, G. F. Bignami, R. Buccheri, N. D'Amico, W. Hermsen, G. Kanback, F. Lebrun, G. G. Lichti, J. L. Masnou, J. A. Paul, K. Pinkau, L. Scarsi, B. N. Swanenberg, and R. D. Wills; COS-B Observation of the Milky Way in High-Energy Gamma Rays, Ninth Texas Symposium on Relativistic Astrophysics, Eds. J. Ehlers, J. J. Perry, and M. Walker, *Annals of the New York Academy of Sciences, 336*, 211 (1980); **Fig. 24–10** Data from W. Becker, Palomar Observatory Photograph; **Fig. 24–11** Agris Kalnajs, Mount Stromlo Observatory; **Fig. 24–12** Courtesy of Humberto Gerola and Philip E. Seiden, IBM Watson Research Center.

CHAPTER 25—**Opener** NOAO/CTIO; **Fig. 25–1, spectrum** David L. Talent, Abilene Christian U.; **Fig. 25–2** Palomar Observatory Photograph; **Fig. 25–4** Harvard U./E. M. Purcell; **Fig. 25–5** Gerrit Verschuur; **Fig. 25–7** Leo Blitz, U. Maryland; **Fig. 25–8** Bart J. Bok; **Fig. 25–9** Contours from Marc L. Kutner; Lick Observatory Photograph; **Fig. 25–10** Model by Ben Zuckerman, U.C.L.A.; **Figs. 25–11***A* and **25–12** NASA/JPL; **Fig. 25–11***B* Meade Instruments Corp.; **Fig. 25–13** NRAO, courtesy of Margaret Weems; **Fig. 25–14** Leo Goldberg; **Fig. 25–15** Jay M. Pasachoff.

PART VI—**Opener** Lick Observatory Archives.

CHAPTER 26—**Opener and Figs. 26–1, 26–2, 26–4, 26–8** Palomar Observatory Photograph; **Fig. 26–3***B* Gerard de Vaucouleurs, U. Texas; **Fig. 26–5** © 1984 Anglo-Australian Telescope Board; **Fig. 26–6** Courtesy of Carlos S. Frenk and Simon White; **Fig. 26–7***A* National Academy of Science; **Fig. 26–7***B* From E. Hubble and M. L. Humason, *Astrophysical Journal 74*, 77, 1931, courtesy of University of Chicago Press; **Fig. 26–11** Radio image from the VLA of NRAO; optical images by Laird Thompson, Institute for Astronomy, U. Hawaii; **Fig. 26–12***A* © 1974 NOAO/Kitt Peak; **Fig. 26–12***B* Halton C. Arp, Palomar Observatory Photograph; **Fig. 26–13** Leftmost photograph from Palomar Observatory; **Fig. 26–18** Jay M. Pasachoff; **Fig. 26–19** Courtesy of Richard G. Strom, George K. Miley, Jan Oort, and *Scientific American;* **Fig. 26–20** Westerbork Synthesis Telescope, George K. Miley, G. C. Perola, P. C. van der Kruit, and H. van der Laan, *Nature, 237,* 269, © 1972 by Macmillan Journals Limited, London; **Fig. 26–21** Courtesy of F. N. Owen, J. O. Burns, and L. Rudnick, NRAO.

CHAPTER 27—Opener and Fig. 27–1 Maarten Schmidt/ Palomar Observatory Photograph; **Fig. 27–2** Palomar Observatory Photograph; **Fig. 27–3** © 1983 Anglo-Australian Telescope Board; **Fig. 27–4** Bruce Margon and E. A. Harlan, Lick Observatory Photograph; **Fig. 27–5** John B. Hutchings and colleagues; Canada-France-Hawaii Telescope; **Fig. 27–6** Halton C. Arp, Palomar Observatory Photograph; **Figs. 27–7 and 27–14** Alan Stockton, Institute for Astronomy, U. Hawaii; **Fig. 27–8A** Palomar Observatory Photographs; montage by W. W. Morgan, Yerkes Observatory; **Fig. 27–8B** Princeton University Observatory; **Fig. 27–8C** R. M. West, A. C. Danks, and G. Alcaino, European Southern Observatory, *Astronomy and Astrophysics, 62,* L113 (1978); **Figs. 27–9 and 27–10A** Observations obtained by Susan Wyckoff, Peter Wehinger, and Thomas Gehren with the 3.6-m telescope of the European Southern Observatory; **Fig. 27–10B** Anthony Tyson, courtesy of AT&T Bell Laboratories; **Fig. 27–11** Stephen Unwin, California Institute of Technology; **Fig. 27–13** Jerome Kristian, James A. Westphal, and Peter Young/Palomar Observatory Photograph.

CHAPTER 28—Opener Lotte Jacobi; **Fig. 28–1B** From the collection of Mr. and Mrs. Paul Mellon; **Fig. 28–3** Wide World Photos; **Fig. 28–6** Robert V. Wagoner, Stanford U.; **Fig. 28–7** Courtesy of AT&T Bell Laboratories; **Fig. 28–9** D. Woody and P. L. Richards; **Fig. 28–10** NASA; **Fig. 28–11** CERN.

EPILOGUE—NASA.

APPENDICES—Appendix 3 The masses and diameters are the values recommended by the International Astronomical Union in 1976, except for new values for Jupiter, Saturn, Uranus, and Pluto. Surface gravities were calculated from these values. The length of the Martian day is from G. de Vaucouleurs (1979). Uranus values from J. Elliot *et al.* Others values from *The Astronomical Almanac, 1984.* **Appendix 4** Based on a table by Joseph Veverka in the *Observer's Handbook 1980* of the Royal Astronomical Society of Canada, but with many updated data. **Appendix 5** Based on a table compiled by Donald A. MacRae in the *Observer's Handbook 1980* of the Royal Astronomical Society of Canada, amended with information from W. Gliese. **Appendix 7** Courtesy of W. Gliese (1979) with new spectral types on a consistent scale by Robert F. Wing and Charles A. Dean (1983), new parallaxes from William van Altena (1983), and additional comments from Dorrit Hoffleit (1983). Transformed with precession and proper motion corrections to 2000.0 coordinates. **Appendix 8** Positions and magnitudes based on a table in the *Observer's Handbook 1980* of the Royal Astronomical Society of Canada. **Appendix 11** Abundances from Edward Anders and Mitsuru Ebihara, *Geochimica and Cosmochimica Acta, 46,* 2363–2380, 1982. Atomic weights are averages for terrestrial abundances.

INDEX

References to illustrations, either photographs or drawings, on pages where there is no related text, are in *italics*. References to color plates are prefaced by CP. References to Appendices are prefaced by A. References to tables are followed by the letter t.

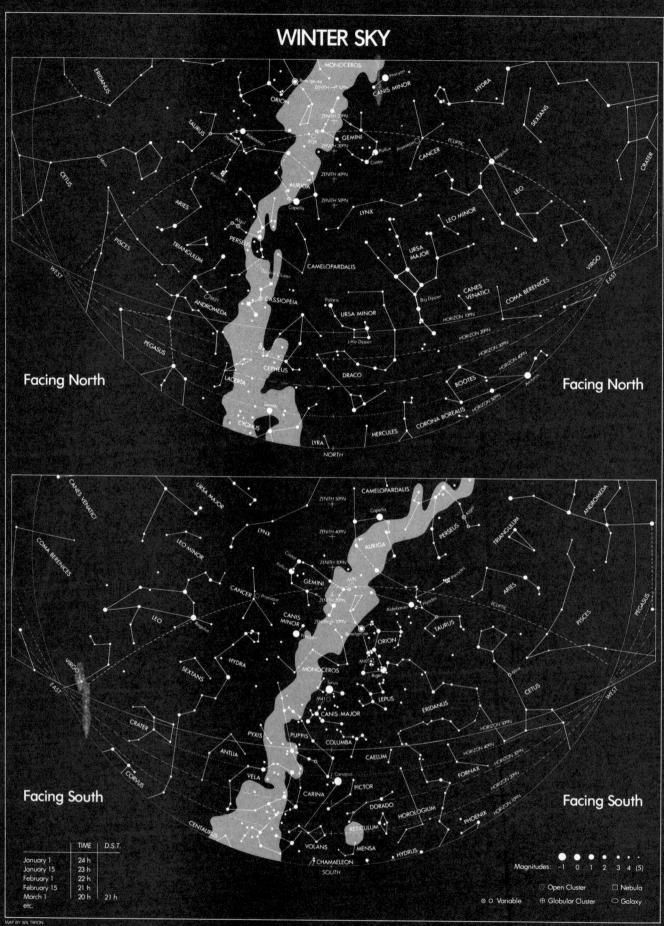

WINTER SKY

SPRING SKY

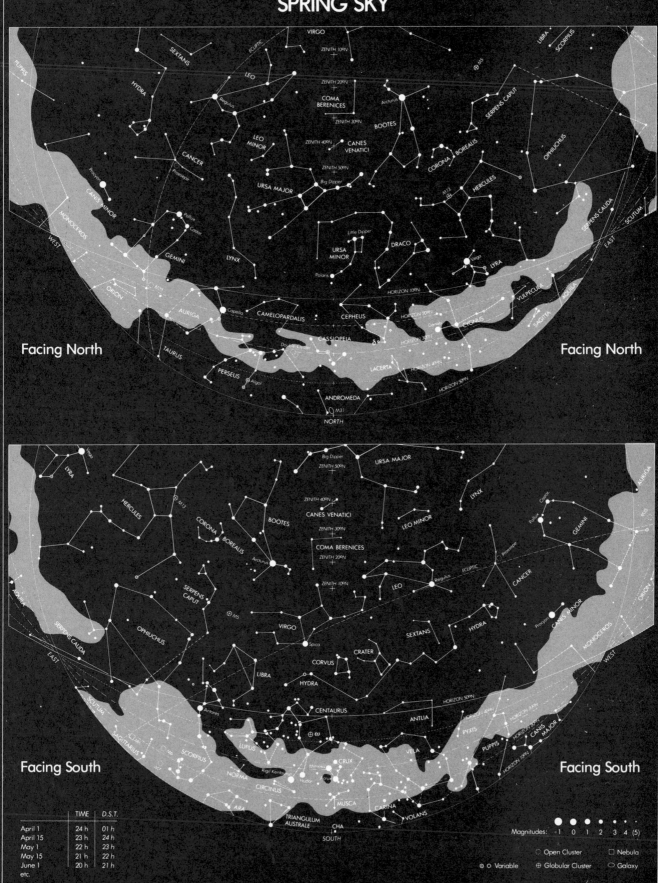

Facing North

Facing North

Facing South

Facing South

	TIME	D.S.T.
April 1	24 h	01 h
April 15	23 h	24 h
May 1	22 h	23 h
May 15	21 h	22 h
June 1	20 h	21 h
etc.		

Magnitudes: -1 0 1 2 3 4 (5)

⊙ ○ Variable   ⚬ Open Cluster   ☐ Nebula
⊕ Globular Cluster   ◯ Galaxy

MAP BY WIL TIRION
FOR JAY M. PASACHOFF

SUMMER SKY

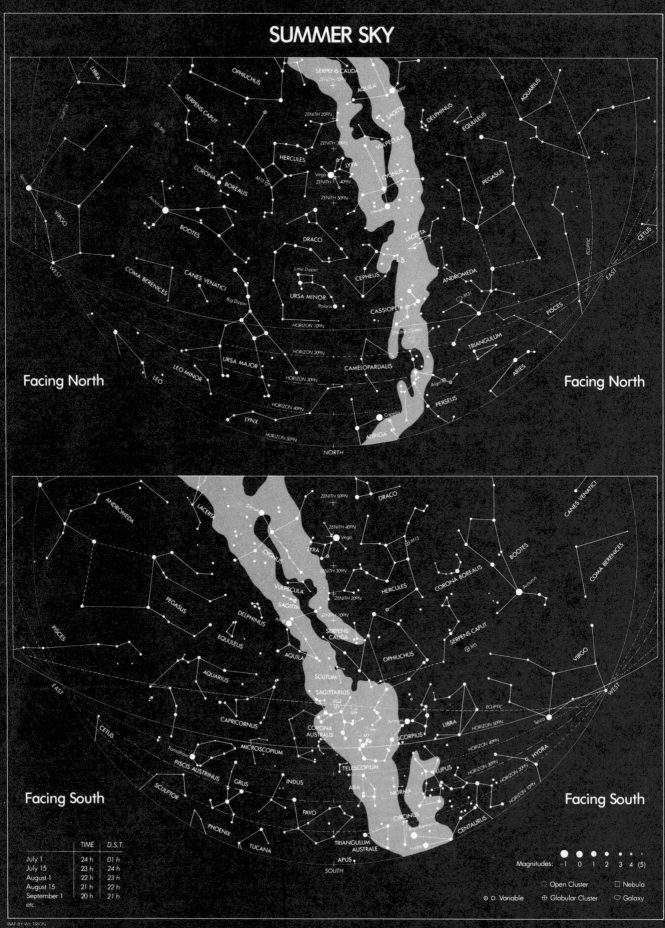

Facing North

Facing North

Facing South

Facing South

	TIME	D.S.T.
July 1	24 h	01 h
July 15	23 h	24 h
August 1	22 h	23 h
August 15	21 h	22 h
September 1	20 h	21 h
etc		

Magnitudes: -1 0 1 2 3 4 (5)

⊙ Open Cluster   ☐ Nebula
⊙ Variable   ⊕ Globular Cluster   ○ Galaxy

MAP BY WIL TIRION
FOR JAY M. PASACHOFF

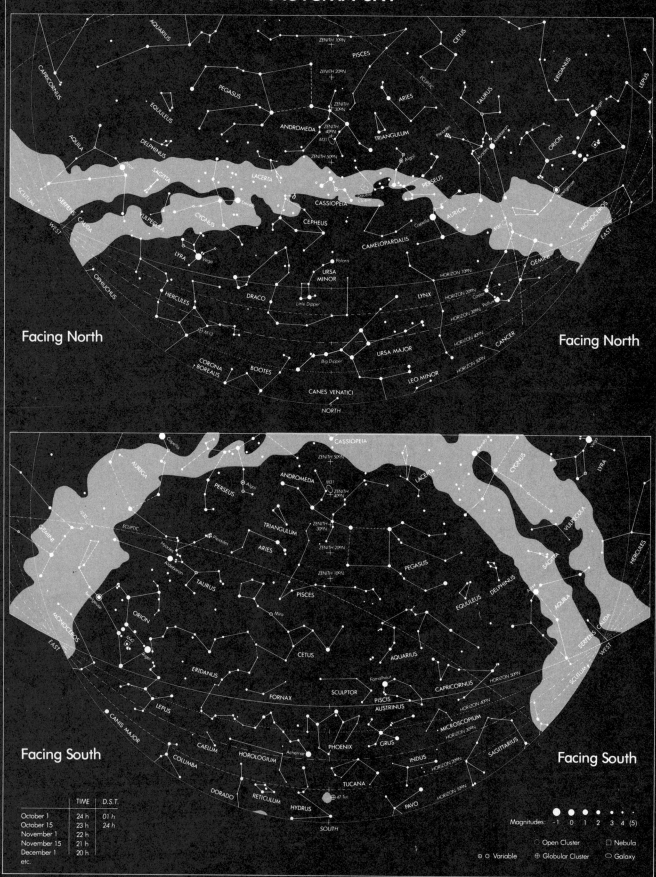

AUTUMN SKY

Facing North

Facing North

Facing South

Facing South

	TIME	D.S.T.
October 1	24 h	01 h
October 15	23 h	24 h
November 1	22 h	
November 15	21 h	
December 1	20 h	
etc.		

Magnitudes: -1 0 1 2 3 4 (5)

○ Open Cluster   □ Nebula

⊙ Variable   ⊕ Globular Cluster   ◯ Galaxy

MAP BY WIL TIRION
FOR JAY M. PASACHOFF